T0257708

IET TELECOMMUNICATIONS SERIES 73

Network as a Service for Next Generation Internet

Other volumes in this series:

Network as a Service for Next Generation Internet

Edited by
Qiang Duan and Shangguang Wang

The Institution of Engineering and Technology

Published by The Institution of Engineering and Technology, London, United Kingdom

The Institution of Engineering and Technology is registered as a Charity in England & Wales (no. 211014) and Scotland (no. SC038698).

The Institution of Engineering and Technology
Michael Faraday House
Six Hills Way, Stevenage
Herts, SG1 2AY, United Kingdom

www.theiet.org

British Library Cataloguing in Publication Data
A catalogue record for this product is available from the British Library

ISBN 978-1-78561-176-6 (hardback)
ISBN 978-1-78561-177-3 (PDF)

Typeset in India by MPS Limited

Contents

**11 Network-as-a-Service in software-defined networking
for end-to-end quality of service provisioning   265**
Qiang Duan

**12 Flow management and orchestration for virtualized network
functions in software-defined networks   287**
Po-Han Huang and Charles H.-P. Wen

Preface

The rapid development of computer networking and information technologies in the past decade has transformed the Internet from a network infrastructure providing a limited number of data communication services to a general service delivery platform for supporting a wide spectrum of highly diverse computing applications. Such transformation has brought significant challenges to the Internet that demand fundamental changes in network architecture and service delivery model. A key strategy taken by the research community of the next-generation Internet comprises two closely related aspects: (i) decoupling service-related network functions from the underlying network infrastructure through virtualization and (ii) encapsulating network resources into self-contained entities that can be exposed and utilized via abstract interfaces. Such a strategy for future Internet design essentially embraces the "as-a-Service" paradigm, which has been widely adopted and successfully deployed for web services and cloud computing, in the field of networking; therefore, it is generally referred to as "Network-as-a-Service" (NaaS) in the literature. By applying various successful IT technologies such as virtualization and the Service-Oriented Architecture (SOA) in networking systems, the NaaS paradigm may greatly enhance network flexibility, simplify network control and improve network service performance. Therefore, NaaS is expected to play a crucial role in the next-generation Internet and has formed an active research area that attracts extensive attention from both industry and academia.

The objective of this book is to present the state-of-the art results about the emerging NaaS paradigm, including its concept, architecture, key technologies and applications. In this book, we attempt to provide readers with a comprehensive reference that reflects the most current technical developments related to NaaS. This book offers a broad coverage of important topics with 16 chapters written by international experts.

The key ideas, principles and benefits of NaaS are first introduced in Chapter 1. In addition, this chapter particularly highlights the impact of NaaS in future Internet that allows cloud technologies to be employed for building network systems and delivering network services. The trend of network cloudification enabled by NaaS offers a promising approach towards convergence of networking and cloud computing.

Methods and infrastructures for NaaS are reviewed in Chapter 2. A dynamic and flexible NaaS architecture for next-generation Internet is presented in Chapter 3. Virtualization plays a key role in realizing the notion of NaaS; therefore, technologies for deploying virtual networks in NaaS are discussed in Chapter 4.

Applying SOA in networking is a key aspect of NaaS, which requires effective mechanisms for network service description, discovery and selection. Chapter 5 introduces a service broker system for discovering and selecting the network services that meet application requirements. Chapter 6 presents a method for service selection and recommendation in integrated network environments.

NaaS allows the uniform service-orientation principle to be employed in both networking and cloud computing thus enabling a convergent holistic view across these two areas that used to be relatively independent. Chapter 7 discusses the evolution of cloud networking technologies with such a holistic view. Chapter 8 introduces the concept and key technologies for cloud mobile networking that employs cloud technologies in mobile networks. Effective and efficient approaches to service composition across the networking and cloud computing domains are presented in Chapter 9.

The flexible and dynamic networking and service provisioning environments enabled by NaaS require sophisticated control and management that cannot be easily provided by traditional networks. The merging Software-Defined Networking (SDN) offers a promising approach to supporting virtualization and SOA in networking; thus may greatly facilitate realizing the NaaS paradigm. On the other hand, NaaS enables a high-level abstraction of network resources and capabilities that offers a promising approach to tackling the challenge of end-to-end service provisioning in SDN.

Technologies for NaaS-enabled service composition in SDN are reviewed in Chapter 10. A NaaS-based service delivery platform for end-to-end QoS provisioning in SDN is presented in Chapter 11. Chapter 12 discusses flow management and orchestration for service chains of virtual network functions in SDN. Chapter 13 presents a survey about the latest development in SDN-based NaaS platforms. Graphics Processing Unit (GPU)-based acceleration of SDN controller for supporting NaaS is reviewed and analysed in Chapter 14.

NaaS may greatly enhance service capability of the next-generation Internet for supporting a wide spectrum of applications. The last two chapters in this book are related to NaaS applications. Chapter 15 introduces an approach called software-defined service network that applies the NaaS model at the application layer. Chapter 16 presents a context-as-a-service platform that adopts NaaS in Internet of Things (IoT) for supporting smart IoT applications.

This book is intended to be accessible to a wide technical audience, including researchers and practitioners in the field of networking and graduate students in computer science and engineering programmes who are familiar with basic networking concepts and technologies.

We would like to thank all the contributing authors who worked diligently to make this book a valuable reference. We also want to thank the editors of IET publisher who helped greatly throughout the contract and publication process and provided valuable comments to improve this book.

Qiang Duan
Shangguang Wang
December 2016

Chapter 1

Network-as-a-Service and network cloudification in the next generation Internet

Qiang Duan[1]

Abstract

A fundamental strategy taken by the research for the next generation Internet lies in application of the service-orientation principle in the field of networking, which enables the *as-a-Service* paradigm that has been widely adopted in cloud computing, for example, Infrastructure-as-a-Service (IaaS) and Software-as-a-Service (SaaS), inside future networking systems. Therefore, such an approach to future network design is referred to as *Network-as-a-Service* (NaaS). NaaS is expected to play a crucial role in the next generation Internet that may introduce significant changes in both network architecture and service delivery model. The main objective of this chapter is to give a high-level overview of the NaaS concept with its key enabling technologies and to present the significant impact of NaaS on the development of Internet architecture and service model. The author hopes to provide readers with a big picture about the NaaS paradigm and its important role in the next generation Internet together with a holistic vision across networking and cloud computing for future service provisioning.

1.1 Introduction

The stunning success of Internet has brought in significant challenges to itself that demand fundamental changes in future network architecture and service models. Rapid development in computer networking in the past decades has transformed the Internet from a network infrastructure providing a limited number of data communication services to a general platform for supporting a wide spectrum of applications with highly diverse service requirements. Such transformation has been further stimulated by the emerging cloud computing, which utilizes networks for remotely delivering various services to end users. However, the current IP-based Internet architecture lacks sufficient capabilities to meet such demands; therefore, various research efforts

[1]Information Sciences and Technology Department, Pennsylvania State University Abington College, Abington, PA, USA

have been made to overcome the ossification of the current Internet in order to support service provisioning required in future networks.

A fundamental strategy taken by the research for the next generation Internet lies in applying the service-orientation principle in the field of networking. This strategy comprises two closely related aspects: (i) network virtualization that decouples the network functions related to service provisioning from network infrastructures for data transport and processing and (ii) service-oriented networking that encapsulates network resources and functionalities into entities that can be exposed and utilized via abstract interfaces. These two aspects of NaaS are interdependent: the decoupling between services and infrastructures enabled by virtualization forms the basis for service-oriented networking while the latter provides a flexible framework for managing the entire life cycle of virtual network functions and services.

The service-oriented strategy for future networking essentially embraces the *as-a-Service* paradigm widely adopted in cloud computing, for example, Infrastructure-as-a-Service (IaaS) and Software-as-a-Service (SaaS), in the networking field; therefore is referred to as *Network-as-a-Service (NaaS)*. The NaaS paradigm is expected to play a crucial role in the next generation Internet with significant changes in both network architecture and service delivery model. The service-orientation principle adopted in both NaaS and cloud computing allows future network designs to fully leverage the successful cloud technologies and enables network services to be delivered through the same model as cloud service provisioning. This evolution trend in networking is called *Network Cloudification*, which then may lead to unification of networking and cloud computing in the next generation Internet.

This chapter gives a high-level overview of the NaaS concept and its key enabling technologies, presents the significant impact of NaaS on the development of Internet architecture and service model, and discusses some possible topics for future research and technology innovation. The author hopes to provide readers with a big picture about the NaaS paradigm and its important role in the next generation Internet together with a holistic vision about networking and cloud computing for future service provisioning.

The rest of the chapter is organized as follows. Section 1.2 provides background information about the Service-Oriented Architecture (SOA). Section 1.3 briefly reviews the evolution path of the NaaS concept in telecommunication and networking technologies. Then the state of the art of two key enabling technologies for NaaS—network virtualization and service-oriented networking—is presented in Sections 1.4 and 1.5. Section 1.6 shows how NaaS may be facilitated by the emerging Software-Defined Networking (SDN) technology. The latest developments in network cloudification and network-cloud unification are presented in Section 1.7, with discussion on some open problems that offer topics for future research and technology innovation in this exciting area.

1.2   The service-oriented architecture

In general, the service-orientation principle advocates that the logic required to solve a large problem can be better constructed, carried out, and managed, if it is decomposed

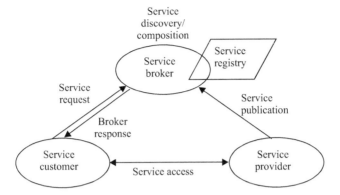

Figure 1.1   Web services-based implementation of SOA

into a collection of smaller units called *services*, which exist autonomously but collaborate with each other. System architecture designed following this principle is referred to the SOA. A service in SOA is a self-contained and platform-independent system module that can be described, published, located, composed, and programmed through a standard interface and messaging protocol. The technologies providing the desired functionality of a service are hidden behind the service interface [1].

A key feature of SOA is loosely coupled interaction among heterogeneous systems in the architecture. The term "coupling" indicates the degree of dependency any two systems have on each other. In loosely coupled interaction, systems need not know how their partners behave or are implemented, which allows systems to connect and interact more freely. Therefore, loose-coupling of heterogeneous systems provides a level of flexibility and interoperability that cannot be matched using traditional approaches for building highly integrated, cross-platform, inter-domain communication environments. Other features of SOA include reusable services, formal contract among services, service abstraction, service autonomy, service discoverability, and service composability. These features make SOA a very effective architecture for heterogeneous system integration with resource virtualization to support diverse application requirements.

Although service-orientation is an architectural principle that may be realized by various technologies, a typical implementation of SOA is based on Web service technologies. A Web service has an interface that describes a collection of operations that are network accessible through standardized XML messaging. Key elements of a Web service-based implementation of SOA include service provider, service broker/registry, and service customer, as shown in Figure 1.1. The basic operations involved in this SOA implementation are service publication, service discovery, service binding/access, and service composition. A service provider makes its service available by publishing a service description at a service registry. The service broker discovers an appropriate service provider in response to a customer request. Then the customer binds with the provider to access the service. Multiple services may be composed into a composite service to meet the customer's requirements.

Representational State Transfer (REST) offers an alternative approach to implementing SOA [2]. REST relies on three main principles: addressability, uniform interface, and statelessness. REST models the datasets to be operated on as resources, which are identified by URIs and accessed via a uniform and standard interface. REST adopts the client-server pattern of the Web and focuses on the concept of resource. Each resource has a representation, which is what the client receives when it sends a request concerning the resource. REST does not restrict client-server communications to a particular protocol, but is most commonly used with HTTP. Each resource is identified by a URI represented by a certain MIME type (such as XML or JSON), and accessed and controlled using POST, GET, PUT, or DELETE http methods. This set of technologies that follow the REST design style for realizing SOA is typically referred to as RESTful Web services.

1.3   Evolution of the NaaS concept in networking technologies

An essential objective of networking is to provide the data communication services needed by upper layer applications for performing their operations. Therefore, telecom systems and computer networks have always been service-driven. The general approach to modeling and specifying a networking system consists of three main steps: starting with description of a service and its capabilities from a high-level user perspective, followed by definition of a functional architecture, and then design of physical elements and corresponding protocols. However, early networking systems were service-specific. Each new service requires development of a corresponding network architecture, describing how functional components interact with each other through certain protocols and structuring the exchange of control and content information to deliver the service to end users [3].

Due to service-specific development of network architecture, each service requires its own infrastructure and functional system that are more or less independent of other services, which leads to a "silo" service mode with minimal capability of reusing and reconfiguring network components. Although such network architecture allows optimal network designs tailored to certain services for providing Quality of Service (QoS), it lacks the ability to offer the services required by various applications and to easily support new services needed by emerging applications. With the rapid development of computer networking into a general platform of data communications for a wide spectrum of computing applications, such service-specific architectural feature becomes an obstacle to the next generation Internet.

Various research efforts have been made in the field of telecom and computer networks in order to enhance service capability for meeting the requirements of future Internet. A key idea behind the various approaches to overcoming the ossification caused by service-specific network architecture and the "silo" service model is to decouple service functions from network infrastructures and encapsulate services through network resource abstraction, which is essentially the concept of NaaS.

Some early attempts toward this direction can be traced back to the 1980s. For example, Intelligent Network (IN) defined an overlay service architecture on top of a physical network and extract service intelligence from switches to centralized service control points. Later on some telecom API standards, such as Parlay, Open Service Architecture (OSA), and Java API for Integrated Network (JAIN), were developed for achieving a similar objective as IN but with easier service development. In early 2000s, a joint group of ETSI and 3GPP developed Paraly X, which employs Web service technology to offer a higher level of abstraction than Parlay/OSA to support service development that is transparent to the details of underlying network implementations.

Next Generation Network (NGN) architecture developed by ITU-T [4] embraces the NaaS concept by separating data transportation and service-related functions into two strata. The service-orientation principle and Web service technologies have been widely employed in NGN, especially in the service stratum to make efficient creation, deployment, execution, orchestration, and management of services [5]. In addition to SOAP-based Web service technologies, RESTful Web service mechanism has also been applied to service provisioning in NGN to enable easy development and deployment of a wide range of services [6].

A Service Delivery Platform (SDP) is a framework that facilitates and optimizes all aspects of service delivery that aggregates network service operation and management functions in a common platform. Representative SDP specifications include OMA Open Service Environment (OSE) [7] and TM Forum Service Delivery Framework (SDF) [8]. Both specifications embrace the NaaS concept by defining a set of service components (called service enablers) that can be composed to form new services. Web service technologies have become a de facto communications standard among system components for supporting service delivery and composition.

The Next Generation Service Overlay Network (NGSON) [9] developed by IEEE aims to bridge the service layer and transport network over IP infrastructure in order to accommodate highly integrated services. Key functional entities of NGSON service control include service discovery and negotiation, service routing, and service composition. Web service technologies have been employed to realize these functional entities [10,11].

Various network research projects followed the service-orientation principle in their network architecture designs. A representative project is UCLPv2 (User Controlled Light Path), in which Web service technologies were employed to expose resources in optical transport networks as SOA-compliant services to enable user control and management of network infrastructure [12]. Such a service-oriented framework of network infrastructure was adopted in a few other projects, including Argia as a commercial implementation for optical network infrastructure services, Ether for developing Ethernet and MPLS infrastructure services, and MANTICORE for supporting logical IP network as services [13]. NaaS has also been adopted by industry in various networking equipment and solution developments. For example, the Service-Oriented Network Architecture (SONA) developed by Cisco based on the IaaS paradigm forms a cornerstone of its architectural approach to network system design [14].

SDN and NFV are two recent significant innovations in the field of networking that are expected to become key components in future networks. Researchers have started integrating the NaaS concept with the SDN and NFV paradigms to address the challenges to service provisioning in next generation Internet. A typical application of NaaS in SDN is to encapsulate SDN controller as a service and expose its control functionalities via an abstract service interface, e.g. RESTful API, to enable various SDN applications to work with heterogeneous controllers. For example, the work on OpenFlow-as-a-Service reported in [15] designs an OpenFlow controller as a plugin for OpenStack's Quantum component to enable network connectivity as a service in a cloud computing environment. A NaaS-based SDN control platform was proposed in [16] to support end-to-end service provisioning in inter-domain SDN through high-level abstraction of SDN controllers enabled by the service-orientation principle. NaaS has also been employed in the NFV architecture for service provisioning in virtualization-based networks. Representative NaaS-based service models of NFV include Network Function Virtualization Infrastructure-as-a-Service (NFVIaaS), Virtual Network Function-as-a-Service (VNFaaS), and Virtual Network Platform-as-a-Service (VNPaaS), which have been identified by NFV-ISG as main NFV use cases [17].

1.4   Virtualization in networking

A key requirement for realizing NaaS is to decouple network service capabilities from the underlying network infrastructures, so that various services may be developed and deployed upon a common infrastructure substrate without being constrained by any specific implementation technology employed in the network infrastructure. Virtualization offers an effective approach to achieving such decoupling thus playing a crucial role in NaaS.

In general, virtualization means separating a service or application from the underlying physical delivery of that service or application. Essentially, virtualization abstracts physical resources as virtual instances and enables multi-tenant access to the resources. Virtualization in computing, including server, storage, and I/O interface virtualization, has been widely employed in data centers as a technical foundation for provisioning cloud services.

1.4.1   Network virtualization

Virtualization technologies have already been employed in networking, for example, in the form of virtual local area network (vLAN) and virtual private network (VPN). Virtualization has also been adopted as an approach to constructing network experiment facilities, for example, Planet Lab [18] and GENI [19]. In order to fully take advantage of virtualization to overcome ossification of the current Internet architecture, the *pluralist* view of network architecture proposed in [20] advocates adoption of virtualization as a key attribute of future networks that allows multi-tenant virtual networks with alternative network architectures to coexist upon shared network infrastructure. Such an architectural vision for future networking is typically referred to as Network Virtualization (NV).

Figure 1.2 depicts a general layered architecture for NV. The layer of network infrastructure comprises physical network resources including nodes and links.

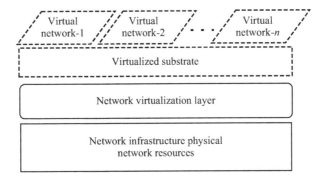

Figure 1.2   Network virtualization architecture

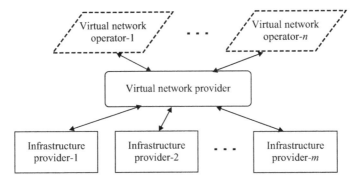

Figure 1.3   Key players in a network virtualization environment

The virtualization layer abstracts physical infrastructure as virtualized substrate, upon which virtual networks can be created and operated. A virtual network is a collection of virtual nodes that are interconnected through virtual links to form a virtual topology. The virtualization layer is responsible for managing the mapping between physical and virtual network resources and providing isolation between multi-tenant virtual networks.

NV decouples upper layer functionalities for service provisioning from the underlying network infrastructures for data transport and process therefore splitting the traditional role of Internet Service Providers (ISPs) to two independent entities: Infrastructure Providers (InPs) who manage the physical infrastructure and Service Providers (SPs) who establish and operate virtual networks by leasing resources from InPs to offer end-to-end services. The role of SPs may be further divided to Virtual Network Providers (VNPs) and Virtual Network Operators (VNOs). A VNP constructs virtual networks for meeting the requests from VNOs while VNOs are responsible for the actual operations of virtual networks for service delivery. The key players in an NV environment and their interaction are shown in Figure 1.3.

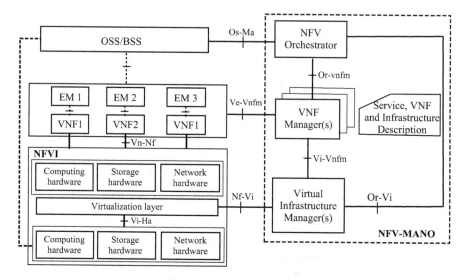

Figure 1.4   ETSI NFV architectural framework

The decoupling between service provisioning and physical infrastructures enabled by NV may greatly facilitate realization of the NaaS model. On the one hand, SPs (including VNPs and VNOs) are able to provision network services upon an abstract view of physical infrastructure; and on the other hand, InPs obtain service-oblivious freedom to implement their infrastructures. NV allows independent evolutions of service functions and infrastructure technologies.

1.4.2   Network function virtualization

The latest innovation in networking that employs virtualization technologies for enhancing network operation and service provisioning is Network Function Virtualization (NFV). The objective of NFV is to address some fundamental challenges to networking by leveraging standard IT virtualization technologies to consolidate many network equipment types onto industry standard servers, switches, and storage [21].

ETSI NFV Industry Specification Group (NFV-ISG) has developed an architectural framework for NFV, which is depicted in Figure 1.4. The NFV architecture comprises NFV Infrastructure (NFVI), Virtualization Network Functions (VNFs), and NFV Management and Orchestration (MANO). The NFVI consists of all infrastructure resources, including both hardware and software components, that build up the environment in which VNFs may be deployed, managed, and executed. VNF is the software implementation of a network function that is capable of running over the NFVI. The MANO component is responsible for management and orchestration of physical and virtual resources in NFVI and lifecycle management of VNFs. The MANO consists of Virtualized Infrastructure Manager (VIM), VNF Manager (VNFM), and NFV Orchestrator (NFVO) [22].

NFVI comprises three functional domains: the compute domain, the hypervisor domain, and the infrastructure network domain. The compute domain provides computational and storage resources for hosting VNFs. The hypervisor domain mediates resources of the compute domain to VNF software. The infrastructure network domain is responsible for providing the required connectivity to support communications among VNFs. A VNF is an abstract entity that allows network function software to be defined and designed, and a VNF Instance (VNFI) is the runtime instantiation of the VNF. Each VNF is managed by an Element Manager (EM). The VIM, VNFM, and NFVO in the MANO component are in charge of the life cycle management and orchestration functions respectively for infrastructure, virtual functions, and network services.

The NFV architecture fully embraces the NV principle, which centers on applying virtualization in networks for decoupling service functions from physical infrastructures; therefore, NFV provides a specific architecture and related mechanisms for realizing the NV notion. On the other hand, NFV and NV focus on different scopes and granularity levels of virtualization in networking. NV focuses on network-level virtualization to enable multiple virtual networks with alternative network architectures for meeting diverse service requirements. NFV focuses on virtualization of individual network functions and provides end-to-end services through orchestration of VNFs [23].

1.5   Service-oriented networking

1.5.1   General architecture for service-oriented networking

The other key aspect of service-orientation in networking focuses on abstraction of network resources and capabilities, encapsulation network functionalities, and composition of network subsystems for service provisioning, which is referred to as Service-Oriented Networking (SON) in this chapter. Virtualization in networking decouples network functions for service provisioning from the underlying infrastructures for data transport and process; thus forming the separated service stratum and infrastructure stratum. Service-oriented abstraction, encapsulation, and composition are applied to both virtual network functions on the service stratum and physical infrastructure resources on the infrastructure stratum. Like in the cloud computing environment, where virtualization and service-orientation of computing resources enable various service models such as IaaS, SaaS, and PaaS, virtualization and service-orientation applied in the networking field introduce multiple service models including NFVIaaS, VNFaaS, VNaaS, and VNPaaS.

A layered architectural framework for service-orientation in networking is given in Figure 1.5. The infrastructure layer is at the bottom of the framework. This layer is essentially the NFVI layer in the NFV architecture that comprises computing resources (including processing capacity, storage space, and hypervisors) as well as network resources, which form an environment in which VNFs can execute. This layer may consist of multiple infrastructure domains owned and operated by different InPs. Network resources in an infrastructure domain can be encapsulated into

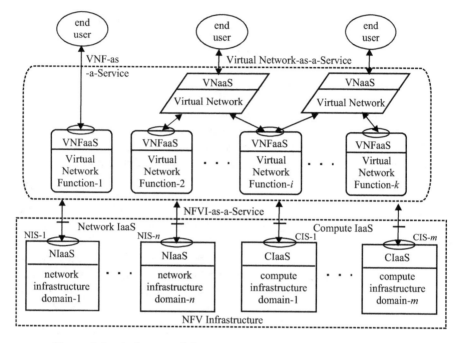

Figure 1.5   Architectural framework for service-oriented networking

Network Infrastructure Services (NISs) and provided by the InP to SPs through a Network Infrastructure-as-a-Service (NIaaS) model. The computing resources in the infrastructure layer can be exposed by following the IaaS model; that is, Compute Infrastructure as a Service (CIaaS). NIaaS plays the same role in network virtualization as IaaS does in a cloud environment, but focusing on networking capabilities instead of general computational resources in the infrastructure. Since NFVI comprise both compute and network domains, NIaaS and CIaaS are combined into NFVI Service (NFVIS) thus realizing a service model NFVI-as-a-Service (NFVIaaS).

The NISs and CISs provided by NFVI can be discovered, selected, and utilized for supporting VNFs. The NFVIaaS model can be used by a network operator for deploying VNFs upon its own NFVI, which is similar to the IaaS model in private cloud deployment mode. In more general cases, a virtual network provider does not own physical network infrastructure, therefore must lease physical resources from an InP to run its VNF instances. NFVIaaS provides a perfect approach to accomplishing this objective and in such cases the NFVIaaS model allows an NFVI to be offered as a public cloud infrastructure.

Applying service-orientation to virtual resources on the service stratum allows VNFs to be encapsulated and provided as SOA-compliant services—VNF-as-a-Service (VNFaaS). The functionality of a VNF may be made available to a customer through VNFaaS, which is comparable to SaaS in clouding computing. For example, with rapid adoption of cloud services, many enterprises have migrated applications to either enterprise data centers or pubic clouds. Instead of asking enterprises to build network infrastructures for deploying some new functions, a network operator may provide the network functions as services that can be utilized by enterprises via the VNFaaS model.

In addition to being offered as a single service, a VNF instance can be selected by a virtual network provider as a building block for constructing a VN, in which a collection of VNF instances are combined for service provisioning. In such cases, VNF services are composed to form a VN, which can then be provided to end users as a service; i.e. Virtual Network-as-a-Service (VNaaS). A VN can be specified first at an abstract level using a VNF Forwarding Graph (VNF-FG), which in general is an order graph that specifies a set of VNFs involved in the VN and the relationship between these VNFs. Then the NFVO in MANO acts as a broker to discover the required VNF services and as an orchestrator to compose these VNFs to instantiate the virtual network service. Essentially, VN construction in the NaaS paradigm takes a service composition process to combine a set of individual VNF service instances into a VNaaS.

In order to facilitate construction of VNs, InPs can make available a suite of infrastructure functions and tools as a platform on which VNs customized for different application requirements can be easily deployed. NFV supports such a use case by providing a toolkit of NFVI with some VNFs as a platform for creating VNs and offers the platform as a service, i.e. VN Platform-as-a-Service (VNPaaS).

1.5.2 Resource description and discovery in NaaS

A key technology for realizing the NaaS architecture depicted in Figure 1.5 is Resource Description and Discovery (RDD), which makes network resources and functionalities available as services that can be selected for hosting VNF/VNs. The multiple-level abstractions in the NaaS architecture required RDD for both physical infrastructure resources (IRDD) and virtual network resources (VRDD).

The main goal of Infrastructure RDD is to provide VNPs with sufficient information about the infrastructure resources for choosing the appropriate NFVIS (or a set of NFVISs) for hosting VNFs/VNs. In general, information about network infrastructure resources includes two main aspects—network topology information that shows how network nodes are interconnected through links; and network capability information such as node processing capacity and link bandwidth. A particular challenge to IRDD in NaaS is to achieve a balance between the exposition and aggregation of network resource information. On the one hand, VNPs need to have enough information about infrastructure resources to make optimal selection of NFVISs. On the other hand, aggregation of resource information is required by NaaS to enable abstraction and encapsulation of network resources as services. Therefore, the virtualization layer should provide an appropriate level of resource abstraction for supporting VNFs/VNs without exposing detailed information of the infrastructure substrate.

Various approaches have been proposed for enabling IRDD. For example, Network Description Language (NDL) [24] is an ontology designed based on Resource Description Framework (RDF) for describing network elements and topologies. In order to facilitate resource abstraction, Network Resource Description Language

(NRDL) proposed in [25] focuses on describing the interactions among network elements rather than individual network objects like switches and links. In [26], the authors developed an approach to describing network resources by first presenting network topology as a full-mesh consisting only of service end nodes and then associating the network connectivity between each pair of end nodes with QoS metrics such as bandwidth and delay. A capability matrix was developed in [27] for describing network capabilities using a capability profile for each pair of ingress and egress nodes. The service curve concept in network calculus theory is employed to make the general service capability profile agnostic to infrastructure implementations. NFVI comprises computing as well as networking domains thus requires RDD applicable to heterogeneous resources. Toward this direction, the authors of [28] proposed extension of NDL with a more powerful ontology defined in Web Ontology Language (OWL) for describing various computing capabilities as well as networking resources.

Research proposals have also been made to support description and discovery of virtual network resources. For example, Virtual eXecution Infrastructure Description Language (VXDL) proposed in [29] is a language for describing virtualized infrastructure resources focusing on computing resources interconnected by network links. The resource description and discovery framework developed in the 4WARD project provides a schema for describing virtual resource properties and their relationship [30].

In NFV architecture, the MANO component is responsible for RDD as part of VNF/VN life cycle management. ETSI NFV-ISG has developed an information model for MANO that facilitates RDD. This information model comprises elements in two main categories: (i) information contained in descriptors, which are deployment templates that provide relatively static information used for deploying VNFs and VNs and (ii) information residing in records, which contain more dynamic run-time data representing VNF and VN instances. The main information elements specified by NFV-ISG in [31] are listed as follows.

Network Service Descriptor (NSD) is a deployment template for the service provided by a VN, which refers all other information elements describing the constituent components of the VN service. VNF Descriptor (VNFD) describes the deployment and operation behavior requirements for a VNF. A VNF FG Descriptor (VNFFGD) describes the topology of a VN by referencing the VNFs involved in the VN and the virtual links connecting these VNFs. A Virtual Link Descriptor (VLD) describes the resource requirements needed for a virtual link. VNF and VN instantiations will create records for representing the newly created instances. Such records include Network Service Record (NSR), VNF Record (VNFR), Virtual Link Record (VLR), and VNF FG Record (VNFFGR). NFV Instances Repository holds information of all VNF instances and network service instances. NFVI Resource Repository holds information about available, reserved, and allocated NFVI resources as abstracted by the VIM across infrastructure domains.

1.5.3 Network service composition in NaaS

Another key enabling technology for the NaaS architecture is network service composition (NSC) that allows VNPs to combine multiple NFVISs to host VNF services and/or compose a collection of VNF instances to form a VN for end-to-end service provisioning.

Network service composition has been actively studied along the evolution of the NaaS concept in networking. For example, OMA OSE supports service composition through a mechanism defined as policy enforcer and adopts WSBPEL as a scheme to express policy for service orchestration. A general framework for dynamic composition of network services for end-to-end information transport was proposed in [32]. Architecture and technologies for realizing dynamic composition of web-based telecom services was developed in [33]. Loosely coupling service composition mechanisms have been adopted in NGSON for integrating communication and data services [34]. Network Functional Composition (NFC) has been proposed as a "clean slate" approach to future network design, which decomposes the layered network stack to functional blocks that can be organized in a composition framework for meeting specific application requirements [35].

Network service orchestration plays a crucial role in the NFV paradigm. The NFVO in the NFV architecture is in charge of orchestration and management of NFV infrastructure resources and virtual functions for realizing network services. NFVO has two main responsibilities: resource orchestration that orchestrates the NFVI resources across multiple VIMs and service orchestration that manages life cycles of network services. These two responsibilities of NFVO respectively focus on orchestration on the physical infrastructure stratum and virtual service stratum in the NaaS architecture. Although both aspects are included in the same NFV-ISG specification [31], they should be realized as two independent entities interacting with each other through a standard abstract interface.

NFVO functionalities for network service orchestration include registering and on-boarding network services; instantiating network services; scaling network services by adjusting service capacities; updating network services by changing service configurations; creating, deleting, querying, and updating VNF-FGs associated to network services; and terminating network services.

NFVO responsibilities for resource orchestration include validation and authorization of requests for NFVI resource allocation; NFVI resource management across infrastructure domains, including distribution, reservation, and allocation of NFVI resources to VNF/VN instances; policy management and enforcement for VNF/VN instances; and collecting information about usage of NFVI resources by VNF instances

Network Service Chaining (NSC), which defines and instantiates an ordered set of network service functions and then performs traffic steering through the chained functions, also offers an effective approach to service composition in NaaS. NSC has received considerable attention in the standardization and research communities lately. IETF has developed an architecture for Service Function Chaining (SFC) that particularly focuses on enabling automatic and dynamic service chaining independently with the underlying infrastructure implementations [36].

It is worth noting that both service orchestration and service chaining may achieve network service composition but subtle difference exists between these two composition approaches. Service orchestration is to arrange and coordinate multiple services to expose their functionalities and capabilities as a single aggregated service to end users. The control logic is specified by a single entity typically called the orchestrator. In contrast, service chaining does not assume a centralized controller but relies on message exchanges between participating service components to realize the composite service. An example of service orchestration in a networking environment is Traffic Engineering (TE) in which a TE application (usually deployed upon a network controller) steers traffic through optimal paths across a network. An example of service chaining is the routing service in which participating routers exchange routing messages and collaborate on packet forwarding from a source to a destination. However, with the significant changes in network architecture introduced by SDN and NFV the subtle difference between service orchestration and service chaining is going away, and the two terms now are often used exchangeably. For example, ONF proposed an SDN-based SFC architecture with an orchestrator serving as a central controller [37].

1.6   Software-defined network control and management for NaaS

The NaaS paradigm will enable much more dynamic and flexible networking environments for service provisioning, which requires sophisticated control and management capabilities that cannot be easily provided by traditional networking technologies. The emerging SDN technology offers a promising approach to supporting virtualization and service-orientation in networking, thus may greatly facilitate realizing NaaS.

SDN is defined by Open Networking Foundation (ONF) as an emerging network architecture where network control is decoupled from data forwarding and is directly programmable [38]. A key objective of SDN is to provide open interfaces that allow software applications to define the data forwarding and processing operations performed by the underlying network. The architectural principles followed by SDN are decoupling network control and management from data forwarding and processing; logically centralized network control and management; and programmability through open interfaces for supporting diverse network services.

A fundamental idea for SDN lies in the notion of resource abstraction, which is an important capability to support network programmability. The centralized SDN controller provides a global abstract view of the underlying network resources, upon which SDN applications may program network behaviors. Figure 1.6 depicts a general architectural framework for SDN, which consists of three planes—the data plane, control plane, and application plane; and two interfaces—the interface between data and control planes (D-CPI) and the interface between application and control planes (A-CPI).

The data plane comprises network resources for performing data forwarding and processing operations. Network devices on the data plane are simply packet forwarding and/or processing engines without complex control logic to make autonomous decisions. The D-CPI, which is also referred to as the Southbound Interface (SBI), allows data plane devices to expose the capabilities and states of their

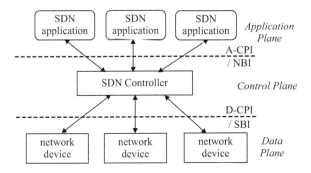

Figure 1.6   A general architectural framework for SDN

resources to the control plane and enables the controller to instruct network devices for their operations. The control plane presents a global view of data plane infrastructure to SDN applications and provide a centralized control platform through which applications may define the operations performed by data plane devices. The A-CPI, which is also called the Northbound Interface (NBI), provides a standard API through which applications can program the underlying network infrastructure.

The network programmability provided by the centralized control platform in SDN may greatly facilitate realization of NaaS by supporting network virtualization and service-oriented networking. As a hypervisor plays a key role in computing virtualization, the network virtualization layer acts as a "network hypervisor" for virtualizing network infrastructure resources. An SDN controller provides a global view of an entire network domain; thus offering a perfect platform upon which a network hypervisor may be realized. Compared to the distributed control in traditional IP networks, the centralized and programmable SDN control platform may greatly simplify implementation of a network hypervisor.

The A-CPI on an SDN controller offers a perfect interface mechanism for encapsulating and exposing network resources as services. RESTful APIs have become a prevalent choice for SDN NBI. For example, Floodlight as one of the most popular open source SDN controller designs provides a built-in virtual network module that exposes a REST API. Meridian, implemented as a module inside Floodlight, provides a REST API for managing virtual networks [39]. The Application Layer Traffic Optimization (ALTO) protocol also supports a REST NBI between an SDN controller and the applications running upon it [40]. The widely available RESTful SDN A-CPI designs provide the basis of implementing service-orientated networking upon the SDN control platform for realizing NaaS.

A framework of SDN-based NFVIaaS implementation is shown in Figure 1.7. In this framework, the NFVI comprises multiple infrastructure network domains as well as compute domains. Each network domain is controlled by an SDN controller that supports a REST NBI. Therefore, the networking capabilities of each infrastructure network domain can be exposed as network infrastructure services (NISs) through the REST NBI on the SDN controller of this domain. Similarly, the compute and storage resources in NFVI can be abstracted as compute infrastructure services

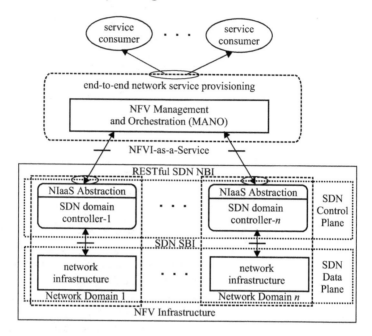

Figure 1.7   SDN-based framework for NFVIaaS implementation

(CISs), for example using OpenStack components such as Nova. NISs together with CISs are provided by NFVI to realize the NFVIaaS model. The service orchestration function provided by NFV MANO can orchestrate the NISs and CISs to form virtual network for end-to-end services provisioning.

1.7   Network cloudification and network-cloud unification

1.7.1   Network cloudification

The NaaS paradigm is expected to make significant changes in the architecture and service delivery model of future networks. Virtualization will become a key architectural attribute in future network designs, where virtual network functions for service provisioning will be decoupled from physical infrastructures for data transport and processing. This change in network architecture allows networking systems to take essentially the same design principle as the virtualization-based design for cloud computing systems. Service-oriented networking enables network resources, including both physical infrastructure capabilities and virtual network functions, to be abstracted and exposed as SOA-compliant services. This change in network service model essentially adopts the cloud IaaS and SaaS models in the network domain thus allowing network services to be delivered as cloud services. Therefore, NaaS enables a trend for implementing networking systems using cloud technologies and delivering network services as cloud services, which can be referred to as *Network Cloudification*.

The notion of network cloudification has attracted extensive research interest from both academia and industry. Stimulated by the rapid development in the fields of NFV, SDN, and cloud computing, researchers are conducting active study on this topic and making exciting progress.

A representative work on implementing networking systems using cloud technologies is Central Office Rearchitected as Data Center (CORD) [41], a joint research project between Open Networking Lab (ONL) and AT&T. This project aims to redesign telecom central office by employing the latest technologies for building data centers, including SDN, NFV, and cloud technologies integrated together. CORD infrastructure consists of a cluster of commodity servers interconnected via an SDN-based fabric network. Network functions of a telecom central office, which were typically provided by purpose-built network appliances, now are realized as more agile VNFs running on VMs hosted by the servers. Resource allocation and management on the servers for hosting VNFs are performed by OpenStack. The fabric network comprises OpenFlow switches controlled by an SDN operating system ONOS. The services provided by OpenStack and ONOS are composed by a service orchestration module called XOS in order to provision various services required by users of the central office.

A research project that fully adopts service-orientation in network architecture design is the EU FP7 project T-NOVA (Network Functions as a Service over Virtualized Infrastructure) [42]. This project aims to design and implement a management/orchestration platform for automated provision, configuration, monitoring, and optimization of Network Functions-as-a-Service (NFaaS) over virtualized network and compute infrastructures, which enables VNFs to be offered on-demand as value-added services to their customers. T-NOVA introduces an innovative *Network Function store* following the paradigm of already successful OS-specific App stores. Through an NF store, T-NOVA establishes an NFV *marketplace* in which network services and functions offered by developers/providers can be published and brokered/traded. Via the marketplace, customers can discover and select the services and virtual functions which best match their needs, as well as negotiate the associated SLAs and be charged under various billing models. The NF store and marketplace in T-NOVA essentially adopts the service registry and broker components of SOA in the NFV architecture thus enabling publication, discovery, selection, and composition of VNF services.

Wireless mobile network is a particular field of networking where cloud technologies have been actively exploited. Research on employing cloud technologies, including virtualization and service-orientation, to enhance mobile network operation and improve mobile service performance began even before ETSI NFV activity and has been making progress since then. For example, in the Cloud-based Radio Access Network (C-RAN) architecture proposed by China Mobile in 2011, baseband process functions are implemented as software applications and consolidated into a cloud data center [43]. The Mobile Cloud Networking (MCN) project develops an end-to-end cloudified infrastructure for mobile network that enables telco capabilities to be deployed and ran on virtualized infrastructure, delivered as services, and composed into a larger service offerings using cloud technologies [44].

1.7.2   Unification of networking and cloud computing

With the emerging trend of network cloudification, the boundary between networking and cloud computing technologies that used to be two relatively independent fields now becomes blurry. The interdependency between clouds and networks can be viewed from two perspectives.

Network-based cloud computing: Networking systems become an indispensable ingredient of cloud infrastructure for supporting cloud service provisioning. Data center networks play a key role in the cloud infrastructure for providing connectivity among servers upon which cloud applications are running. Research results have indicated that in many cases networking is a key impact factor for service performance of a cloud data center. The emergence of inter-cloud federation requires wide area networks to be part of cloud infrastructures that comprising multiple interconnected data centers located at different geographical sites.

Cloud-based networking: As reviewed in the previous subsection, NaaS essentially introduces key cloud technologies—virtualization and service-orientation—into networking; therefore, cloud computing is expected to play a crucial role in future networks. The infrastructure stratum of the future Internet architecture will consists of computing resources as well as networking devices. Network functions will be virtualized as software applications running on VMs as cloud applications and delivered to end users via the same service model as for cloud services.

Therefore, network cloudification will enable unification of networking and cloud computing in the next generation Internet, which will provide functions for data communication, processing, and storage with unified architecture and deliver composite network-cloud services to end users.

A layered architectural framework proposed in [45] for NaaS-based unification of networking and cloud computing is depicted in Figure 1.8. In this framework, resources in both networking and computing infrastructures are virtualized into services by following the same service-orientation principle, which offers a uniform mechanism for coordinating networking and computing systems for end-to-end service provisioning. Service-oriented virtualization in NaaS enables a holistic vision of both networking and computing resources as a single collection of virtualized and dynamically provisioned resources, which allows federated control, management, and optimization of resources across the networking and computing domains.

NaaS-based service composition allows network and cloud services, which used to be offered separately by different providers, to be unified into composite network-cloud services. Composition of network and compute services greatly expands the spectrum of Internet services that can be offered to customers. In addition, network-cloud unification enables a new service delivery model in which the roles of traditional Internet service providers and cloud service providers merge together into one role of composite network-cloud service providers. This new service model may greatly stimulate innovations in service development and provide a wide variety of new business opportunities.

Inspired by the notion of NaaS-based network-cloud unification, researchers recently started various projects toward this direction. Unifying Cloud and Carrier

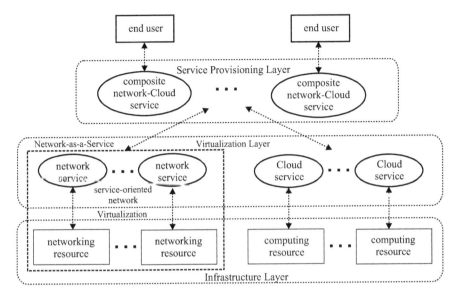

Figure 1.8   NaaS-based unification of network and Cloud services

Network (UNIFY) is an EU FP7 project for realizing network-cloud unification in order to achieve much more flexible service provisioning and efficient operation in future Internet [46]. UNIFY envisions a unified network-cloud service environment that spans from home/enterprise networks through aggregation/core networks up to cloud data centers. In order to achieve this objective, the UNIFY architecture combines service-oriented abstraction of compute, storage, and network resources to realize logically centralized, automated, and recursive resource orchestration across networking and computing domains.

Service composition across the networking and cloud computing domains plays a crucial role in NaaS-based network-cloud unification; thus has formed an active research area where encouraging progress has been made recently. For example, an approach to network-aware cloud service composition was proposed in [47]. This approach optimizes service composition decisions by taking into account network resource consumption and end-to-end QoS constraints among other decision-making factors. In [48], the authors formulate QoS-aware composition of network and cloud services as a variant of multi-constrained optimal path problem and propose an efficient approximate algorithm to solve the problem for obtaining the composite network-cloud service that meets multiple constraints for end-to-end QoS performance.

Although researchers have started making exciting progress toward network-cloud unification, this area is still at an infant stage with a lot of open problems that must be thoroughly studied before the notion can be fully realized. Some of the most challenging issues are briefly discussed in the following paragraphs, which provide some interesting topics for future research and technology development.

Implementing network functions as software instances running on cloud infrastructure is challenging. Network functions are often complex functions with highly specialized operations and very stringent capacity and performance requirements. Traditional network function designs assume purpose-built network appliances instead of general-purpose commodity servers as their execution platform. Therefore, network cloudification requires redesign of such non-cloud-native network functions by considering the key characteristics of cloud infrastructures, such as loose-coupled architecture and asynchronous non-blocking communication patterns. In addition, how to realize network functions as virtualized software instances and still achieve a comparable level of performance as what dedicated network appliances guarantee is also a challenging problem.

Resource management in unified network-cloud service environments, especially allocation of physical resources for hosting virtual functions, is an open issue that deserves more thorough investigation. Network resource allocation for supporting virtual networks and computing resources management in data centers have been widely studied respectively in the contexts of virtual network embedding and cloud control/management. However, network-cloud unification demands federated management of both networking and computing resources to achieve an optimal balance between resource utilization and service performance guarantee. Therefore, resource management for network-cloud unification brings in new challenges that must be fully addressed by future research.

NaaS-based network-cloud unification requires effective mechanisms for service publication, discovery, and selection. Although various technologies have been developed for implementing the SOA, including both conventional Web services and RESTful schemes, applicability of such technologies to a unified network-cloud service environment needs to be fully examined. For example, light-weight RESTful API seems to be a good choice for meeting the real-time requirements of many network services; however, lack of automatic service discovery and composition mechanisms may significantly limit application of RESTful schemes in the large-scale dynamic networking environment required for network-cloud unification.

Service composition across networking and cloud computing domains plays a crucial role in NaaS-based network-cloud unification. Although service composition has been extensively studied in the field of Web services and cloud computing, Web/cloud service composition approaches must be applied to network services with caution due to distinction between the two types of services; for example, network services are highly event-driven and often require stateful implementations. New challenges brought in by network-cloud unification for unified composition across network and cloud services include scalability for supporting numerous service consumers in large scale networks , flexibility for meeting requirements of a wide spectrum of existing and emerging services, and agility for being adaptive to fluctuation in both resource availability and traffic loads.

1.8   Conclusion

The Network-as-a-Service (NaaS) paradigm essentially applies the service-orientation principle, which has been widely adopted in Web and cloud areas, in the field of networking in order to enable some fundamental changes in network architecture and service delivery model for meeting the requirements of the next generation Internet. Key enabling technologies for NaaS include: (i) network virtualization that decouples service-related network functions from physical network infrastructures and (ii) service-oriented networking that encapsulates network resources and exposes network functionalities as SOA-compliant services. Although the idea of NaaS has been explored since earlier stage of telecom and networking research, recent innovations in networking technologies such as Network Function Virtualization (NFV) make NaaS a key component for future networks. The flexible and agile networking environment enabled by NaaS calls for more advanced network control and management, which may be greatly facilitated by the emerging SDN technology. The NaaS paradigm leads to a trend of network cloudification, which allows networking systems to be implemented using cloud technologies and network services to be delivered as cloud services. NaaS-based network cloudification offers a promising approach to enabling unification of networking and cloud computing, which may significantly enhance service provisioning in the next generation Internet. Although exciting progress has been made toward network cloudification and network-cloud unification, this area is still on an infant stage. Many challenging problems must be thoroughly studied before this vision can be fully realized; thus offering a lot of interesting topics for future research and technology innovation.

References

[1]   T. Erl, *Service-Oriented Architecture—Concepts, Technology, and Design.* Englewood Cliffs, NJ: Prentice Hall, 2005.

[2]   M. Lanthaler and C. Gutl, "Toward a RESTful service ecosystem," in *Proceedings of the Fourth IEEE International Conference on Digital Ecosystems and Technologies*, pp. 209–214, Dubai, UAE, Apr. 2010.

[3]   T. Magedanz, N. Blum, and S. Dutkowski, "Evolution of SOA concepts in telecommunications," *IEEE Computer Magazine*, vol. 40, no. 11, pp. 46–50, 2007.

[4]   ITU-T, "Recommendation Y.2012: Functional Requirements and Architecture of the NGN Release 1," Sep. 2006.

[5]   G. Branca and L. Atzori, "A survey of SOA technologies in NGN network architectures," *IEEE Communications Surveys & Tutorials*, vol. 14, no. 3, pp. 644–661, 2012.

[6]   F. Belqasmi, R. Glitho, and C. Fu, "RESTful Web services for service provisioning in next-generation networks: a survey," *IEEE Communications Magazine*, vol. 49, no. 12, pp. 66–73, 2011.

[7]   OMA, "Open Service Environment version 1.0," available at http://www.openmobilealliance.org/technical, Oct. 2009.

[8]   TMForum, "TMF061 Service Delivery Framework (SDF) Reference Architecture," Jul. 2009.

[9]   IEEE, "Standard 1903: Functional Architecture of the Next Generation Service Overlay Networks," Oct. 2011.

[10]   C. Makaya, A. Dutta, B. Falchuk, *et al.*, "Enhanced next-generation service overlay networks architecture," in *Proceedings of the 2010 IEEE International Conference on Internet Multimedia Systems Architecture and Application*, pp. 1–6, Bangalore, India, Dec. 2010.

[11]   C. Makaya, B. Falchuk, D. Chee, *et al.*, "Service composition based on next-generation service overlay network architecture," in *Proceedings of the 2011 IFIP International Conference on New Technologies, Mobility, and Security*, pp. 1–6, Paris, France, Feb. 2011.

[12]   E. Grasa, G. Junyent, S. Figuerola, A. Lopez, and M. Savoie, "UCLPv2: a network virtualization framework built on Web services," *IEEE Communications Magazine*, vol. 46, no. 3, pp. 126–134, 2008.

[13]   S. Figuerola and M. Lemay, "Infrastructure services for optical networks," *IEEE/OSA Journal of Optical Communications and Networks*, vol. 1, no. 2, pp. A247–A257, 2009.

[14]   Cisco White Paper, "Using infrastructure service orchestration to enable a service-oriented architecture," 2009. http://www.cisco.com/c/en_ae/solutions/enterprise/sona.html

[15]   F. Hsu, S. Malik, and S. Ghorbani, "OpenFlow as a Service," *Technical Report of University of Illinois at Urbana-Champaign*, vol. 21, 2014.

[16]   Q. Duan, "Network-as-a-Service in software-defined networks for end-to-end QoS provisioning," in *2014 23rd Wireless and Optical Communication Conference (WOCC)*, pp. 1–5, IEEE, Newark, NJ, 2014.

[17]   NFV-ISG, "Network Function Virtualization: Use Cases v1.1.1," Oct. 2013.

[18]   T. Anderson, L. Peterson, S. Shenker, and J. Turner, "Overcoming the Internet impasses through virtualization," *IEEE Computer Magazine*, vol. 38, no. 4, pp. 34–41, 2005.

[19]   GENI-Planning-Group, "GENI design principles," *IEEE Computer Magazine*, vol. 39, no. 9, pp. 102–105, 2006.

[20]   J. Turner and D. E. Taylor, "Diversifying the Internet," in *Proceedings of the 2005 IEEE Global Communications Conference*, pp. 755–760, St. Louis, MO, Nov. 2005.

[21]   NFV-ISG, "Network Functions Virtualization – an introduction, benefits, enablers, challenges and call for action," in *Proceedings of SDN and OpenFlow World Congress*, Darmstadt, Germany, Oct. 2012.

[22]   NFV-ISG, "NFV 001: Network Function Virtualization (NFV) Architectural Framework v1.2.1," Dec. 2014.

[23]   Q. Duan and M. Toy, *Virtualized Software-Defined Networks and Services*. Norwood, MA: Artech House, 2017.

[24]   J. Ham, P. Grosso, R. Pol, A. Toonk, and C. Laat, "Using the network description language in optical networks," in *Proceedings of the 10th IFIP/IEEE International Symposium on Integrated Network Management*, pp. 199–205, Munich, Germany, May 2007.

[25]   A. Campi and F. Callegai, "Network resource description language," in *Proceedings of the 2009 IEEE Global Communication Conference (GLOBECOM'09)*, Honolulu, HI, Dec. 2009.

[26] C. E. Abosi, R. Nejabati, and D. Simeonidou, "A novel service composition mechanism for the future optical internet," *Journal of Optical Communications and Networking*, vol. 1, no. 2, pp. A106–A120, 2009.

[27] Q. Duan, "Network service description and discovery for high-performance ubiquitous and pervasive grids," *ACM Transactions on Autonomous and Adaptive Systems*, vol. 6, no. 1, pp. 3:1–3:17, 2011.

[28] I. Baldine, Y. Xin, A. Mandal, *et al.*, "Networked Cloud orchestration: a GENI perspective," in *Proceedings of the Second IEEE Workshop on Management of Emerging Networks and Services*, pp. 573–578, Miami, FL, Dec. 2010.

[29] G. P. Koslovski, P. V.-B. Primet, and A. S. Charao, "VXDL: virtual resources and interconnection networks description language," in *Proceedings of the Second International Conference on Networks for Grid Applications*, Beijing, China, Oct. 2008.

[30] I. Houidi, W. Louati, D. Zeghlache, and S. Baucke, "Virtual resource description and clustering for virtual network discovery," in *Proceedings of the 2009 IEEE International Conference on Communications (ICC'09)*, Dresden, Germany, Jun. 2009.

[31] NFV-ISG, "NFV-MAN 001: Network Function Virtualization Management and Orchestration v1.1.1," Dec. 2014.

[32] C. Fortuna and M. Mohorcic, "Dynamic composition of service for end-to-end information transport," *IEEE Wireless Communications*, vol. 16, no. 4, pp. 56–62, 2009.

[33] R. Karunamurthy, F. Khendek, and R. H. Glitho, "A novel architecture for Web service composition," *Elsevier Journal of Networks and Computer Applications*, vol. 35, no. 2, pp. 787–802, 2011.

[34] S. Komorita, M. Ito, H. Yokota, *et al.*, "Loosely coupled service composition for deployment of next generation service overlay networks," *IEEE Communications Magazine*, vol. 50, no. 1, pp. 62–72, 2012.

[35] C. Henke, A. Siddiqui, and R. Khondoker, "Network functional composition: State of the art," in *Proceedings of the 2010 IEEE Australasian Telecommunication Networks and Applications Conference*, pp. 43–48, Auckland, New Zealand, Nov. 2010.

[36] J. Halpern and C. Pignataro, "IETF RFC 7665: Service Function Chaining (SFC) Architecture," Oct. 2015.

[37] ONF, "TS-027: L4-L7 Service Function Chaining Solution Architecture," Jun. 2015.

[38] ONF, "Software-Defined Networking: The New Norm of Networks," White Paper, Apr. 2012.

[39] M. Banikazemi, D. Olshfski, A. Shaikh, J. Tracey, and G. Wang, "Meridian: an SDN platform for cloud network services," *IEEE Communications Magazine*, vol. 51, pp. 120–127, Feb. 2013.

[40] IETF, "RFC 7285: Application Layer Traffic Optimization (ALTO) Protocol," Sep. 2014.

[41] L. Peterson, "CORD: central office re-architectured as a datacenter," in *IEEE Software Defined Networks White Paper*, Nov. 2015.

[42]   G. Xilouris, E. Trouva, F. Lobillo, *et al.*, "T-NOVA: a marketplace for virtualized network functions," in *Networks and Communications (EuCNC), 2014 European Conference on*, pp. 1–5, IEEE, Piscataway, NJ, 2014.

[43]   C. Mobile, "C-Ran: The Road Towards Green RAN," *White Paper*, 2011.

[44]   B. Sousa, L. Cordeiro, P. Simoes, *et al.*, "Towards a fully cloudified mobile network infrastructure," *IEEE Transactions on Network and Service Management*, vol. 13, no. 3, pp. 547–563, 2016.

[45]   Q. Duan, Y. Yan, and A. V. Vasilakos, "A survey on service-oriented network virtualization toward convergence of networking and cloud computing," *IEEE Transactions on Network and Service Management*, vol. 9, no. 4, pp. 373–392, 2012.

[46]   A. Császár, W. John, M. Kind, *et al.*, "Unifying cloud and carrier network: EU FP7 project UNIFY," in *Proceedings of the Sixth IEEE/ACM International Conference on Utility and Cloud Computing (UCC2013)*, pp. 452–457, IEEE, Dresden, Germany, 2013.

[47]   S. Wang, A. Zhou, F. Yang, and R. N. Chang, "Towards network-aware service composition in the cloud," *IEEE Transactions on Cloud Computing*, in press.

[48]   J. Huang, Q. Duan, S. Guo, Y. Yan, and S. Yu, "Converged network–cloud service composition with end-to-end performance guarantee," *IEEE Transactions on Cloud Computing*, in press.

Chapter 2
Methods and Infrastructure of Network-as-a-Service

Wu Chou[1] and Li Li[2]

Abstract

Network-as-a-Service (NaaS) is fast advancing and it abstracts the heterogeneous and dynamic network resources and their relationships as connected and uniform REST services. This chapter covers the REST service infrastructure for NaaS as well as methods and techniques in REST service modeling and service design for applications in data networking.

Keywords

NaaS, RaaS, REST Chart, REST service patterns, hypertext-driven navigation, REST client design, SDN, NFV

2.1 Introduction

Network-as-a-Service (NaaS) allows network operators and users to provision, manage, and control data networks through high-level operations, which can be relatively opaque from the underlying network complexity and uncertainty. NaaS abstracts a data network as a collection of services that can be described, published, discovered, and invoked through standard interfaces. On one hand, these services are loosely coupled with network elements that perform the actual operations and data handlings, so that the locations and implementations of the network elements can change without affecting the operations they provide. On the other hand, NaaS-based services decouple the clients from the network operations, such that both of them can evolve independently. Compared to traditional network architectures where such separation and decoupling are difficult, NaaS can provide more flexible and efficient network

---

[1] Network and Enterprise Communications, Huawei.
[2] Shannon Cognitive Computing Lab, Huawei.

operations on-demand to the clients, and at the same time, increase the network utilization, and reduce management cost for network operators. NaaS is made possible by the recent advances in distributed computing architecture, namely, Software-Defined Networking (SDN), Network Function Virtualization (NFV), Container technologies, and Resource-as-a-Service (RaaS). It can also be argued that the rapid architectural changes brought by these advances call for a service-oriented architecture to make data networking more agile, open, and service oriented to support existing and new applications from various domains including large scare cloud data centers and Internet-of-things (IoT).

SDN [1–4] decouples the data plane and control plane that were tightly coupled in the traditional network architecture, such that the control plane can evolve without changing the data plane. Furthermore, SDN logically centralizes the distributed control plane in the traditional network through southbound application programming interfaces (APIs), and makes it easier to expose a network as services to the clients. Instead of having to configure a collection of network elements separately, sometimes using different protocols, SDN allows a client to create, configure, and control a network through a single northbound API.

NFV [5] aims to replace proprietary hardware middle boxes, such as network address translation (NAT) and Firewall (FW), by software components running on commodity servers. NFV not only reduces the facility, energy, and operating cost of network operators, but also increases the operator's flexibility on where and when the network functions can be deployed. By treating the network functions as services, a network operator can easily choose and combine the functions from different vendors into a coherent service, without worrying about their implementations.

Container-based virtualization is accelerated by the advance of Docker [6,7] technologies that make the containers easy to build, deploy, update, and control. Processes in a container run much faster than in Virtual Machines, because the process share the same Linux kernel as the host processes, while isolated by namespaces. Containers are themselves services that can be managed through REST APIs. Containers make it easy for network operators to package and deploy middle boxes and expose them as services without overhead. Furthermore, the fine-grained resources of a container, including CPU and memory, can be dynamically changed through API while the container is running, allowing a network operator to scale the containers horizontally and vertically based on the workloads.

RaaS [8] is a cloud computing paradigm to capture the current trends from renting virtual machines with prebuilt resources toward renting resources to compose elastic machines with variable capacities. In RaaS clouds, fine-grained resources, like memory, CPU, network, and storage, will become the basic services that a client can request. For example, a client can request 1 GB memory for 10 s at $0.05, and 100 MB bandwidth to transfer data to a public database at $0.04 per second. The interplay of NaaS and RaaS will further transform distributed computing toward service-oriented paradigm. On one hand, the success of NaaS will bring RaaS closer to reality. On the other hand, RaaS can significantly increase resource utilization for a network operator, and at the same time, decrease the rental cost for the clients without compromising the performance of the services.

To fully realize the potentials of NaaS based on the above-mentioned technologies, the service layer needs to address the following challenges, in order to abstract the network elements as services that provide certain network functions:

1. Heterogeneity: how to deal with variations in space, that is, abstract the wide range of network elements (e.g., switches, routers, middle boxes, and SDN controllers) that differ in many dimensions, including types, functions, sizes, access methods, and communication protocols, into some uniform services for the clients.
2. Dynamism: how to deal with variations in time, that is, hide from clients the changes to network elements due to hardware, firmware, software, and configuration changes.
3. Composability: how to maintain and represent various and dynamic organizations of network elements, for example, a Service Function Chain, by a uniform, extensible, and easy to understand way for the clients.
4. Overhead: how to minimize the potential overhead introduced by the service layer to the communications between the network elements, when they have strict latency requirements.
5. Performance: how to improve the performance within the service layer by taking advantage of the REST cache mechanisms and the properties of network services.

Heterogeneity of network elements can come from two sources: intrinsic and configuration variations. Network elements can differ in types, functions, interfaces, performances, and costs, and a network element, for example NAT, can be configured to perform the same function in different ways. An extensible service description and flexible service discovery mechanism are needed to address this issue.

Dynamism of network elements can also come from two sources: evolutions by design and changes by configurations. Existing network elements can be reconfigured to perform different functions and this change must be republished at the service layer. New types of network elements will become available that provide new functions not anticipated before. Such changes may require a new service description to be agreed upon before it is published. To address this issue, we believe a bottom-up approach, where the individual services control the service description, publication, and discovery process, is more suitable than the top-down approach, where a central authority controls the entire process for all the services.

To address these challenges, an extensible and flexible service framework becomes critical. The REST service framework is well suited to fulfill this role. The REST service framework consists of three basic planes: data, REST resources, and hypertext representations. This framework transforms a network into REST services, based on a clear separation of three network planes: data, control, and management, as shown in Figure 2.1.

In the REST service framework, the data plane of the network remains unchanged, to avoid introducing service overhead to the data plane, while the control and management planes are reorganized as interconnected REST resources. The REST resources are identified by URI and they expose uniform interfaces that clients can access

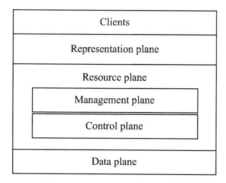

Figure 2.1   Relationship between REST service framework and network planes

over a variety of network protocols. The dynamic relationships between the network functions are maintained as connections between the REST resources.

The clients do not manipulate the states of the REST resource directly or know the resource connections at design time. Instead, the clients exchange hypertext representations with the REST resources and the REST resources provide dynamic service publication, discovery, and invocation such that the REST resources and clients can evolve independently based on a common representation plane. To improve service performance, layered caches can be used between the REST clients and the REST resources by taking advantages of special properties of network elements.

2.2   REST API for SDN

SDN decouples the data and control planes of a network to provide enhanced flexibility, programmability, and openness to applications. This decoupling allows a logically centralized controller to control the network based on a global view of the network resources. As shown in Figure 2.2, at the center of the control plane is an SDN controller, which controls the behaviors of the underlying data forwarding elements through some southbound APIs, e.g., OpenFlow [9,10]. The SDN controller also provides a programmable interface for applications to observe and change the network states dynamically. This interface is called the northbound API of SDN, shorthanded as NBI (North Bound Interface).

A network is a complex distributed system whose structure, function, resource, and behavior can change dynamically. As the network structure evolves and the application paradigm shifts, the NBI of the network has to be flexible enough to accommodate these changes. As such, RESTful NBI has become a prevalent choice for the NBI in SDN, where a well-designed REST API can be highly extensible and maintainable for managing distributed resources, such as data networking. Extensibility in REST means that a REST API can provide different functions at the same time and it can also make certain changes to those functions over time without breaking its clients.

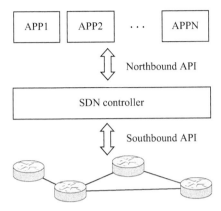

Figure 2.2   The architecture of software-defined network

There are plenty of cases that the REST APIs need to undergo rapid changes and updates as evidenced by many open source development projects. One important mechanism of REST APIs is hypertext-driven navigation, also known as "hypertext as the engine of application state" in [11,12], which offers an effective means for clients to cope with certain changes automatically. To facilitate hypertext-driven navigation, we need to design the REST API following the REST principles [13] and convey the REST API design to the users precisely, so that they can create the well-designed clients accordingly. Furthermore, it needs to address the overhead associated with hypertext-driven navigation to balance the flexibility and efficiency of a REST API.

2.3   Hypertext-driven navigation principle of REST API

A primary goal of hypertext-driven navigation is to cope with REST API changes by reducing the design time dependency between the clients and a REST API, such that a client can navigate to the target resources as driven by the hypertext from interacting with those resources. Hypertext-driven navigation relies on five layers of a REST API described below:

1. Connection: links of connected resources implemented in any programming languages and running on any devices.
2. Interactions: methods/protocols (e.g., HTTP) to interact with those resources.
3. Identification: identifiers (e.g., URI) that identify the resources.
4. Representation: hypertexts (e.g., XML) sent to and obtained from the interactions.
5. Description: descriptions (e.g., REST Chart) about the possible representations.

At the design time, a client is programmed against a description of a REST API, without knowing how the resources are identified and connected. At the runtime, the client becomes an automaton that starts from the entry point and navigates to the target resource driven by hypertext representation. During the navigation, the client moves from layer 4 down to layer 1 and back up in cycles. In each move, the client uses representations to determine identifications, identifications to determine interactions, interactions to determine connections and representations.

Besides following hyperlinks, content negotiation and redirection can also be regarded as special mechanisms for hypertext-driven navigation. In content negotiation, the URI does not reveal the media type [14] of the representation, but determines an interaction (e.g., HTTP) that can negotiate and determine the appropriate representation. As a result, the representation of a resource can change without changing its identification—a loosely coupled architecture for distributed resources. In HTTP redirect, the URI does not need to identify the target resource, but it determines an interaction that in turn determines the correct connection to the target resource. As a result, a resource can change its connections without changing its identification. As URI resolution depends on the context, it is also possible to change the interaction with a resource without changing its identification. Hypertext-driven navigation, therefore, makes it possible to allow certain independent changes in these layers, as long as those changes are permitted by the description layer.

On the basis of hypertext-driven navigation principle, defining media types in URI, such as /network/port/json and /network/port/xml, would not be a good practice because it makes representation dependent on and tightly coupled with the identification. Consequently, it is difficult to evolve without changing the identification, and the client is deprived of the right to negotiate for the optimal media type that best fits its capabilities. For example, an SDN controller can be accessed over a variety of network resources, including switches, routers, and devices; separating identification from representation allows a client to select the most efficient representation for the desired services.

Moreover, due to the similar reason, exposing a set of fixed URIs in the REST API is not a good practice either, as it ties identification with connection, and any changes in the connections can invalidate the URIs. For example, URI /networks/1/ports/2/ implies there is a connection from the networks resource to the ports resource. If this connection changes, then the URI becomes invalid. In SDN, such resource reorganization is common and critical. For instance, SDN controllers are expected to provide a more dynamic, open, and programmable network, where the network can be dynamically configured and planned against the network traffics to support various applications and quality of service (QoS) requirements [1].

2.4   Flexible REST API design framework

To provide a flexible model for hypertext-driven navigation in REST API, we can adopt REST Chart [15]; a Petri-Net based framework and language as the foundation for REST API modeling. In particular, REST Chart models a REST API as a special type of Colored Petri Net [16] that consists of places and transitions connected by arcs. Resource representations are modeled as tokens stored in the places. The colors

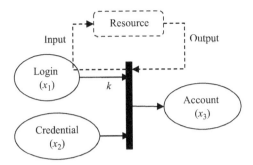

Figure 2.3   Illustration of a basic REST Chart

of a place designate the schemas that define the representations (tokens) accepted by the place. A transition binds a hyperlink between two places to a protocol that interacts with the hyperlink. Under this modeling framework, a REST Chart describes both the representations and interactions of a REST API as well as the hypertext-driven navigation of its client.

Figure 2.3 illustrates a basic REST Chart with three places (schemas) and one transition that describe a REST API to login an online account with username and password. To use this REST API, a client selects the login hyperlink k from token x_1 in the login place, and deposits token x_2 (username and password) in the credential place. With these two tokens in the input places, the transition will fire as defined by Petri-Net semantics, where firing a transition here models the interaction with the remote resource using the protocol defined by the transition, as being shown here as the dotted arrows between the transition and the remote resource, which is not part of, but is implied by, the REST Chart. If the interaction is successful, the resource will return the account information in a response, which is modeled as token x_3 in the account place.

Hypertext-driven navigation of a REST API is determined by the reachability of the corresponding Petri-Net that models it, as client states are modeled as token markings of the Petri-Net. In this example, the initial state of the client is $(x_1, 0, 0)$, meaning that it has token x_1 in the place login, but no tokens in the credential and account places. From this initial state, the REST Chart indicates that the final state $(0, 0, x_3)$ is reachable as follows:

$$(x_1, 0, 0) \rightarrow (x_1, x_2, 0) \rightarrow (0, 0, x_3).$$

Not all possible states are reachable from the initial state, and such unreachable states include $(x_1, 0, x_3)$ and $(0, x_2, x_3)$. State $(x_1, 0, x_3)$ means that a client can access the account without valid credential, and state $(0, x_2, x_3)$ means that the client can access the account without following the hyperlink. An interesting property of REST Chart is that the set of reachable states defines the possible resource connections taken by the clients following hypertext-driven navigation. If a client can enter an

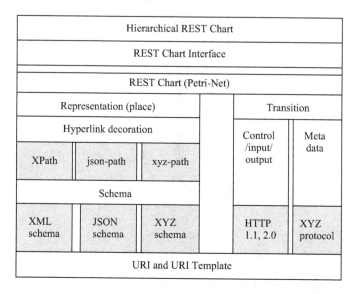

Figure 2.4   Layered modular architecture of REST Chart

unreachable state, it means that the REST API description violates the hypertext-driven navigation principle [17].

REST Chart is based on a layered and modular architecture to model the extensibility of a REST API. As shown in Figure 2.4, a REST Chart allows extensions at several layers with local changes, including a local change at a module only expands the layers below it without modifying its interface to the upper layer. For example, a REST Chart can make local changes to the schemas of a place without affecting the places or the transitions. Similarly, a REST Chart can make local changes to the network protocols of a transition without affecting the transitions or the places.

The *hyperlink decoration* mechanism in REST Chart separates hyperlinks (resource identification) from schemas (resource representation), such that we can define hyperlinks independently from schemas. A hyperlink decoration has two parts: (1) a hyperlink, and (2) a mapping from the hyperlink to the schema. A hyperlink also has two parts: (1) a URI that identifies the service provided by the resource, and (2) a URI template [18] that describes the locations of the resource.

For example, a hyperlink decoration *k* in Figure 2.3 could be defined as an XML element, where we use short service names to illustrate the idea:

```
<link id= "k">
<rel value= "urn:login" />
<href value= "http://bank.com/{path}/{userid}" />
</link>
```

The <rel> element defines the service and the <href> element defines the actual location of the service. Two hyperlinks are *compatible* if they have the same service

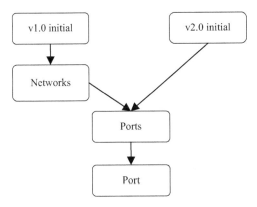

Figure 2.5   Version 1.0 (left) and version 2.0 (right) connections

URIs. A hyperlink path is a sequence of place-hyperlink pairs, where hyperlink is a [service, location] pair as defined above. Two hyperlink paths are *compatible* if for each corresponding place-hyperlink pairs, the place is identical and the decoration is *compatible* as defined above.

For example, the hyperlink path for the REST Chart in Figure 2.3 could be:

```
login-[urn:login, http://bank.com/users/1]
account-[]
```

The first place-hyperlink pair corresponds to token markings $(x_1, 0, 0) \rightarrow (x_1, x_2, 0)$, while the second one corresponds to token marking $(0, 0, x_3)$, that is, the final state. If a REST API changes the location URI but keeps the service URI, it produces a compatible hyperlink path:

```
login-[urn:login, http://bank.com/accounts/1]
account-[]
```

A client that navigates a REST API based on compatible hyperlinks can follow compatible hyperlink paths without any change using the decision rule: in place x, follow hyperlink with service y from x, regardless of location z. In other words, as long as the relations between x and y in pairs x-$[y, z]$ remain the same, a REST API is free to change z anywhere along a path without breaking any hypertext-driven clients.

Not all connection changes produce compatible hyperlink paths. Figure 2.5 shows two versions of REST Chart for network APIs in OpenStack [19,20]: Neutron version 1.0 and 2.0 that contain an incompatible path between them. These hierarchical REST Charts [21] have places that contain nested REST Charts, so that we can organize a complex REST API as modular REST Charts. For example, the REST Chart in Figure 2.6 contains the REST Chart in Figure 2.8. We refer readers interested in hierarchical REST Chart to [22].

To facilitate machine processing, REST Charts are encoded as machine readable XML. For example, the networks-ports connection of version 1.0 REST Chart in

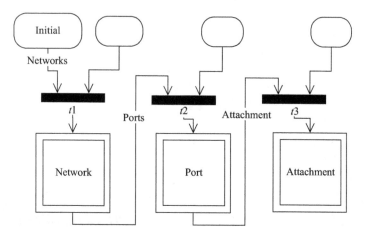

Figure 2.6   Part of REST Chart for Neutron REST API v1.0

Figure 2.6 is represented as follows (all places are represented by <representation> elements in REST Chart):

```
<representation id= "networks">
<link id= "ports_v1.0">...</link>
<schema>...</schema>
</representation>
<transition>
<input>
<representation ref= "network" link= "ports_v1.0" />
</input>
<output>
<representation ref= "ports" />
</output>
</transition>
```

Similarly, the initial-ports connection of version 2.0 REST Chart in Figure 2.7 is a follows:

```
<representation id= "initial">
<link id= "ports_v2.0">...</link>
<schema>...</schema>
</representation>
<transition>
<input>
<representation ref= "initial" link= "ports_v2.0" />
</input>
<output>
```

```
<representation ref= "ports" />
</output>
</transition>
```

Table 2.1 lists the main URI namespaces of Neutron REST API versions 1.0 and 2.0, which contain both compatible and incompatible hyperlink paths. Neutron REST API Namespace changes.

To identify the invariants and changes between two versions of a REST API, such as the one in Figure 2.5, we developed a Petri-Net based algorithm and tool [23] to

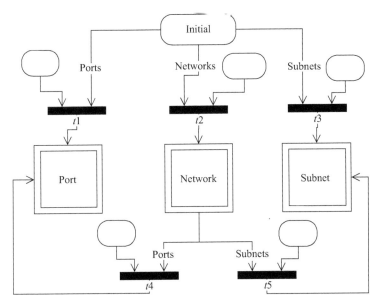

Figure 2.7   Part of REST Chart for Neutron REST API v2.0

*Table 2.1   Namespaces of two versions of neu-
tron API*

Core resources of Neutron REST API v1.0

/networks/{network-id}
/networks/{network-id}/ports/{port-id}
/networks/{network-id}/ports/{port-id}/attachment

Core resources of Neutron REST API v2.0

/networks/{network-id}
/ports/{port-id}
/subnets/{subnet-id}
/devices/{device-id}

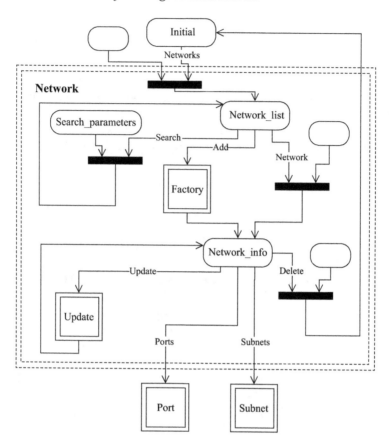

Figure 2.8   REST Chart nested inside the network place

compare two REST Charts, for example, Figures 2.6 and 2.7. The algorithm emulates a hypertext-driven client navigating through the REST Chart of Figure 2.7 with only the knowledge about the REST Chart defined by Figure 2.6, and it outputs all the places it can reach in Figure 2.7 when assisted by the places in Figure 2.6. In other words, the places it can reach in Figure 2.7 are the services that are compatible to those in Figure 2.6, while the places it cannot reach in Figure 2.7 are the new services that have no counterparts in Figure 2.6.

In this case, it identifies that to migrate a v1.0 client to v2.0 REST API, the minimum change to a v1.0 client is to modify the decision rule at the *initial* place to reach the *port* place because if we align place *initial* in v2.0 with place *networks* in v1.0, then the following two paths become compatible:

```
v1.0: initial-[networks,_], networks-[ports,_], ports-[port,_]
v2.0: initial-[ports, _], ports-[port, _].
```

To cope with compatible and incompatible changes to a REST API, it can use a more structured way to break a REST client into two cooperative functional

Table 2.2   *REST API and client changes*

Client changes		API message changes	
		Compatible	Incompatible
API path changes	Compatible	No	Agent
	Incompatible	Oracle	Oracle, agent

components, that is, client oracle and client agent. As elaborated in the next section, the client oracle is responsible for selecting hyperlinks and the client agent is responsible for exchanging protocol messages based on the selected hyperlinks. By this structural separation, we can reuse the client agent and/or the client oracle. Table 2.2 summarizes the possible changes to a REST API and the corresponding changes required to the client components.

2.5   Flexible REST client design framework

As pointed in the previous sections, REST is based on the principle that *any client of a REST API should be driven by nothing but hypertext*. This principle is easy to understand if we treat a REST API as a distributed finite state machine, where the states are resource representations and the transitions are the links between the representations. In this model, hypertext-driven means that a client can enter a REST API from an entry point, and then be guided by the hypertext from the resources to reach a final representation.

Intuitively, this principle makes it possible for a user without any technical background to use the Web and the distributed services in general, in which a user can enter a URI into a Web browser (or the user agent) to get the initial page, and then follows the links on the pages until the desired page is retrieved. In this process, the user makes decisions on which links to visit based on the information on the page, and the user agent carries out these decisions by interacting with the resources identified by the links.

To mirror this process on machine clients of a REST API, it is structurally desirable that a machine client can be organized by such modular functional components according to the REST framework as well, that is, a client oracle responsible to select links to follow from the resource representations, and a client agent responsible to interact with the resources based on the selected links. This decomposition of a REST service client has several advantages over a monolithic REST client where these two functions are intertwined and inseparable:

- Client oracle can be reused with different versions of a REST API, especially if a REST API version updates changes some resource representations and identifications, but does not change the link relations in the representations;

- Client oracle can be reused across different service description languages of the same REST API;
- Client oracle can drive client agents in different programming languages to achieve consistent behavior;
- A client agent can be driven by different client oracle to accomplish different tasks; and
- Client oracle can significantly reduce the size of a client agent if only a small portion of the resource representations are selected by the client oracle.

Given a REST API, a client oracle and client agent can be programmed manually. But this can be difficult, time consuming, and error prone. However, if a REST service client can be generated automatically from a machine-readable service description of a REST API, such as REST Chart [18], the manual programming process can become unnecessary or greatly reduced. Therefore, in addition to the advantages mentioned above, automated REST service client generation can drastically speed up the REST API development process, and significantly improves the consistencies between REST API description, documentation, and system implementation.

In case that a REST API is described by REST Chart, the goal of a REST service client can be defined in terms of the resource representations (typed places), where an optimal client oracle for a goal can be defined as the shortest path from the initial place to the goal place. This is due to the process by which a client tries to reach a goal in a REST API can be modeled as the client navigates a REST Chart, following the path of smallest number of transitions between the places, to reach a place corresponding to the goal in the REST Chart. As such, REST Chart makes it possible to find such shortest paths, as it specifies all possible paths between all possible resource representations.

For instance, when a client wants to deposit a check to an account guided by a REST Chart, it must reach the login place first, then the place to deposit a check, then the place to scan the check, then the place to upload the check, then the place to verify the information, until it reaches a place corresponding to the positive acknowledgement of the entire process. If the client takes detours, it will take many more steps than necessary to accomplish the transaction.

On one hand, it is possible to find such shortest paths in REST Chart based on Petri-Net reachability algorithms [24] or cover ability algorithms [24]. However, these algorithms are usually complex or may take exponential space as they compute all possible token markings in arbitrary Petri-Net. For this reason, we decide to use graph search algorithm whose time complexity is not dependent on token markings and is polynomial to the number of places and transitions of a Petri-Net.

To apply this approach, we convert the REST Chart to a nested directed graph. The server places become the vertices, and each edge is labeled with corresponding transition including the client place and a cost. This process can be recursively applied to the nested REST Charts.

In addition, the cost introduced on the transitions is a positive real number that quantifies the expense of the resource interaction and it is usually not part of a REST Chart. The costs depend on the implementation and runtime environment of a REST

```
1.   Client_Oracle(C, A, Pi, Pj): Oracle
2.     C: REST Chart
3.     A: adjacency matrix for C
4.     Pi: source place
5.     Pj: target place
6.     IN = {Pi}
7.     For each Pk in P of C do d[Pk] = A[Pi,Pk]
8.     While Pj not in IN do
9.       Pk = a place X in P–IN with minimum d[X]
10.      s[Pk] = C.transition(Pi)
11.      IN +=Pk
12.      For each place X in P–IN do
13.        dist = d[X]
14.        d[X] = min(d[X], d[Pk] + A[Pk, X])
15.        if (d[X] < dist) then s[X] = C.transition(Pk)
16.      End
17.    End
18.    T = s[Pj]
19.    Oracle = (C.server_place(T), T, C.client_place(Y))
20.    While C.server_place(T) ≠ Pi do
21.      T = s[T]
22.      Oracle +=(C.server_place(T), T, C.client_place(T))
23.    End
24.    Return reverse(Oracle)
25. End
```

Figure 2.9   Client Oracle algorithm adapted from Dijkstra's Shortest Path algorithm

API, and they can be obtained by various performance measuring methods of the REST API.

Moreover, it is often useful to direct a REST client to not just one, but to a series of goal places when interacting with a workflow. To realize this requirement, we can adopt Dijkstra's Shortest Path algorithm [25] to the nested directed graph to find the shortest path from the initial place to the first goal place, and then repeat the algorithm from current goal place to the next goal place in the series, until the last goal place is reached. If a goal place is not reachable, then the process will stop. The output of the process is an oracle for the reached goals.

The Client Oracle algorithm is outlined in Figure 2.9. The core of the algorithm (lines 2–17) uses Dijkstra's Shortest Path algorithm on the directed graph which is converted from the REST Chart to find a shortest path from an initial place Pi to a final place Pj and record the transitions on the path (e.g., s[X] = transition(Pk)). The rest of the algorithm (lines 18–24) reconstructs from the recorded transitions of the oracle as a sequence of triples (Server_Place, Transition, Client_Place).

This structured approach to design REST service client can be applied to process very complex REST services, in which a REST client is decomposed into two reusable functional modules: a client oracle that decides the resources to interact based on the resource representations, and a client agent that carries out the resource interactions instructed by the client oracle. More about this approach and compact modeling representation based on hierarchical REST Chart are discussed in [26].

2.6   Hypertext-driven client-resource interaction patterns

In order to effectively use the REST Chart framework to support hyperlink-driven navigation in large scale REST APIs, a REST API should be designed and implemented following certain patterns, each addressing a common problem in REST APIs. In addition to the five basic patterns in [27], this section considers five new practical patterns to facilitate hypertext-driven navigation. As each design pattern has advantages and disadvantages, a REST API can combine them in a more meaningful ways.

2.6.1   Tree pattern

If we have 10^4 types of resources, we can organize them as an *n*-ary balanced tree to reduce the cost of hypertext-driven navigation. In a 10-ary balanced tree, the number of messages to access any resource is bounded by $O(\log N)$, which is 4 for $N = 10^4$.

The advantage of this approach is that we can reduce the overhead of hypertext-driven navigation. The disadvantage is that we may not find a balanced tree for all domains.

2.6.2   Backtracking pattern

Once a client navigates to a resource following a hypertext path, it can remember the hyperlinks in caches and use them to backtrack a REST API when the connections change. For example, after a client navigates a hyperlink path:

$$\text{initial-[networks, \_], networks-[ports,\_], ports-[port, \_],}$$

It can cache the places {initial, networks, ports} and the hyperlinks {networks, ports, port} separately to allow them to be recombined in different ways. When getting stuck, the client can backtrack the path as follows: at place x, if hyperlink h is not available, go back to x's parent and select hyperlink h again. If the REST API reorganizes the resource as follows:

$$\text{initial-[ports,\_], ports-[port, \_],}$$

Then the client can no longer find the ports hyperlinks from the networks place. However, the client can backtrack to place initial, the parent place of networks, and rediscover the hyperlinks for ports.

In order to support backtracking, the meaning of service URI of hyperlinks should be independent of places, despite that its location URI can be dependent on the place. Furthermore, REST API should avoid HTTP cookie as it may make the client's interactions nonrepeatable. With HTTP cookie, a client may get different tokens when it backtracks to the same place. An approach that avoids the side effect of HTTP cookie is to use session resources [28].

The advantage of backtracking is that it can rediscover the relocated resource. Its disadvantage is that a client can only find relocated resource with compatible hyperlinks reachable from the hyperlink path it has visited. For example, if an SDN REST API changes the original hyperlink path to:

initial-[subnets, _], subnets-[ports,_], ports-[port, _],

Then the client will not be able to find the ports hyperlinks because subnets-[ports, _] is a new place and the client has never visited it before, and consequently, the backtracking will fail.

2.6.3 Redirection pattern

HTTP 1.1 redirection [29] can be used to inform a client the existence of new service and/or new location of a resource. For example, if a client follows a hyperlink path: network-[urn:update, u1], and receives from u1 a HTTP 3xx response with new hyperlink pair [urn:set, u2], then the client can learn that this change is not specified in REST Chart:

network-[urn:update, u1] → network-[urn:set, u2].

One primary advantage of redirection is that a REST API does not have to change its REST Chart. The disadvantage is that the REST API cannot release the previous resource as it is used for redirection. It is possible to release the previous resources after a certain transition period, so that clients who use the REST API frequently can learn the new locations from redirection, and the casual clients can learn from a new REST Chart published later on.

2.6.4 Search pattern

```
GET /neutron/v2.0/tenants/t100/portsHTTP/1.1
Host: localhost:8080
Accept: text/xml, application/json

HTTP/1.1 200 OK
Content-Type: text/xml
<ports>
  <network>
  <name>myNet</name>
  <id>net1</id>
<link rel="network"href="/net1"/>
  </network>
<link rel="add"href="/factory" />
<link rel="search"
href="/search?{k1} = {v1}...{kN} = {vN}"/>
</ports>
```

The advantage of this approach is that a client can reach any resources with two interacts: search and visit. The disadvantage is that a client can only search resources that exist, and the client has to know some information about the resources it searches for.

2.6.5   Generator pattern

If a REST API wants to return a large number of resources to a client, it can return a small generator to create those hyperlinks, instead of the hyperlinks themselves. A generator defines a list of hyperlinks with a URI template and functions $f_1 - f_n$ that instantiate the variables $v_1 - v_n$ in the URI template [30]:

$$[\text{URI\_template}(v_1, \ldots, v_n)|v_1 \leftarrow f_1, \ldots, v_n \leftarrow f_n].$$

For example, the following is a generator that generates the hyperlinks to network ports for a tenant tid:

$$[/\text{networks}/\{\text{nid}\}/\text{ports}/\{\text{pid}\}|[\text{nid, pid}] \leftarrow f(\text{tid})].$$

If $f(1234) = \{[1,a], [2,b], [3,c]\}$, then the generated list is [/networks/1/ports/a, /networks/2/ports/b, /networks/3/ports/c]. A REST API can send the generator functions to a client using a mechanism called Code On-Demand [11], in the similar way that Web servers deliver JavaScript code to Web browsers.

The advantage of this approach is that the REST API can convey a large number of items with a small amount of message exchanges, and a client has the freedom to select the items to fit the need. The disadvantage is that the client has to execute code from a REST API, which increases the security risks in the REST API.

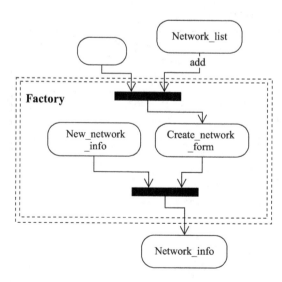

Figure 2.10   REST Chart model of creating a new network

2.6.6 *Factory pattern*

In many situations, hypertext-driven navigation can be the way to access resources that have not been created in a REST API. The hyperlink path in this case can be a chain of factory resources, where one factory resource creates another factory resource in the chain, and the final factory creates the target resources. Figure 2.10 shows the REST Chart model of a factory resource that accepts a form and creates a network, from which a client creates subnets, and from subnets the ports. The example HTTP request and response to interact with the factory resource is shown in Figure 2.10. Here the form marks the required attributes by making attribute required = true, while it provides default values for some other elements. In particular, the attribute method of the form element indicates that the client should submit the filled-out form to the factory resource by the HTTP command POST.

```
GET /neutron/v2.0/tenants/t100/networks/factory HTTP/1.1
Host: localhost:8080
Accept: text/xml, application/json

HTTP/1.1 200 OK
Content-Type: text/xml
<form method="POST">
  <network>
    <id required="true" />
    <name/>
    <admin_state_up>true</admin_state_up>
    <shared>true</shared>
    <tenant_id required="true" />
  </network>
</form>
```

A chain of factories defines a hyperlink path that can be navigated by a client only once and in the fixed order. To support backtracking along this hyperlink path, each factory resource should also support a delete operation, such that a client can always go back to the parent and recreate a child resource that is not working.

The advantage of this approach is that a REST API has the freedom to reorganize a factory chain, such as inserting and removing factories, in the same way as it reorganizes any resources, but with the additional semantics that a creation operation must be executed at most once. For this reason, a client can deal with changes to factory chains using any patterns discussed herein. The disadvantage of this approach is that it does not support transaction that can automatically abort all the previous creations if the last one fails. REST transaction is still a research area [12].

2.7   Efficient cache mechanisms

Clients visiting the hypertext-driven REST API for the first time should start from the entry URI and then follow the returned hyperlinks to access the resources. Compared to the fixed-URI scenario where the clients access the target resource directly, this could introduce considerable performance overhead if the target resources take many rounds of interactions to reach from the initial resource. This section presents a client cache mechanism to reduce the performance overhead when performing the hypertext-driven RESTful interactions. Consequently, a REST API and its clients can take advantages of the extensibility in REST APIs through the hypertext-driven navigation with the much reduced performance overhead.

2.7.1   Issues of caching for REST service clients

The basic idea of caching for REST service clients is to save clients from additional rounds of interactions in future accesses. As the current REST APIs are commonly implemented over HTTP, we started by considering deploying a client-side classical Web cache based on HTTP 1.1 cache controls [30], which stores the complete HTTP response, including all the attributes, entities, and hyperlinks contained in the representation.

Figure 2.11(a) shows an example of such Web cache. Within the cache, the initial representation contains the hyperlinks only, while the network_list representation contains not only hyperlinks, but also a set of network entities. Each network entity has its ID attribute and a network hyperlink.

Such a Web cache could effectively improve the performance of clients for retrieving representations regarding the application state. However, many hypertext-driven interactions with a REST API are for retrieving the hyperlinks rather than the application state. These hyperlinks are more stable than the resource states, especially in SDN. Therefore, caching stable hyperlinks with unstable elements in the same cache entry would cause unnecessary cache invalidations. For example, as shown in Figure 2.11(a), to add a new network, clients follow the create hyperlink in the network network_list representation. Although this hyperlink remains constant, the entire representation may become invalid when a network resource is added or removed.

A closer examination of SDN REST API shows that a resource representation typically contains three types of disjoint content:

1. Entity hyperlinks that point to unstable entity resources;
2. Action hyperlinks that point to stable action resources;
3. Elements describing the state of the resource.

For example, as shown in Figure 2.8, the network hyperlink is an entity hyperlink pointing at a network resource, while other hyperlinks are action hyperlinks providing various services related to the network. In general, the action hyperlinks are defined at design time and tend to be stable. In contrast, the entity hyperlinks tend to change with the resource state. For example, the create action hyperlink in Figure 2.8 is always present in the network_list representation, whereas the network entity hyperlink will change as networks are added or removed.

If we apply the same cache control to the representation, we are likely to cache the entity hyperlinks too long but the action hyperlinks too short. A cache saved too long will not only waste storage, but also give clients incorrect states about the system, whereas a cache saved too short will cause undue misses and decrease the performance.

These observations suggest that we should cache entity hyperlinks, action hyperlinks, and elements differently to maximize the client performance while reducing the rate of cache invalidation.

Saving the complete hyperlinks in cache may waste a lot of spaces if a large number of hyperlinks are instances of a few URI templates. Applying the idea of Generator Pattern discussed previously to caches, we only save the distinct URI templates and the variable values. Since in our system, a common URI template and a distinct resource ID will be sufficient to identify most resources, we can save cache space by maintaining a mapping from the distinct IDs to URI templates.

2.7.2   Differential and layered cache at clients

Our solution to cache entity hyperlinks, action hyperlinks, and elements differently is a *differential and layered cache* mechanism. The differential cache control is a fine-grained cache control that allows a REST API to specify which elements of a representation can be cached for how long, which is in addition to representation level HTTP 1.1 cache control. The proposed differential cache control takes the meaning of HTTP 1.1 max-age cache directive [30] but applies it to individual elements of a representation as a special attribute. For example, the `cache-control` attribute in the link element `<link cache-control="max-age:{delta}"/>` indicates that the link element can be cached for {delta} seconds, which overwrites any same cache control at the representation level. When applied to an element with children, the attribute indicates that the element and its children share the same cache control, unless a child element has its own cache control.

This approach is domain and language independent as it can specify cache controls in different mark-up languages at different granularity. For this reason, the differential caches can be maintained by clients or trusted proxies who have access to the representations.

Based on the differential cache control, the action hyperlinks are cached in the hyperlink cache, which contains *rel* → (*hyperlink, expiry*) mappings, where the hyperlink points to entries in the representation cache, which contains *hyperlink* → (*representation, expiry*) mappings.

The entity hyperlinks are saved in a bookmark style cache, called Entity Index. The Entity Index table saves (*rel, ID*) → (*hyperlink, expiry*) mappings, where the *rel, ID*, and *hyperlink* are extracted from the *rel, id,* and *href* attributes of entity hyperlinks, respectively. The Entity Index acts like bookmarks that offer clients shortcuts to access the resources. Combined with the hyperlink cache, clients can perform operations on the entity resources with much fewer interactions.

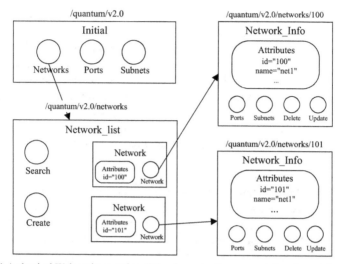

(a) A classical Web cache contains all the elements of a representation, i.e., attributes, objects and hyperlinks.

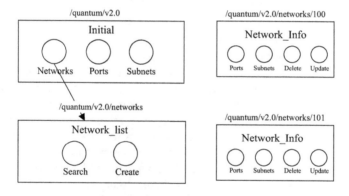

(b) Our hyperlink cache contains only the static hyperlinks.

Figure 2.11   The differences between (a) a classical Web cache and (b) our hyperlink cache

These caches are populated and updated at clients according to the differential cache controls and the regular HTTP 1.1 cache controls, whenever a fresh representation is retrieved from the origin REST API server.

Figure 2.12 shows an example of the Entity Index with three entries extracted from the representations of the previous hypertext-driven interactions. To update the network of ID 100, the client first constructs the resource URI from the Entity Index and then uses this URI to lookup the updated hyperlink from the hyperlink cache. As all these operation happens at the caches, the client can find the URI to the target resource without any interaction with the REST API.

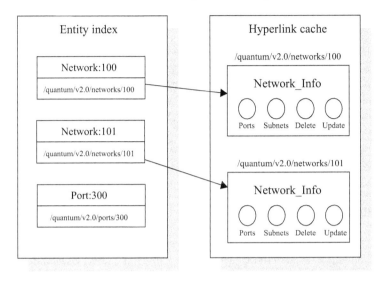

Figure 2.12   The layered caches: bookmark, hyperlink, and representation

2.7.3   Cache replacement

To prevent the cache overflow, new entries should replace old entries when the cache reaches the size limit. The problem of cache replacement has been well studied. The efficiency of the replacement algorithm depends heavily on the workload. Finding the most efficient cache replacement algorithm is out of the scope of this chapter, we consider that the classical LRU (least recently used) algorithm should match the data locality well for many cases.

Furthermore, the sizes of entries in the Entity Index are mostly very similar. This allows the lightweight LRU type algorithm to perform well rather than having to adopt more complex nonuniform size cache replacement algorithms, such as Greedy Dual-Size (GDS) algorithm [31]. This observation also applies to the hyperlink cache, as its entries contain the static hyperlinks only, of which the size differs much less comparing to the case of including all the attributes and entities in the cached representations.

The hyperlink cache and Entity Index carry out cache replacement independently rather than sharing the same buffer. This is because the size of an entry in the Entity Index is typically several times smaller than the size of an entry in the hyperlink cache.

2.8   Summary

NaaS is fast advancing as it provides the critical service layer abstraction for network services and makes network RaaS. It is well suited in the SDN/NFV-based network architecture with enhanced flexibility, manageability, and composability for data networking. In particular, REST services have become the major enabler in NaaS.

It abstracts the heterogeneous and dynamic network resources and their relationships as connected and uniform REST services.

In this chapter, NaaS was investigated from multiple angles, including data networking in cloud computing, relation to RaaS architecture, and REST Chart for REST service modeling. Moreover, we described the mechanism of *hypertext-driven navigation* in REST, structure and semantics of service representation, identification and connection in REST services, hypertext-driven REST client design framework, REST service patterns, service caching for efficient service interaction, and other related methods and techniques. Examples and use cases were described with references for further studies.

References

[1]   S. Wang, A. Zhou, F. Yang, R. Chang. "Towards network-aware service composition in the cloud," *IEEE Transactions on Cloud Computing*, PP(99), pp. 1–14. doi: 10.1109/TCC.2016.2603504, 2016. http://ieeexplore.ieee.org/document/7553440/.

[2]   M. Luo, Y. Tian, Q. Li, J. Wang, W Chou. "SOX – a generalized and extensible smart network openflow controller," *The First SDN World Summit*, Germany, October 2012.

[3]   N. Foster, M. J. Freedman, R. Harrison, J. Rexford, M. L. Meola, D. Walker. "Frenetic: a high-level language for OpenFlow networks," *Proceedings of the Workshop on Programmable Routers for Extensible Services of Tomorrow*, Philadelphia, Pennsylvania, Article No. 6, 2010.

[4]   OpenDaylight Platform: https://www.opendaylight.org/. Last accessed on 13 April 2017.

[5]   NFV (2015). https://portal.etsi.org/NFV/NFV_White_Paper.pdf, retrieved February 2015.

[6]   Docker (2015). Docker website. https://www.docker.com/, retrieved February 2015.

[7]   Docker Remote REST API: https://docs.docker.com/reference/api/docker_remote_api/. Last accessed on 13 April 2017.

[8]   O. A. Ben-Yehuda, M. Ben-Yehuda, A. Schuster, D. Tsafrir "The rise of RaaS: the Resource-as-a-Service cloud", *Communications of the ACM* 57(7), pp. 76–84, 2014

[9]   N. N. McKeown, T. Anderson, H. Balakrishnan, *et al* "OpenFlow: enabling innovation in campus networks," *SIGCOMM Computing and Communications Review*, 38(2), pp. 69–74, 2008.

[10]   Open Networking Foundation, OpenFlow Switch Specification, Version 1.5.0, December 19, 2014, ONF TS-020.

[11]   R. T. Fielding "Architectural styles and the design of network-based software architectures," Ph.D. Dissertation, University of California, Irvine, CA, 2000, http://www.ics.uci.edu/~fielding/pubs/dissertation/top.htm. Last accessed on 13 April 2017.

[12]   L. Richardson, S. Ruby *RESTful Web Services*, O'Reilly, 2007.

[13]   R. T. Fielding, "REST API must be hypertext driven," October 28, 2008, http://roy.gbiv.com/untangled/2008/rest-apis-must-be-hypertext-driven. Last accessed on 13 April 2017.

[14]   N. Freed, N. Borenstein. RFC2046: Multipurpose Internet Mail Extensions (MIME) part two: media types, 1996, http://www.ietf.org/rfc/rfc2046.txt. Last accessed on 13 April 2017.

[15]   L. Li, T. Tang, W. Chou "Automated creation of navigable REST services based on REST chart," *Journal of Advanced Management Science*, 4(5), pp. 385–392, September 2016.

[16]   T. Murata "Petri nets: properties, analysis and applications," *Proceedings of the IEEE*, 77(4), pp. 541–580, April 1989.

[17]   L. Li, W. Chou "Design and describe REST API without violating REST: a Petri net based approach," *Proceedings of the 2011 IEEE International Conference on Web Services*, Washington DC, USA, pp. 508–515, 2011.

[18]   J. Gregorio, R. Fielding, M. Hadley, M. Nottingham, D. Orchard (eds.). *URI Template*, RFC6570, 2012, https://tools.ietf.org/html/rfc6570. Last accessed on 13 April 2017.

[19]   R. Fehling "A concept of hierarchical Petri nets with building blocks," *12th International Conference on Applications and Theory of Petri Nets: Advances in Petri Nets*, Gjern, Denmark, pp. 148–168, 1993.

[20]   OpenStack Foundation. *OpenStack Networking Administration Guide*, February 2013, http://docs.openstack.org/trunk/openstack-network/admin/content/index.html. Last accessed on 13 April 2017.

[21]   OpenStack Networking API: http://developer.openstack.org/api-ref-networking-v2.html. Last accessed on 13 April 2017.

[22]   L. Li, T. Tang, W. Chou "Designing large scale REST APIs based on REST chart", *IEEE ICWS* 2015, pp. 631-638, New York, USA, June 27–July 2, 2015.

[23]   L. Li, W. Chou "Compatibility modeling and testing of REST API based on REST chart," *WEBIST 2015*, pp. 194–201, Lisbon, Portugal, May 20–22, 2015.

[24]   C. G. Cassandras, S. Lafortune *Introduction to Discrete Event Systems*, Springer, New York, 2008.

[25]   J. L. Gersting *Mathematical Structures for Computer Science*, third edition, Computer Science Press, New York, 1993, pp. 422–423.

[26]   L. Li, W. Chou "Finding optimal REST service oracle based on hierarchical REST chart," *Service Computation 2015*, pp. 21–26, Nice, France, March 22–27, 2015.

[27]   W. Zhou, L. Li, M. Luo, W. Chou "REST API design patterns for SDN Northbound API," *The 28th IEEE International Conference on Advanced Information Networking and Applications Workshops (AINA-2014)*, pp. 358–365, Victoria, BC, Canada, May 13–16, 2014.

[28]   L. Li, W. Chou "Design patterns for RESTful communication web services," *ICWS 2010*, pp. 512–519, Miami, July 5–10, 2010.

[29]   R. Fielding, J. Gettys, J. Mogul, *et al. Hypertext Transfer Protocol – HTTP/1.1*, RFC 2616, section 12, IETF (Internet Engineering Task Force), June 1999.

[30]   L. Li, T. Tang, W. Chou "A XML based monadic framework for REST service compositions," *IEEE ICWS* 2015, pp. 487–494, New York, USA, June 27–July 2, 2015.

[31]   L. Cherkasova "Improving WWW proxies performance with greedy-dual-size-frequency caching policy," *Computer Systems Laboratory HPL-98-69 (R.1)*, November 1998.

Chapter 3

Flexible and dynamic Network-as-a-Service for next generation Internet

Amina Boubendir[1,2], Emmanuel Bertin[3], Gladys Diaz[2,4] and Noëmie Simoni[2]

Abstract

This chapter presents Network-as-a-Service (NaaS) architecture leveraged with flexibility and dynamicity. Convergence and stronger collaboration between network planes while keeping separation is the driving notion for this chapter. We first analyze the convergence of the needs and the convergence of conceptual and technological solutions in the new telecommunication ecosystem.

We have considered solutions such as Network Functions Virtualization (NFV), Software-Defined Networking (SDN), Cloud and Service-Oriented Architecture (SOA). The advantages brought by these paradigms contribute to the pillars of NaaS. Thus, their convergence and collaboration is necessary for realizing NaaS. However, they present challenges that yet need to be addressed. Therefore, we study the evolution that is crucial for a flexible and dynamic NaaS.

We describe an approach for designing Virtual Network Functions (VNFs) as service components. This "as-a-Service" design presents a service component architecture model with functional and non-functional aspects, and a set of properties to be respected for the structure, interconnection and for the management of VNFs.

We also describe the importance of adopting and integrating dynamic Application Programming Interfaces (APIs) in the relational dimension of NaaS for more agility in contracts. These new schemes allow us to introduce flexibility then dynamicity of NaaS. These two NaaS features rely on the interactions of network planes (data, control and management plane).

Flexibility is achieved through the offers of network services that can be customized through a network exposition layer. This layer offers discovery, selection and composition of VNFs. These VNFs are first described with quality of service (QoS)

[1] Network Control Architectures Department, Orange Labs, Châtillon, France
[2] Networking and Computer Science Department, Télécom ParisTech, Paris, France
[3] Unified Communications and Advanced Services Department, Orange Labs, Caen, France
[4] L2TI Department, Paris 13 University, Sorbonne Paris Cité, Villetaneuse, France

information for a QoS-based selection according to the required or desired QoS. The customization is possible thanks to service composition.

Dynamicity is achieved through an automated global orchestration of network services. We define the functions of this orchestration after an integration of SDN and NFV, then an incremental integration of SDN-enabled NFV with "as-a-Service," dynamic APIs and a new network virtualization layer, all in the same architecture. The global orchestration is responsible for automatic life-cycle management of network services for dynamic Service Level Agreements (SLAs). Indeed, this automation aims at ensuring compliance of QoS with SLAs and to react as dynamically as the changes in user requests or preferences.

3.1   Introduction

Over the last decade, many elements have drastically changed the telecommunication ecosystem. At the service level, no doubt the development of an entirely new type of service providers commonly known as "Over-The-Top (OTT) players" has been a real game-changer. The emergence and continuous evolution of OTT applications as well as Cloud applications is fostering this trend. OTT applications are driven by web actors. These applications push for more versatile audio and video communications with dynamic behaviors with more constraints on the network, like mobility. However, traditional Telco architectures are ill-suited to meet the dynamic requirements of today's application services. Current monolithic networks hinder Telcos and OTTs from choosing the most suitable network behavior needed by applications and even less to perform it dynamically and "on-the-fly."

Nonetheless, with the dynamicity of service applications, it becomes essential for networks to provide dynamic reactions. For that, Telcos need to transform their networks for such variations in customer demands. Within such an ever-changing telecommunication ecosystem, agility in delivering services is a must for Telcos. Network architectures will be transformed as Telcos adopt virtualization and softwarization at different layers of networks.

These transformations aim to achieve higher agility in network management and operations and in network services delivery. In the context of network openness and Network-as-a-Service (NaaS), Telcos are more and more seeking to expose value-added network functions to monetize their networks.

The ultimate objective is to enable network operators to expose Virtual Network Functions (VNFs) toward third-party actors through an NaaS architecture that allows Telcos to build customized network services in a dynamic way. Our direction is the notion of convergence. We present hereafter the convergence of needs (Section 3.1.1), then the convergence of solutions (Section 3.1.2) and the different views of convergence (Section 3.1.3).

3.1.1   Convergence of needs

With OTT services such as Web real-time communications (rich instant messaging, audio and video conferencing), cloud services and content services (live and on-demand video), users expect ubiquitous access to network services, from any device, via any access network with service continuity and a QoS in accordance with application constraints. In addition, user mobility, network mobility and session mobility are expected as given since devices become network agnostic capable of connecting to any available network. Offloading an ongoing media session from an access network or a device to another in a transparent way becomes a strong quality of experience requirement.

These application-layer evolutions push forward the user-centric models. These changes raise challenges in Telco network infrastructures. Indeed, Telcos still rely on mobile services designed for and deployed on dedicated mobile-enabled infrastructures. They reshape application-to-network communication models. Thus, new deployment models, new networking paradigms and new management approaches are required. For example, mobility requires a smooth multi-access continuity which itself requires adequate network mechanisms and automatic orchestration and management operations.

Also, as services and devices impose a huge throughput demand pushing networks and IT infrastructures to their limits, Telcos need to be able to dynamically scale their networks. Scaling alone would not be enough though. The new communication models impose new network control solutions. This requires developing and deploying new network services and control mechanisms.

In addition, OTT services that use Telco networks have diverse requirements in terms of network services that vary with their profiles and core business (a service provider, an application developer or an enterprise). Operator networks cannot address this diversity of demand today as network reaction to changing networking needs of applications is currently very slow. Indeed, vendor-dependency imposes to acquire new equipment and to develop new network services. Then, the traditional way of implementing services in hardware is a long process: on-site human intervention, physical installation and manual configurations through Command Line Interfaces are needed, which increases service deployment cycles and thus Time-To-Market.

Moreover, networking devices and mechanisms are vertically integrated and implement numerous standard and vendor-specific protocols. This makes management operations complex and results in static and rigid networks unable to respond as applications requirements rapidly change. This static and rigid nature of networks makes it difficult for operators to offer adapted solutions.

Furthermore, increasingly service customization is presented as the desired property to be integrated in future network services. Indeed, network solutions offered nowadays are monolithic and are being delivered in a standard way without any service-level or network-level customization. Therefore, service providers are seeking to achieve adaptation and differentiation of their service offerings to accurately answer applications needs and target the largest range of users' demands.

3.1.2   Convergence of solutions

On the one hand, in Cloud Computing, virtualization and SOA play important roles in the construction of service offerings. Virtualization enables SOA which in turn enables a structured interaction of services in heterogeneous and distributed systems. It presents tools for service exposition, composition and orchestration. Also, virtualization of computing resources enables dynamic placement of services.

On the other hand, networking technologies like SDN, which breaks the vertical integration of network architectures, allow a network provider to achieve network adaptability through programming and provides advantages in network configuration processes automation, network control and orchestration. Moreover, NFV applies virtualization of computing resources to network functions and brings dynamic network function deployment which is essential within NaaS.

In order to expose such an architecture, the need is tremendous for sustainable network tools providing network service description, discovery, invocation, composition as well as network service management and orchestration. Therefore, SDN and NFV converge with SOA-like models to consider the principles of service architectures for the design of VNFs. This makes NaaS inherit SOA and Cloud characteristics.

These enablers, together, offer ways to take advantage of the virtualization, programmability and service architecture assets. The convergence of these networking and virtualization technologies is a strong foundation for an adaptable network. In such converged environments, the structure of services or components is absolutely important. In order to be able to adapt services to all organizations, services need to be as autonomous as possible, easy to be used and to be managed in all contexts (Cloud, edge or fog computing).

In front of the needs for flexibility and dynamicity, how would these solutions provide service differentiation? That is to allow a service provider to build, on-demand, customized network services based on users' profiles. Then, as soon as requests change, how the service provider would modify these services on-the-fly? And what do flexibility and dynamicity really mean in this new ecosystem?

Our objective is to propose an NaaS architecture with capability of network service differentiation considering both computing and networking technologies. We propose in this chapter an NaaS architecture achieved through the convergence of emerging virtualization and networking technologies.

3.1.3   Convergence views

The first convergence view is the "user-centric" aspect. It is the most encompassing view because the user, with its strong needs and requirements regarding mobility and network service ubiquity, is in the center of Cloud and Network architectures and will be the heart of NaaS architecture in the future. It is quite equivalent to the Cloud Software-as-a-Service (SaaS).

This convergent view needs to rely on two other convergence views with service continuity requirements:

- Convergence of networks and
- Convergence of services and applications (IT, Telco and Web).

However, for these two convergence views to be efficient, an important convergence view is needed. It is the convergence of network planes: management plane, control plane and data plane. Indeed, services in general and network services in particular should be accessible by the three planes. This implies a convergence of interfaces or Application Programming Interfaces (APIs). We further explain how our contribution relies on this convergence to introduce flexibility and dynamicity features for NaaS.

We propose in this chapter an NaaS architecture achieved through the convergence of emerging virtualization and networking technologies. In Section 3.2, we survey and analyze the strengths of NFV, SDN, Cloud along with component and service-oriented models which have the potential to enable achieving NaaS. We then highlight the challenges and the research issues that yet need to be tackled to achieve the flexible and dynamic NaaS objective. Next, we present the main schemes that are crucial for NaaS. We focus on the meaning of "as-a-Service" through service component structure for VNFs and the properties guaranteeing the delivery of personalized network services, the importance of dynamic APIs for NaaS and finally the customization of virtual networks (VNs) with respect to SLAs in NaaS. Later, we propose NaaS architecture that integrates flexibility and dynamicity features. We explain in depth these features step-by-step through different network architectures where we detail the architectural, functional and relational dimensions of our proposition. We finally present our conclusions and future work.

3.2 Technologies enabling Network-as-a-Service

We survey in this section the paradigms and technologies that are strongly considered as enabling technologies for achieving NaaS. We study NFV (Section 3.2.1), SDN (Section 3.2.2), Cloud along with component and service-oriented models (Section 3.2.3). We then discuss the challenges that need to be leveraged (Section 3.2.4).

3.2.1 Network functions virtualization

Virtualization has had a major impact in the computing world, and is now impacting the networking world as well. NFV is an initiative to move the network functions and services that are now being carried out by proprietary, closed, dedicated hardware to virtual containers in a form of VNFs on Commercial-Off-The-Shelf (COTS) hardware. It promises a decrease in the amount of proprietary hardware that is needed to launch and operate network services within the operator network infrastructure. Therefore, it should overcome network complexity and facilitates network management. The main benefit of NFV is the dynamic deployment of network services over a virtualized infrastructure. It then reduces service deployment cycles and thus Time-To-Market. In ETSI NFV architecture [1] (Figure 3.1), the NFV Infrastructure (NFVI) is virtualized and comprises a set of virtual resources providing computing, storage and network facilities. NFVI resources are controlled and managed by the Virtualized Infrastructure Manager (VIM). At the intermediate level, VNFs run on top of

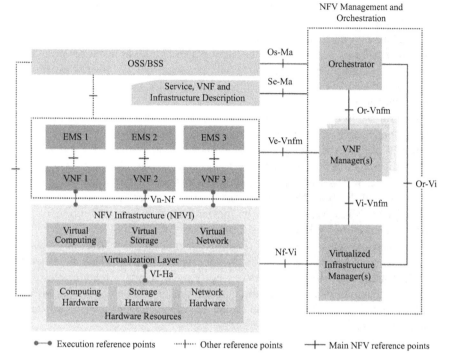

NFV REFERENCE ARCHITECTURAL FRAMEWORK

Figure 3.1   NFV architecture framework

one or several NFVI resources and the VNF Manager (VNFM) and Element Management System (EMS) are responsible for controlling and managing VNF resources. NFV handles VNF lifecycles (on-boarding, instantiating, monitoring, terminating and deleting). An NFV service is built based on a combination of VNFs using VNF Forwarding Graphs (VNF FGs). The NFV Orchestrator (NFVO) controls and manages infrastructure services. It coordinates the resource allocation to infrastructure services and VNFs, either by a direct interaction with the VIM or via the VNF Manager. It takes into account deployment policies based on various criteria including affinity rules, location constraints and performance criteria. The orchestrator brings more intelligence to the whole system comparing to the operations performed by the OSS/BSS which perform traditional management operations. NFV covers mainly service deployment issues which make it a resilient technological opportunity for Telcos.

3.2.2   Software-defined networks (SDN)

SDN [2] breaks the vertical integration of Telco networks. It proposes decoupling the network control logic (control plane) from the underlying traffic forwarding devices (data plane). All the intelligence is held by the control logic implemented in a logically centralized controller, also called Network Operating System (NOS), which maintains a global view of the network. This structural design eases the introduction of programmable interfaces through which the network becomes directly programmable. The control plane is programmable through northbound (NB) APIs by software network applications running on top of the controller. The forwarding devices are programmable by the controller through southbound (SB) APIs. The Management plane in the SDN architecture defines the infrastructure support tasks and operations that application, control and data planes cannot do for policy reasons. However, orchestration is a strong support for the OSS/BSS. It adds intelligence through new management operations (like on-demand deployment) comparing to classical operations performed by the OSS/BSS.

It is important to mention the difference between the different interfaces (APIs) in the SDN architecture.

- SB APIs allow the SDN controller to program network devices,
- Northbound APIS allow the Orchestration layer to program functions included in the controller,
- The APIs between the management plane (OSS/BSS) and the virtualized infrastructure are one-way APIs that allow OSS/BSS to directly program the networking resources and components,
- The APIs between the management plane (OSS/BSS) and the control plane are two-ways APIs that allow, from one side, OSS/BSS to program functions included in the controller, and from another side, the controller to notify OSS/BSS about changes or events either in the control plane or in the data plane.
- Another one-way API allows OSS/BSS to program Orchestration and network applications.

Figure 3.2 shows an SDN architecture with physical and virtualized network infrastructures.

These SDN principles simplify networks greatly. First, network devices no longer need to interface and process many standards and protocols but merely accept instructions from SDN controllers. Then, the centralized control simplifies policy enforcement and modification of the network. More globally, programmability brings dynamicity and automation. Indeed, network administrators manage and control network devices, network topologies, traffic paths and packet handling policies using high-level languages. This facilitates the design, delivery and operation of network services.

3.2.3   Cloud computing and service and component-based models

With the objective to extract the requirement features and study their relevance to be integrated in an "as-a-Service" architectural model for VNFs, we present a landscape of Component-based Models issued from application development technologies in the software development discipline. We survey architectural models from component-oriented models and service-oriented model to self-controlled component model.

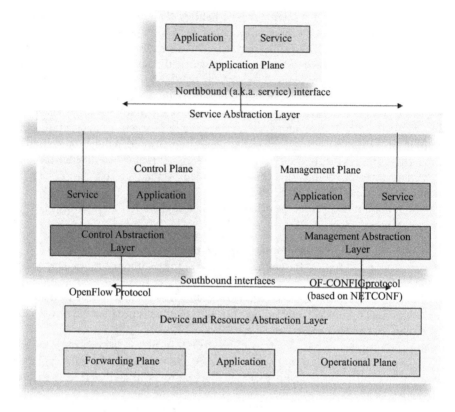

Figure 3.2   SDN architecture

We highlight and analyze the strengths and shortcomings of each model as well as the associated management approaches.

In Component-Based Software Engineering (CBSE), component models offer a structured programming paradigm that allows developers to reuse software components. A component is a software module offering predefined services, and able to communicate with other components. The functionality that a component provides is defined with its dependencies using offered and requested interfaces in a structured, usually hierarchical, form where components can either be primitives or composites (i.e., containing one or many inner components). This is adapted for the specification of large-scale distributed systems which will be the case of future virtualized network operators' NFV environments. This structure makes component models good candidates for the architecture design of an NFV network service and consequently the composed VNFs.

Cloud computing strongly relies on user-centric models. Cloud model and applications are based on service- and component-based models, called also CBSE. We usually talk about SOA for Cloud service models. Figure 3.3 shows the SOA-Cloud

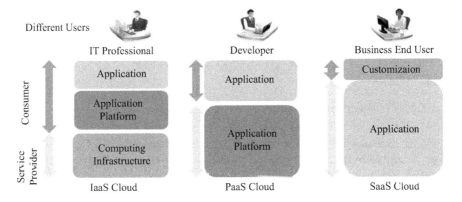

Figure 3.3   Cloud architecture model

architecture model. We distinguish three layers as defined by NIST [3]: Software-as-a-Service (SaaS), Platform-as-a-Service (PaaS) and Infrastructure-as-a-Service (IaaS).

Let us then analyze the architecture components and services in different initiatives. We focus on approaches that achieve user-centric services.

3.2.3.1   Fractal component model

Fractal [4–6] defines a general conceptual component model. It is defined in OW2 Consortium as a modular and extensible component model that can be used to design, implement, deploy, reconfigure and manage complex software systems and applications, from operating systems and middle-ware platforms to graphical user interfaces. *Fractale* stands for objects that have a regular invariant structure even at different scales. This scalability and elasticity feature of Fractal is important for distributed systems. In Fractal, dynamicity aspects are presented through the possibility to add or remove components to or from an already deployed application thanks to reconfiguration of the bindings between components (at run-time scaling) and to the introspection of composite components.

The strength of Fractal resides in its interfaces. It has three interfaces:

* Usage interfaces: client and server interfaces for functional bindings, where the component defines what it needs and provides.
* Control interfaces: controllers or membrane, useful to control and manage the component life cycle and provides methods to start or stop the component.
* Management interfaces: also through the membrane for component configuration.

In addition to that, Fractal defines an Architecture Description Language (ADL). The ADL uses an XML syntax and is a way to describe a component based system without having to worry about the implementation code.

However, Fractal component server interfaces may be functional interfaces or non-functional interfaces (controllers or membrane). The functional connections are well-defined and composition is possible but the non-functional interfaces (the membrane) are not explicit where control and management are not separated. There is no way for non-functional bindings or compositions.

To address these shortcomings, the Grid Component Model (GCM) was proposed as we explain next.

3.2.3.2   Grid component model

GCM [7] has extended Fractal toward autonomous distributed systems. Specified in ETSI [8], it defines well-structured non-functional aspects defined with all necessary management elements and interfaces in the membrane. Thus, non-functional bindings or compositions are possible similarly to the functional bindings in Fractal. GCM architecture strength resides in this separation of concerns in designing separately:

- Business Content: functional aspects of a primitive component responsible for business logic, and
- Membrane: responsible for non-functional aspects.

In GCM, the membrane structure is a strong asset for component management and control (non-functional aspects). For the functional aspects, communication is performed on interfaces and follows component bindings. GCM components management approach uses an autonomic approach in hierarchical architectures implementing Monitoring-Analysis-Planning-Execution (MAPE) loop [9] to manage this hierarchy where *Execution* is used to change the bindings. Figure 3.4 shows a GCM component architecture with a MAPE loop.

However, the main shortcoming of GCM is this hierarchical nature of the bindings between components which adds functional coupling and makes primitive components hard to be composed and recomposed. Bindings are needed to be less tight in order to have a dynamic composition. SOA has addressed this rigidity in bindings as we present hereafter.

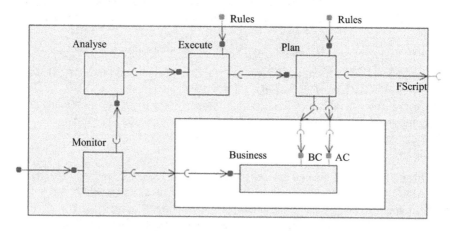

Figure 3.4   GCM component architecture autonomic control (MAPE loop)

3.2.3.3 Service-oriented architecture

Quite differently, SOA [10] is an architectural model, where a component is a *service* and is the unit of work done by a service provider to achieve desired service for a service consumer. It proposes a composition model with loose coupling and reduced functional coupling among interacting service components. These features are a must for building customized services.

Composition consists in building a global service composed of a set of elementary service components. Flexible composition [11] is customizable by adding, replacing and removing service components based on user needs.

SOA proposes a structure: Service Component Architecture (SCA) and a technology: Web Services (WS).

Component Architecture

SCA [12] is a component model adapted to SOA features and enables creation of service components and modeling of service compositions. The strong features of SCA are re-usability, reference points and loose coupling. Reference points allow the service component to have all the needed connections, while loose coupling brings flexible compositions (any combination) of service components through loose bindings to eliminate all functional couplings. The coordination of service components is achieved through Business Process Execution Language (BPEL).

Web Services

WS [13] implement SOA requirements and thus support flexible composition and methods of service integration in Cloud applications. The strong asset of WS is the Client–Provider relationship based on service publication and description in catalogs by the service provider, and service discovery and invocation by the client. For this, they support the WS Description Language (WSDL), Universal Description Discovery and Integration (UDDI), EXtensible Markup Language (XML) and APIs such as Simple Object Access Protocol (SOAP) and Representational State Transfer (REST).

However, in practice, coupling in SCA is not loose as the reusability feature in BPEL is strong and relies on existent functional coupling. Thus, SCA model is not dynamic and it is not possible to add, replace or remove service components or bindings dynamically (at run-time). Also, web services do not integrate the notion of contract or SLA and do not support QoS management as GCM does.

3.2.3.4 Self-controlled service component model: SCC model

Because of the shortcomings of SOA implementations exposed above, self-controlled service component (SCC) Model [14] combines the strong features of GCM and SOA. SCC components are designed based on the resilient GCM membrane structure, explicit non-functional composition and autonomic management of GCM; and on loose coupling, highly dynamic composition and service discovery capacities of service-oriented models. SCC has adopted GCM and to a certain extent MAPE loop. SCC approach has relied on initial SOA features for the design of SCC components. Thus, SCC components inherit the ability to achieve dynamic and at run-time service composition. Further, SCC model has defined new service component features beyond the ones advocated by existing models. SCC has proposed a dynamicity in handling components through auto-control approach for the management of non-functional aspects (QoS) in flat architectures enabled by the membrane structure. Moreover, SCC components include, starting from the design step, functional and non-functional specifications. Furthermore, self-control mechanisms are attached automatically to SCCs to enable at run-time autonomic management.

This analysis has allowed us to study the principles and evaluate the relevance (strengths and shortcomings) of component and service models that targeted user-centric services in Cloud, virtualized and distributed environments, which makes them good candidates for adoption within NFV environments.

Based on the stakes and awaited objectives, we have presented earlier, regarding the adoption of component models for the design of network function in dynamic Telco NaaS architecture, the SCC model seems to gather the most important features and answer the requirements for modeling VNFs as services.

Indeed, the service-oriented features of SCC are the bases for the exposition of VNF: description, discovery, selection and dynamic composition thanks to the "loose coupling" feature. It makes it possible to build different compositions of network functions and thus customize network services based on users' needs. Moreover, the SCC functional and non-functional specifications will be a plus for the deployment, invocation and life cycle management of network functions in NFV. Once the VNFs deployed and invoked, the QoS monitoring will allow a network provider to always monitor VNFs behavior and react if needed to respect the QoS SLA. The "at run-time" dynamic composition feature will allow a network provider to consider a changing in the network service request of a user (OTT, application developer, etc.).

So, in order to have such exposition and dynamic composition of network functions to build customized network service offerings, we propose to adopt an "as-a-Service" architectural model and design properties for VNFs based on the SCC model as we present in Section 3.3.

3.2.4 Challenges to undertake for achieving NaaS

3.2.4.1 NFV and Cloud-related challenges

In today's user-centric ecosystem, network providers are working to take advantage from NFV to deliver network services with high dynamicity to meet the continuously changing network users' requests. However, adoption of NFV by network operators is still at a first step. Indeed, NFV implementations are Cloud-based. They correspond to an Infrastructure-as-a-Service (IaaS) model. VNFs are VMs corresponding exactly to the network functions that network operators have in legacy network architectures. Basically, the operation of VNF deployment corresponds to a VM placement over the NFVI. Thus, it is clear that to date, NFV consists in implementing legacy network functions logic (performed today by physical proprietary dedicated hardware) in software systems over VMs of the NFVI keeping the same service logic as defined by 3GPP specifications.

Actually, this way of implementing network functions logic in VMs over virtualized infrastructures is so similar to implementing applications in VMs in a Cloud

Computing environment. So, NFV today is a way to "cloudify" the operator network by virtualizing legacy physical network functions. So, exposing APIs over legacy network functions, even if they are value-added functions, is not enough to achieve resilient network openness. Neither is enough for the deployment of the VNFs with NFV. Simply virtualizing existing network functions (e.g. gateways, CSCFs) may make them cheaper to implement but will not reduce the network complexity and does not eliminate the monolithic nature of network services and architectures. Indeed, the authors in [15] present a critical overview of the architectures proposed so far for cloudifying IP Multimedia Subsystem (IMS) and give research directions including reconsidering granularity levels of network functions. That means to separate the functional entities' logic and data or state, to consider decomposing the IMS network functional entities' logic into smaller sub-functional entities, which should lead to finer control over the distinct functions. These are major research directions we have considered in our work. We believe that for greater flexibility and to be able to offer differentiated network services, network operators should reconsider their legacy functional architectures to rethink the services' logic to make them less monolithic. The challenge here is to push further the actual VNF developer/provider decomposition into a functional decomposition of VNFs. Moreover, NFV promises the ability to provide tailored network services based on compositions of VNFs offered as services similar to cloud services. However, Cloud services are based on well-defined *Service and Component based Models* like SOA. So, we strongly believe that NFV should apply moreover the Cloud Computing service models as we have exposed above. Of course, the best approach or model for designing network functions to leverage the benefits of NFV is not immediately obvious, but there are several enablers: Network-to-Cloud convergence [16], service and component-based models (see Section 3.2.3) and micro-services. Apart from evident benefits like resource utilization that NFV surely brings through this cloudification of network functions, how can we take more advantage of advances in Cloud computing and Cloud application architectures within this network-to-Cloud convergence to find the best way to architect VNFs to provide more flexible NFV network service offerings?

3.2.4.2 SDN-related challenges

The separation of control and data planes that SDN brings is an advance for scenarios of network policies enforcement or networking operations like reconfiguration that do not consider dynamicity of application services. SDN-based network virtualization approach today is more or less static. However, while users have continuously changing requests, this approach is not adapted to the user-centric era. Mobility constraints, SLA requirements and user preferences (temporal and spatial) should be taken into consideration at the SDN networking level. For that, the challenge is to conceive virtual control planes or Virtual Network Operating System (vNOS) that would be customized based on users' needs and requirements at the network level. This customization should be dynamic and each would be dedicated to an application service, a session or a user. Virtualization of control plane is crucial for the customization of network services. VNs need to answer SLA and QoS requirements. For that, there is a need for dedicated virtual control plane for each VN. The role of the actual control plane in the SDN architecture would become a Network Hypervisor supporting multiple virtual control planes.

3.3   Schemes of transformation: the way toward NaaS

The convergence required to design and build the new network ecosystem becomes possible thanks to the paradigms of "as-a-Service," virtualization and programmability.

We need to correctly understand their scope and accurately refine their use. Our perception of these three paradigms leads us to the following refinements:

- An "as-a-Service" VNF component (Section 3.3.1) should comply with all the properties that we define (Section 3.3.2) in order to facilitate its use, its interactions and its management. In addition, we advocate a self-control mechanism based on QoS management and control leveraged with a QoS-aware description of VNFaaS components.
- Virtualization has been a real technology push for the agility that was always needed in IT. It is therefore urgently crucial to make OSS evolve to support this agility and introduce multiple levels of orchestration. This would be a solution to the current practice of explicit configurations through provisioning management applications (between OSS and EMS in NFV). For that, the nature of APIs must adapt to these new structures and become more dynamic thanks to programmability (Section 3.3.3).

3.3.1   Modeling VNFs as service components

The architecture of an "as-a-Service" component is depicted in Figure 3.5. We strongly rely on SCC model so that the architecture of an "as-a-Service" component inherits SCC structure, interfaces and membrane. An "as-a-Service" component has two functional interfaces: a Server interface that includes the service methods (processing functions) that are performed by the VNF service component, and a Client interface which invokes the following service in a composition transmitting the result of its processing to another component or for exploitation of the final result. Note that the functions of these interfaces have no return value, according to the SCC specification.

The triptych In Monitor, Out Monitor and QoS Control, associated to each service component, introduces an homogeneous autonomic management approach. Here, the idea is to position the Monitoring mechanism of the MAPE loop and integrate QoS monitors at the nearest level from components which are in fact services. This brings the most pertinent monitoring information about a component. The QoS measurements help notify in case of any QoS requirements violation. If any, a dynamic reaction to repair or replace the component is performed. The structure of the membrane allows components to have a generic characteristic and thus components' reuse.

IConfigMnitor (In and Out), IConfigQoS and IActivate are non-functional server interfaces representing the management interfaces.

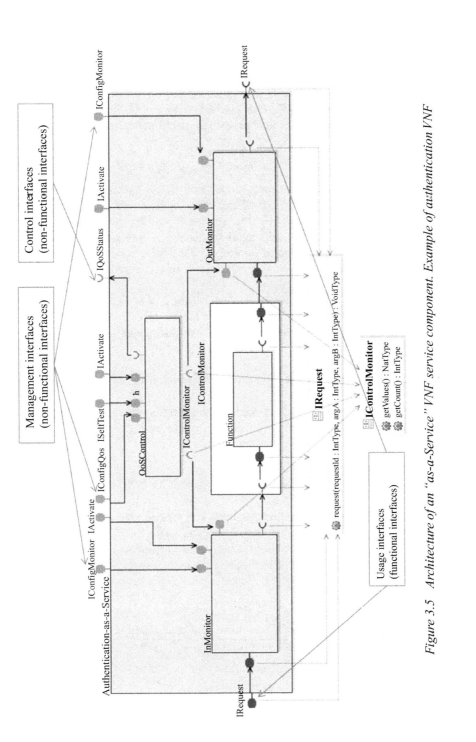

Figure 3.5   Architecture of an "as-a-Service" VNF service component. Example of authentication VNF

They hold the necessary mechanisms to manage the configuration of non-functional components (the monitoring and QoS control components) in the membrane.

The control interface QoSStatus is a client non-functional interface. It contains mechanisms to enforce the self-control information to the administrator in charge of reacting to QoS violation events. It outputs InContract notifications as long as the behavior is compliant to the SLA, otherwise it triggers OutContract. Absence of InContract events can be used to detect severe failures from the service component.

3.3.2 Properties of a VNF "as-a-Service" component

For a component, a VNF in our case, to be "as-a-Service" component, it needs to gather all the following properties. The service components that feature these properties make it possible to benefit from service exposition and dynamic service composition advantages. SCC components gather the interesting properties from Fractal and GCM, as well as properties expressed in SOA methodology but that have not been implemented in SCA and Web Services. It has proposed SOA+ properties as described in [14].

We propose that "as-a-Service" components verify new properties in addition to the SCC ones. All the features are tightly linked as the non-respect of one of them may impact the verification of others. We distinguish through properties related to:

- The definition of the structure and the formal descriptions of service components, i.e., the nodes themselves,
- The definition or design of the service logic and functional architecture of service components, i.e., the interactions or links between service components,
- The management of the service components.

3.3.2.1 Properties related to the structure of VNF service components

Cohesion

VNF service components should be consistent. The service logic offered should be relevant and recognized as a meaningful business service for potential customers. The service rendered by the component should find all its functionalities in a logical way internally. We also call this feature: self-sufficiency, autonomy or even functional decoupling.

Reuse

A VNF service logic in a service component should be reusable to build different services, in different compositions and different environments.

Abstraction

VNF service components must dissimulate the service logic from the outside world and act as black boxes beyond service descriptions that should appear on service catalogs and SLAs.

Invariance

A VNF service component should have an identical structure, that would not vary from a level to another in a hierarchy of service components.

That means that the structure is invariant when scalability and elasticity are needed. This property is expressed in components construction, derived from the Fractal core principles.

Statelessness

A service is stateless if it processes each received request as an independent transaction without any relationship with previous ones. A VNF service component should then neither keep information regarding its state or its processing state nor handle information about previous requests. If it maintains its state for a long period, it will lose the "loose coupling" feature, its availability for other incoming requests and even its possibility to scale. Each component should handle data coming only from outside its area of responsibility i.e., from other service components so that its functional behavior does not use data received from previous invocations. For that, we need to rely on transactions in unconnected mode that define well-specified formats of in-requests and out-responses. Here, the SCC structure with its interfaces would help. We also need to delegate information handling and state management to external entities. This feature is crucial as it impacts the independency of a service component and thus the possibility to include it in compositions that need to be dynamic.

Mutualization

By mutulization we mean multi-tenancy. VNF service components should support multi-tenancy in order to be invoked by multiple users requiring the offered service either simultaneously or not. This reinforces the statelessness feature. Mutualization requirement calls for a need for loose bindings or connections between VNF service components to have the capacity to provision multiple users and answer multiple service requests autonomously. Thus, Mutualization will help realizing minimum functional coupling and loose coupling between functions. We can check the respect of loose coupling A network provider would offer the same VNF service component instance as-a-Service to multiple users.

3.3.2.2 Properties related to the interactions or links between VNF service components

Loose Coupling

VNF service components should have no predefined sequence between them and should maintain relationship with minimized functional coupling.

Invocation

A VNF service component should be accessible and invoked based on service requirements in SLAs (invocation interface, function or service and QoS level). We distinguish three types of service contracts: syntactical contract (service interface, function, service or process name, input/output parameters and structural constraints), semantic contract (informal description of the function or service with service use rules and constraints) and service-level (QoS and SLA) contract (defines the service commitments (time to response, process or access)

Composition

Multiple VNF service components should be able to be chained as elementary entities (primitive or composite components) to create an NFV network service. They should be effective service composition participants, regardless of the size and complexity of the composition. This composition requirement feature is verified only if all of the features described are verified.

3.3.2.3   Properties related to the management of VNF service components

Description

VNF service components should be describable based on meta-data in an independent manner from their implementation specificity. The formal description should have a logical and meaningful structure.

Registration

VNF service components should be able to be registered in a Domain Registry. This registration can be made through a publication (publish) of its service offering, QoS level and state. It should also be able to discover its environment through service discovery.

Exposition

This feature includes cohesion, description, registration and invocation. Exposition is providing the functional and non-functional description of VNF service components as well as their inherent QoS level offered through catalogs on service portals to allow a third-party actor to select and/or build (based on his profile and competences) an NFV network service.

Auto-management

VNF service components should be able to monitor and control their behaviors (non-functional aspects) using autonomic management approaches like MAPE loop. Placing the monitoring of QoS very close around each service component and business logic helps to detect exactly the malfunctioning component.

Ubiquity

Ubiquity is the high equivalence between service components. VNF service components should be defined and described based on their core function and the level of QoS (the values of QoS parameters) they offer. According to this definition, VNF service components may be grouped into communities of ubiquitous or identical service components where service components of community provide the same function or service even if their business codes or algorithms are different, with the same offered QoS level. This feature goes with scalability issues, as the VNF service provider may decide to scale the NFV service by adding ubiquitous VNF service components. It also enables higher service availability as this gathering in ubiquitous component

Table 3.1   Evolution of dynamic APIs

	Interface	Implementation	Change characteristics
Dynamic API Level 1	Static contract include metadata	Tightly coupled syntactic integration	Not programmable
Dynamic API Level 2	Static contract include metadata and links	Tightly coupled syntactic integration	Not programmable
Dynamic API Level 3	Static contract include metadata, links and constraints	Loosely coupled, semantic interoperability	Programmable

community is an approach to set redundancy schemes (i.e., more chances to find the requested service with the desired QoS level).

3.3.3   Dynamic APIs

The concept of dynamic APIs provides a path to a future mode of operations. Today's APIs provide interfaces to access to services with predefined and static contracts (SLAs) between the parties. The objective behind dynamic APIs is to bring more agility to the contracts, more coherency and respect of QoS requirements.

Introducing dynamicity in APIs would mean a closer cooperation between the actors or parties of a system with faster reactions in case of changes. Indeed, dynamicity allows an API to advertise new/modified features in its payload so that a consuming system or a collaborating party can be notified of the changes and react to them. We distinguish three levels in the evolution of dynamic APIs. Table 3.1 synthesizes the evolution of dynamic APIs.

3.3.3.1   Level 1 dynamic APIs

Dynamic APIs of level 1 provide for simple notification. They are semi-automatic and disruptive. They have an interface that presents static predefined contract between parties with ability to include metadata. Their implementations are tightly coupled and they rely on syntactic integration. The change to interface is not programmatically supported. This means that the contract is broken after a change. However, unlike convential broadly practiced APIs, level 1 dynamic APIs can automate notification of change. This level of dynamic APIs 1 is a known practice with limited capabilities and implementations.

3.3.3.2   Level 2 dynamic APIs

Dynamic APIs of level 2 provide notification and documentation to improve human intervention. They are also semi-automatic and disruptive. They have an interface that presents static predefined contract between parties with ability to include metadata and links. Like level 1 APIs, level 2 APIs are tightly coupled and they rely on syntactic integration. Also, change to interface is not programmatically supported. However, in addition to automation of notification, level 2 APIs can facilitate changed mappings between systems. Level 2 builds incrementally on Level 1 but still does not address the fundamental problem of agility.

3.3.3.3 Level 3 dynamic APIs

Level 3 builds incrementally on level 2 but enables higher dynamicity. Indeed, level 3 APIs have an interface that supports static predefined contract but include metadata, links and constraints (like QoS constraints). Unlike levels 1 and 2, level 3 APIs are loosely coupled and rely on semantic interoperability. Thus, change to interface is programmatically supported and the contract is adaptable. In addition to automation of notification and changed mappings, level 3 APIs can fully automate mappings between systems to maintain synchronization. Level 3 APIs add constraints to dynamically maintain synchronization between systems. They are fully automatic and non-disruptive.

Dynamism is represented by actual behavior as a property of an actor. What makes an API dynamic is expanding the scope of its contract to define an interaction-model between the engaged parties that is designed to cope with change either in the request or in the behavior. This definition should be semantically understood. In effect, it supports the ability for the actors to negotiate change during the life-cycle of the agreement. For these reasons, and thanks to the properties of level 3, dynamic APIs, the use of dynamic APIs is crucial for NaaS. Indeed, the new usages call for faster reactions and faster integration of changes in front of the very dynamic requests coming from the application layer and from users with constraints on QoS that are so likely to change during mobility while seeking for service continuity. We further highlight the importance of using dynamic APIs. Dynamic APIS are used by the orchestrator and between the building blocks of NaaS architecture in Section 3.4.

3.4 Flexible and dynamic NaaS

We present in this section the flexibility and dynamicity as features of NaaS architecture. These features result from different convergence. We first introduce the actors of the convergence in NaaS where we describe the roles of network planes and the interactions needed within the new telecommunication environment (Section 3.4.1). Then, we introduce flexibility of NaaS through network exposition (Section 3.4.2). Finally, we present the dynamicity of NaaS step-by-step through different network architectures with a focus on convergence of solutions and the introduction of automatic orchestration of network services (Section 3.4.3).

3.4.1 Introduction: the convergence of NaaS actors

In the new telecommunication environment, new usages with new users' needs have risen in the form of converged needs. In parallel, a convergence of network and IT solutions has emerged and is widely supported. In front of this convergence of needs (new usages through Cloud, IoT, digitization, etc.), a convergence of solutions is necessary.

Indeed, the new needs with their very changing service requests require a "flexibility" regarding building and choosing network service offerings. The possibility of choosing services imposes the possibility of making different compositions of network functions' components. This implies achieving differentiation and customization of network service offerings through composition of network functions. It is more and more becoming a required property to be integrated in future networks.

The convergence of needs require also "dynamicity" from the network. The network reaction to requests should be as dynamic as the requests and behaviors themselves. Dynamicity would make it possible to deploy network services on-demand and to change their behaviors on-the-fly. Moreover, there are needs for "adaptability" in order to make network services meet the service requirements and SLAs.

All these features are vital in realizing Network-as-a-Service (NaaS). The responses and reactions of the network in front of application service needs are achieved through multiple invocations of network functions' components. These invocations are made over the management plane, the control plane and the data plane. The management plane, the control plane and the data plane are the three integral components of a telecommunications architecture. Therefore, the three network planes are the three main actors to realize these features and their convergence is necessary. But what would a convergence of these planes brings?

Let us first describe the role of each network plane.

Management plane
The management plane is the set of tools to coordinate and monitor the different deployed resources and capabilities. The management plane carries administrative traffic. Its role is built around FCAPS (Fault, Configurations, Accounting, Performance and Security) functions. On a daily-basis, it ensures operational management and fault management. In a continuous manner, it performs configurations, accounting and performances management operations.

Control plane
The control plane is responsible for provisioning and end-to-end resources allocation and reservation. It aims at optimizing the use of resources by reducing the times of processing in network nodes. Control plane operations are faster than management plane operations.

Data plane
The control plane and management plane serve the data plane. The data plane, also known as the user plane, ensures the transfer of user traffic over the network between users through multiple protocols. Service or processes are executed in the infrastructure at the data plane.

Based on these roles, we identify and define new functionality where these planes collaborate more closely to provide the required capacities for the convergence of needs. The collaboration makes the convergence of planes necessary. The new functionality nest under NaaS features. Each feature of NaaS is the result of the interaction of two planes, as we show in Figure 3.6 and define hereafter.

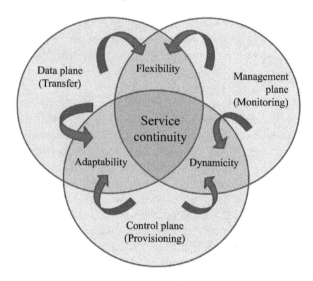

Figure 3.6   Convergence of NaaS actors: convergence of planes

3.4.1.1   Capacity of flexibility

The interaction and collaboration of the data (or user) plane and the management plane corresponds to the capacity of flexibility in order to take into account the temporal and spatial preferences of users. Indeed, if these two planes interact, users would always benefit from the best connection with the appropriate charging. Users would be reachable using any of their identifiers. For that, users' profiles needs to be well structured and with accurate information. To realize flexibility, it is necessary to describe and expose services.

3.4.1.2   Capacity of dynamicity

The interaction and collaboration of the control plane and the management plane corresponds to the capacity of dynamicity in order to react as fast as possible by merging relevant management information (reactive monitoring) and automation of the control plane (for re-routing).

Indeed, dynamicity enables to perform changes during operation (at run-time during exploitation or execution) to maintain compliance with SLAs. In this case, we talk about a dynamic QoS. The life cycle of QoS becomes: procurement at the beginning of the transfer and reactive monitoring during transfer inducing a renegotiation and reallocation if necessary.

3.4.1.3   Capacity of adaptability

The different mobilities lead, among other things, to changes in access networks and core networks that may have different QoS levels while we must maintain continuity of connectivity and continuity of service. The interaction and collaboration of user plane and control plane correspond to the capacity of adaptability in order to achieve interoperability in heterogeneous networks and reservation of resources in adequation with the QoS of each sub-network.

We focus in this chapter on flexibility and dynamicity capacities which are the features we propose for NaaS. We present next the flexibility of NaaS through network exposition.

3.4.2   Flexibility of NaaS: flexibility in service offerings through exposition

Achieving flexible network service offerings is enabled by service components composition. This implies a possibility of choice and a very accurate selection of components of network functions. Service components would be composed based on this selection. This flexibility relies on modeling network functions as service components (to become as-a-Service), as we have presented in Section 3.3. Service components would be invoked through dynamic APIs and compositions would be recomposed on-the-fly according to events, requests or user preferences. In order to achieve network flexibility, network function components should also be efficiently exposed.

We present here the three main modules of a network exposition layer in an NaaS architecture. First, the service Description, then the service Discovery and Selection, and finally service Composition.

3.4.2.1   Service description

The description of services in general and network services in particular is a fundamental tool in NaaS. The description holds information about network services that are exposed. Indeed, VNF service components should be describable based on meta-data in order to be accurately identified and used in a virtualized environment. Description languages should enable a description of network services in an independent manner from their implementation specificities.

Description of Web Services
W3C (World Wide Web Consortium) and OASIS are the two leading standardization bodies in the description of Web services. Descriptions can be based on standards like:

- Web Service Description Language (WSDL) or eXtensible Markup Language (XML) which focuses on syntactic (not semantic) description of interfaces which makes it restrictive to support the loose-coupling feature required by SOA.
- Service Component Description language (SCDL) is an XML-based language used to describe Service Component Architecture (SCA) elements such as modules, components, references, imports and exports.
- Web Application Description Language (WADL) for service description of RESTful Web services. WADL is closely related to WSDL.
- HTML for RESTful Services (hRESTS) is a promising approach to describe RESTful services.

- Web Ontology Language (OWL) is a W3C standard for formal ontology description with more semantics to describe business-oriented aspects.

Non-functional service aspects known as QoS in general are also an important aspect of service description in Web services description. WSQM and WS-QDL are two examples of description language with QoS parameters:

- Web Service Quality Model (WSQM) is an OASIS standard for Web service QoS specification.
- Web Service Quality Description Language (WS-QDL) is a description language to represent QoS using WSQM model.

Description of Telecom Network Services

Open Service Access (OSA) Parlay X and Open Mobile Alliance (OMA) Service Environment (OSE) are two standards for exposing telecom infrastructure to upper layer applications through a service abstraction layer. Also, the European Computer Manufacturers Association (ECMA) published a service description of computer supported telecommunication applications. These standards rely on WSDL for describing network functions as Web services. This helps with the creation and the deployment of Web services-based telecom applications. It has become obvious that adopting WSDL for the description of network functions limits capabilities of describing rich semantic and mainly QoS information. They also need to be extended to multimedia network services in virtualized environments. In contrast to functional service features, there is less agreement regarding the definition and specification of QoS attributes.

The integration of QoS in the description of services becomes essential.

QoS-aware description of VNF service components

For service components that feature the architecture and properties exposed in Sections 3.3.1 and 3.3.2, composition and adaptation are made easier. We here present another property to as-a-Service components. Each service component should have a definition of the QoS level offered: "offered QoS." It is the QoS of the component at the design: design QoS. This introduces a QoS-based description of service components in order to choose a service component on the basis of its functionality and behavior (QoS). The triad input monitor, output monitor and QoS control of self-controlled service components present generic QoS components guaranteeing that the service composition will provide the predefined functionality and QoS.

The QoS control component is associated with the business component to ensures compliance with the service contract. It should reflect the behavior defined at design time and proposed in the offer as the offered QoS level. So it controls the QoS level at run-time and compares it to the QoS at design time. This guarantees the QoS requested by the user when choosing the service.

Service components would be gathered into communities of components that are equivalent in functionality and QoS. Service components are defined equivalent if they provide the same services with the same QoS even if their algorithms are different.

3.4.2.2   Service discovery and selection

Service discovery plays a key role in flexible NaaS. It allows NaaS users (applications, developers or service providers) to discover the network services that the network provide offers. Then, based on discovery, the network services that best meet requirements are selected. Service discovery has been widely studied in the area of Web services. The main approach of Web service discovery is the OASIS standard Universal Description, Discovery, and Integration (UDDI), which specifies a data-model to organize service information with APIs for publishing and querying service descriptions. UDDI has been applied in telecom systems, for example, Parlay X and OMA OSE specifications adopted UDDI as the technology for publishing and discovering network services. Although UDDI served as the defacto standard for service registry during a certain period of time, its data model and query mechanism lack support for semantic and QoS information, thus significantly limiting its application in network service discovery. QoS-based network service discovery is particularly important for NaaS in order to meet service requirements. NaaS requires further development on network service discovery. Due to the wide variety of services, network service discovery for realizing the NaaS paradigm must be able to cope with the heterogeneity in services. A possible approach to addressing heterogeneity in service information is to develop a general purpose model that can be used for mapping different service descriptions. Dynamic and adaptive network service discovery for NaaS is important to support extensibility and elasticity of Cloud service provisioning. Dynamic and adaptive network service discovery relies on the latest service information to find the most appropriate services. However, keeping the latest network service descriptions precise becomes particularly challenging due to the dynamic nature of network states and capacities.

3.4.2.3   Service composition

Service composition is essential for realizing flexible NaaS. Service composition has also been widely studied in Web services. Web Service business Process Execution Language (WS-BPEL) is an OASIS standard that has been widely accepted by industry for modeling Web service composition. Web Service Choreography Description Language (WS-CDL) is a W3C candidate recommendation for Web service composition specification. For RESTful Web services, elementary building blocks can be composed into more complex services through mashups. This enables data, presentations or functionalities from two or more resources to be combined for creating new services. Web service composition with respect to QoS requirements is particularly important to network services. QoS-aware composition has been studied in the Web services area under different assumptions.

3.4.3   Dynamic NaaS: orchestration

In order to achieve the desired dynamicity in NaaS, the convergence of solutions is necessary. Solutions are the emerging network and virtualization technologies, mainly SDN and NFV as we have described in Section 3.2.

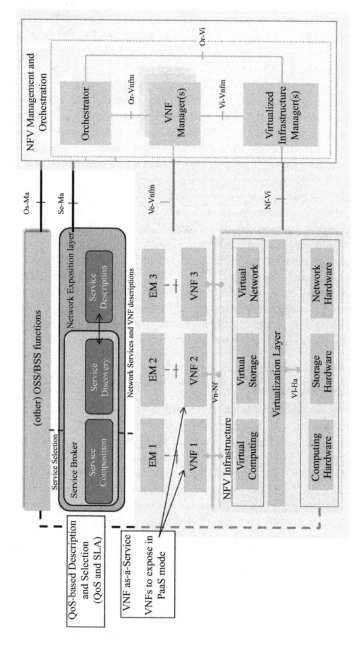

Figure 3.7   Flexible NaaS: network exposition layer for service customization

Even though SDN and NFV are independent from each other, each of them brings sets of advantages that are complementary to each other. So, SDN may benefit from NFV and NFV may benefit from SDN.

The integration of SDN and NFV is widely advocated. We also advocate this integration which is crucial in the NaaS context. Indeed, on the one hand, on the networking point of view, SDN separates control and data planes for more efficient cooperation between these two planes. This separation itself eases the use of APIs which provide advantages for automating network configuration processes. This helps to achieve network adaptability.

On the other hand, NFV virtualizes network functions by virtualizing their hardware resources and introduces a three-level management and orchestration plane, from where VNFs, network services and NFV infrastructure are managed. Therefore, the control in SDN and the management in NFV are key elements that worth being integrated to achieve NaaS. Moreover, it is necessary to integrate the elements we have presented in Section 3.3. Indeed, the strong VNF as-a-Service model with the triptych structure, properties and dynamic APIs are differentiating elements to be integrated with SDN and NFV into NaaS to provide it with the desired flexibility and dynamicity features.

Furthermore, driven by mobility, user-centric, IoT and the different possible organizations (Cloud, Fog, Dew, etc.), the new usages impose to consider a new level of abstraction for the virtualization at the network level in addition to the virtualization of network equipment which corresponds to a hardware virtualization. Thus, this integration should rely on the virtualization of the control plane (VNOS) for the customization of VNs as we detail further. This enables to build VNs that take into account SLA requests and adapt dynamically to the changes in SLAs using monitoring. On top of these integrations, the dynamicity feature of NaaS is introduced through our definition of global orchestration to automate the network service life-cycle management.

We present the dynamic feature of NaaS through three dimensions. First, the architectural dimension, highlighting the building blocks of NaaS. Then, the relational dimension to show the interactions between these building blocks, and finally the functional dimension to describe the functions of each element of the architecture.

3.4.3.1 Architectural dimension

We present here the architectural dimension of NaaS. We explain the NaaS architecture through different phases of integration. We consider that the network exposition layer described in Section 3.4.2 is already integrated. The first phase consists in integrating SDN and NFV. Figure 3.8 shows the building blocks of an SDN-enabled NFV architecture. The integration of SDN and NFV in this figure describes the positioning of SDN in the NFV architecture.

At the lowest level, we find the NFV infrastructure. It corresponds to the IaaS level of the Cloud. It is managed by an NFV IaaS manager (ex: OpenStack). It comprises COTS hardware resources abstracted using hypervisors (abstraction layer) to host virtual resources on top of them. In order to integrate SDN at this level, we introduce the SDN control and the SDN management to the manager of the NFV

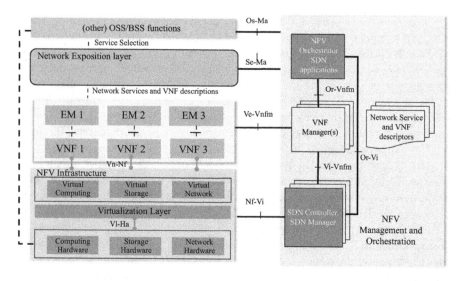

Figure 3.8   NaaS phase 1: SDN-enabled NFV architecture

infrastructure (VIM). We here integrate SDN in a virtualized environment since the VN in the NFVI may be a WAN. VN resources in the NFVI become virtual SDN network resources. The integration of SDN control, SDN management and NFVI management to NFV results in defining an orchestration at the NFVI level. This would achieve engineering of the infrastructure, which means that all dynamic and automatic operations would be achieved by the orchestrator. This integration implies also the introduction of northbound APIs on top of the SDN controller and SDN manager building blocks towards the SDN applications introduced to the NFV orchestrator.

The second phase consists in integrating as-a-Service to NFV, and integrating network virtualization through virtualization of the control plane (VNOS) with VN control and management to NFV. Figure 3.9 shows the integration of VNF-as-a-Service and network virtualization based on VNOS and QoS-based descriptions.

So, at the intermediate level, we find the VNFs and their associated EMS for management. It corresponds somehow to the PaaS level in the Cloud. At this level, we integrate the VNF as-a-Service model we have presented in Section 3.3. We apply the structure and properties proposed by the model to the software structure of VNFs to model them as composable service components. This makes VNFs become autonomous services. It also allows to achieve automatic control and provisioning thanks to the architecture of VNF-as-a-Service components and their self-control (see Section 3.3). In addition, the recursive nature of VNF-as-a-Service components—i.e., the possibility to reproduce the structure and the triptych at multiple levels—allows to perform auto-scaling and auto-elasticity.

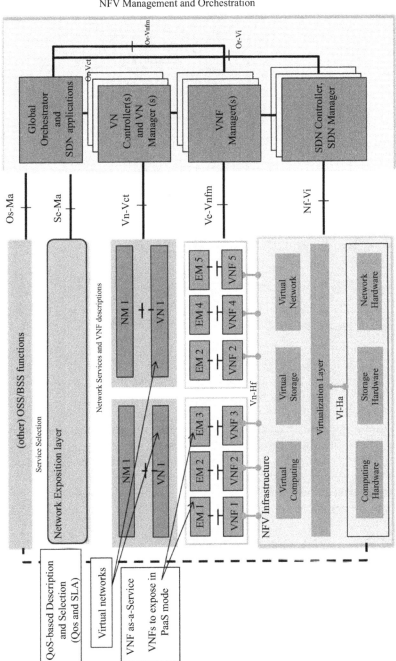

Figure 3.9   NaaS phase 2: SDN-enabled NFV, VNF as-a-Service and virtual networks control

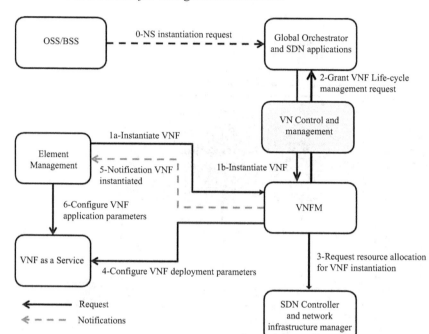

Figure 3.10   NaaS relational dimension

At a higher level, on top of the VNF-EMS level, we introduce a virtualized level that considers the building of VNs. This implies considering a VN control plane (VN controller or Virtual Network Operating System: VNOS) for each VN. The VN control is integrated within NFV management and orchestration plane. This latter should be defined and designed by the orchestrator according to the user SLA request. This virtualization level allows the orchestrator to consider changes or requests in the application layer and in SLAs through an event-based approach. These changes would be considered on-the-fly, whether they were top-down events (i.e., related to user preferences and contexts) or bottom-up (i.e., reactions based on monitoring).

The last phase consists in introducing global orchestration for automatic life-cycle management of network services with service descriptions and dynamic APIs. Figure 3.11 shows the automatic life-cycle management of network services through the global orchestrator. Automatic orchestration of network services is enabled by the high-level descriptions and the use of dynamic APIs.

3.4.3.2   Relational dimension

We describe here the relational dimension of NaaS. The relational dimension corresponds to the interfaces that are used between the building blocks of Naas communicate. Figure 3.10 synthesizes the main exchanges.

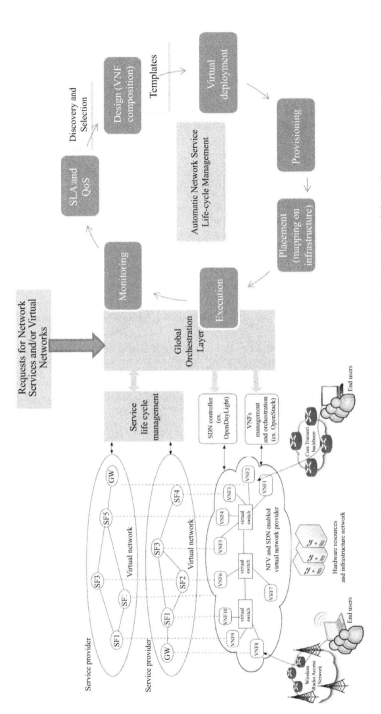

Figure 3.11 NaaS phase 3 Global orchestration: automatic network service life-cycle management

Os-Ma-nfvo reference point
This reference point is used for communication between OSS/BSS and NFV Orchestrator. It supports the following:

- Network Service Descriptor and VNF package management
- Network Service instance life-cycle management:
  - Network Service instantiation
  - Network Service instance update (update VNF instance comprised in the Network Service instance)
  - Network Service instance query (e.g. retrieving summarized information about NFVI resources associated to the Network Service instance, or to a VNF instance within the Network Service instance)
  - Network Service instance scaling
  - Network Service instance termination
- VNF lifecycle management: for VNF lifecycle management, the NFV Orchestrator identifies the VNF Manager and forwards such requests
- Policy management and/or enforcement for Network Service instances, VNF instances and NFVI resources (for authorization/access control, resource reservation/placement/allocation, etc.)
- Querying relevant Network Service instance and VNF instance information from the OSS/BSS
- Forwarding of events, accounting and usage records and performance measurement results regarding Network Service instances, VNF instances, and NFVI resources to OSS/BSS, as well as information about the associations between those instances and NFVI resources, e.g. number of VMs used by a certain VNF instance.

Ve-Vnfm-em reference point
This reference point is used for exchanges between EM and VNF Manager. It supports the following:

- VNF instantiation
- VNF instance query (e.g. retrieve any run-time information)
- VNF instance update (e.g. update configuration)
- VNF instance scaling out/in, and up/down
- VNF instance termination
- Forwarding of configuration and events from the EM to the VNFM
- Forwarding of configuration and events regarding the VNF from the VNFM to the EM.

Ve-Vnfm-vnf reference point
This reference point is used for exchanges between VNF and VNF Manager. It supports the following:

- VNF instantiation
- VNF instance query (e.g. retrieve any run-time information)

- VNF instance update (e.g. update configuration), scaling (out/in, and up/down) and termination
- Forwarding of configuration and events from the VNF to the VNFM
- Forwarding of configuration, events, etc. regarding VNF, from the VNFM to the VNF
- Verification that the VNF is still alive/functional.

Nf-Vi reference point

This reference point is used for exchanges between Virtualization Infrastructure Manager and NFV Infrastructure. It supports the following:

- Allocate VM with indication of compute/storage resource, Migrate and terminate VM
- Update VM resources allocation
- Create, configure or remove connection between VMs
- Forwarding of configuration information, failure events, measurement results and usage records regarding NFVI (physical, software, and virtualized resources) to the SDN controller.

Or-Vi reference point

Or-Vi is a reference point used for exchanges between NFVO and the SDN controller and manager. It supports the following:

- NFVI resource reservation/release
- NFVI resource allocation/release/update
- VNF software image addition/deletion/update
- Forwarding of configuration information, events, measurement results, usage records regarding NFVI resources to the NFV Orchestrator

NFVO may request Virtual Links based on network connectivity services through the Or-Vi reference point. The Resource Orchestration functions of the NFVO interfaces the SDN controller and manager through the Or-Vi reference point for requesting creation and management of VNs.

Or-Vnfm reference point

This reference point is used for exchanges between NFV Orchestrator and VNF Manager. It supports the following:

- NFVI resources authorization/validation/reservation/release for a VNF.
- NFVI resources allocation/release request for a VNF.
- VNF instantiation.
- VNF instance query (e.g. retrieve any run-time information).
- VNF instance update (e.g. update configuration).
- VNF instance scaling out/in, and up/down.
- VNF instance termination.
- VNF package query.
- Forwarding of events, other state information about the VNF that may impact also the Network Service instance.

Vi-Vnfm reference point

This reference point is used for exchanges between the VNF Manager and the VIM. It supports the following:

- NFVI resources reservation information retrieval
- NFVI resources allocation/release
- Exchanges of configuration information between reference point peers, and forwarding to the VNF Manager such information for which the VNFM has subscribed to (e.g. events, measurement results and usage records regarding NFVI resources used by a VNF).

3.4.3.3   Functional dimension

The functional dimension of NaaS relies on the global orchestration. Let us describe briefly the functions of the known building blocks in order to focus on the functions of the orchestrator in NaaS.

The SDN controller and the SDN manager, at the NFVI management level, perform control and management of the NFVI resources. The NFVI resources in the NaaS include virtual SDN network resources. The VNF Manager is responsible for the life-cycle management of VNF instances which are designed as "as-a-Service components" in the NaaS architecture. The VNF manager is supported by EMSs for the management of VNFs. EMs perform FCAPS functions. The VN control and management is dedicated to the control and management of the life-cycle of VNs. VN control is supported by Network Management systems (NMs) for the management of VNs. Like EMs, are responsible for FCAPS management for VNs.

The global orchestration relies on these control and management blocks for the management of the life-cycle of network services. As Figure 3.11 shows, requests for customized network services and VNs are addressed to the global orchestrator. These requests are accompanied by SLAs and QoS constraints and requests. The objective of the orchestrator is to build the adequate network service over a dedicated VN respecting the SLA and QoS at all levels. The network service is composed of an accurate selection of VNF components. Indeed, the orchestrator uses the information about requested QoS in the SLA in order to select the VNF components that best corresponds to the request. This is possible since VNF components are exposed with the functional and non-functional aspects and described on the basis of the QoS they offer.

Then, the selected VNFs are composed at the design phase. This phase is the construction of the network service of the requested VN. It is the definition of the service logic.

It is important to mention that this selection is preceded by a discovery phase in the case where the composition is performed by a third-party actor that has a profile that allows him to perform service composition. The discovery and the selection are made through the network exposition layer.

The orchestrator uses then the templates of VNFs, the QoS constraints to build a template for the VN. We define this phase as a virtual deployment. It corresponds to a construction of the VN at a logical level (over the new abstraction level) and the definition of the VNOS that controls it.

The next phase is the provisioning of the network service and the VN. The orchestrator relies on an automatic provisioning where the QoS constraints are taken into consideration for the resources allocation which enables dynamicity. The description of the VN is sent to the next phase, the placement phase. It consists in mapping the components of the VN service over the NFVI hardware resources. The provisioning phase makes it possible to consider QoS constraints for the placement phase. Thus, the selected virtual nodes enable to distribute the components of the VN service (defined in the virtual deployment phase) over the resources that fulfill the service QoS requirements at the physical equipment level where the components of the VN services are executed.

In order to maintain the requested QoS level, the management plane through the orchestrator performs a continuous monitoring. The collected information of the control plane will perform a re-provisioning either to respect user preferences and thus ensure a compliance with the SLA or to respond to new requests or events that are very likely to occur and that would change the SLA or the QoS. These changes may be related to mobility with service continuity for example.

3.5   Conclusions

The new telecommunication ecosystem reveals several convergence aspects. The new application services push for new usages while Telcos do not have the required endorsed network evolutions. These changes impose huge needs for flexibility and dynamicity in Telco networks.

We have presented in this chapter all the elements required to achieve a resilient, but over all, a flexible and dynamic NaaS architecture based on the convergence of new paradigms and technologies and the convergence of network planes.

Among the proposed architectures, flexible NaaS architecture shows that flexibility is achieved through an efficient network exposition layer with very accurate service description, service discovery and service selection for flexible service compositions. Flexibility in service composition is the foundation of the customization of network services.

Considering VNFs as "as-a-Service" components and adopting dynamic APIs are foundations for realizing such a network exposition. The rest of the architectures show dynamicity of NaaS. We first rely for the definition of these architectures on the integration of the emerging NFV and SDN technologies. Then on the use of "as-a-Service" component model for VNFs. We have also showed the importance of introducing an additional virtualization level for the definition VNs and the virtualization of the control plane. Finally we have presented the role of an automated global orchestration for the life-cycle management of network services. The NaaS architecture has been presented according to the architecture dimension which describes the building blocks of NaaS. Then, the relational dimension highlighting the interfaces between these building blocks and finally the functional dimension that presents the functions of the orchestration.

References

[1] *ETSI NFV ISG, Network Functions Virtualisation (NFV); Architectural Frame-work. Group Specification*, Dec 2014.

[2] ONF, "SDN architecture overview, onf tr-504," tech. rep., Nov 2014.

[3] P. M. Mell and T. Grance, "Sp 800 145. the NIST definition of cloud computing," tech. rep., Gaithersburg, MD, United States, 2011.

[4] G. Blair, T. Coupaye, and J.-B. Stefani, "Component-based architecture: the fractal initiative," *Annals of Telecommunications – annales des télécommunications*, vol. 64, no. 1, pp. 1–4, 2009.

[5] OW2 (ObjectWeb, Orientware) Consortium. The Fractal Project. Available at: http://fractal.ow2.org/.

[6] E. Bruneton, T. Coupaye, M. Leclercq, V. Quéma, and J.-B. Stefani, "The fractal component model and its support in Java: experiences with auto-adaptive and reconfigurable systems," *Software Practice Experiments*, vol. 36, pp. 1257–1284, Sep 2006.

[7] F. Baude, D. Caromel, C. Dalmasso, *et al.*, "GCM: a grid extension to fractal for autonomous distributed components," *Annals of Telecommunications – annales des télécommunications*, vol. 64, no. 1–2, pp. 5–24, 2009.

[8] *ETSI GRID; Grid Component Model (GCM); GCM Fractal Management API. Technical Specification, ETSI TS 102 830 V1.1.1*, Mar 2010.

[9] M. Maurer, I. Breskovic, V. C. Emeakaroha, and I. Brandic, "Revealing the MAPE loop for the autonomic management of cloud infrastructures," in *2011 IEEE Symposium on Computers and Communications (ISCC)*, pp. 147–152, June 2011.

[10] T. Erl, *Service-Oriented Architecture – Concepts, Technology and Design.* Prentice-Hall: Upper Saddle River, NJ, 2005.

[11] H. Zhao and P. Doshi, "A hierarchical framework for logical composition of web services," *Service Oriented Computing and Applications*, vol. 3, no. 4, pp. 285–306, 2009.

[12] *Service Component Architecture.* Prentice-Hall: Upper Saddle River, NJ, http://www.osoa.org, Nov 2007.

[13] D. Guinard, V. Trifa, S. Karnouskos, P. Spiess, and D. Savio, "Interacting with the SOA-based internet of things: discovery, query, selection, and on-demand provisioning of web services," *IEEE Transactions on Services Computing*, vol. 3, pp. 223–235, Jul 2010.

[14] T. Aubonnet, L. Henrio, S. Kessal, *et al.*, "Management of service composition based on self-controlled components," *Journal of Internet Services and Applications*, vol. 6, no. 1, pp. 1–17, 2015.

[15] M. Abu-Lebdeh, J. Sahoo, R. Glitho, and C. W. Tchouati, "Cloudifying the 3GPP IP multimedia subsystem for 4G and beyond: a survey," *IEEE Communications Magazine*, vol. 54, pp. 91–97, Jan 2016.

[16] Q. Duan, Y. Yan, and A. V. Vasilakos, "A survey on service-oriented network virtualization toward convergence of networking and cloud computing," *IEEE Transactions on Network and Service Management*, vol. 9, pp. 373–392, Dec 2012.

Chapter 4

Virtual deployment of virtual networks in Network as a Service

Gladys Diaz[1,3], Amina Boubendir[2,3] and Noëmie Simoni[3]

Abstract

While telecommunication and networking are driven by massive adoption of virtualization and the emergence of network programmability, the process of deployment remains an important phase in virtual networks (VNs) life-cycle.

We raise through this chapter the question about the importance and opportuneness of redesigning VN deployment process in this new ecosystem. For that, we have first raised the impacts and advantages brought by Cloud Computing, Software-Defined Networks (SDNs) and Network Function Virtualization (NFV) over network architectures and operations. We have then focused on the notion of Network-as-a-Service (NaaS) which integrates these new technologies. We have come to define the role and the characteristics of NaaS. This way, we have been led to distinguish two issues for the implementation of VNs. The first one is the actual process of deployment of VNs at a network service delivery level that takes into account the applicative flows, and the second one is the placement process at the physical network infrastructure.

Our objective is to present the "Virtual Deployment" as a phase within the VN deployment process. We define the virtual deployment phase using NaaS architecture. Our virtual deployment proposition takes into account the properties of flexibility, adaptability and dynamicity that are vital for the NaaS, where Cloud, SDN and NFV represent major building components in operators' network architectures. Thus, the virtual deployment of a VN is meant to integrate a response to a Service Level Agreement (SLA) request. Indeed, it supports the adaptation of a VN to integrate, starting from the deployment phase, network-level Quality of Service (QoS) constraints as a response to the service level agreement (SLA) request. To automate the virtual deployment, we propose to introduce the QoS constraints at the design phase. Therefore, through a Network Application Programming Interface (API), a network orchestrator

[1]L2TI Department, Paris 13 University, Sorbonne Paris Cité, Villetaneuse, France
[2]Network Control Architectures Department, Orange Labs, Chatillon, France
[3]Networking and Computer Science Department, Télécom ParisTech, Paris, France

will be able to consider "On Demand Services" and "User-Controlled services." Also, we rely on the abstraction of the Network Operating System (NOS) to address the need for an ability to choose and adopt the most efficient network control functions. Thus a personalized VN is consolidated, which takes into account all network constraints coming like requests from SLA and from usage.

We end this chapter by presenting technical tools that can be used for this purpose but they still have some limitations. The conclusion highlights the strengths of our proposal and introduces the perspectives.

4.1   Introduction

Today's networks are having a hard time meeting the demands of new usages and applications driven by Cloud computing, the development of M2M and more generally those of the Internet of Things (IoT). As the complexity of network architectures increases, their maintenance has become a challenge; the need to enrich their functionalities has indeed become a challenge when it comes to accommodating all these new requirements without making operation and maintenance overly complex.

As a result, networks must evolve both in terms of capacity and scaling, these networks and the services they support have to be easier to deploy, whilst operating them must remain simple and flexible. Promises have been made to (and by) network operators and vendors that technologies such as softwarization (or network programmability) and virtualization through emerging approaches known as Software Defined Networking (SDN) [1] (Section 4.1.1) and Network Function Virtualization (NFV) [2] (Section 4.1.2) would make it possible to design "simpler, more agile and flexible" network architectures, and ultimately provide a "full Network as a Service" (Section 4.1.3). This section examines these promises and considers the potential of these technologies to radically change today's approach of network design and operation.

4.1.1   SDN: changes and benefits

As a paradigm, SDN offers a layer of abstraction between hardware and software coupled with a central interface to instantiate and manage network elements dynamically to meet demand at any point in time.

In this respect, the constraints relative to the hardware specifics or interfaces are expected to be hidden by the SDN layer [3,4]. With SDN, Network administrators can control and adapt the network in real time, regardless of the physical constraints of the underlying hardware, which is made possible through an abstract layer called a controller, which centralizes the network logic (including routing).

This network programmability makes it possible to adapt network architectures in a more dynamic and simple manner than with traditional architectures. Network applications and services would then become independent from the physical network infrastructure, and network environment becomes more agile and flexible as network resources can be solicited on demand. A number of network maintenance procedures can also be automated through the abstraction of the network layer and access to network resources via APIs that help achieve efficient enforcement of network policies, such as network forwarding, for network elements (re)configuration and adaptation. In this respect, SDN would help operators to move away from the vendor-specific features and dependencies in their infrastructures that are generally counter to the quest for full interoperability of their systems.

Therefore, in the general context of network virtualization, SDN, through the SDN-based network virtualization, provides an efficient tool for building virtual networks (VNs) with abstract hardware and network elements organized around an SDN controller. Whilst having a global reach on the physical infrastructure, the SDN controller offers separate logical network topologies for distinct users (i.e., tenants) programming networks defining flows and paths for each user (application or tenant).

However, although virtualization is achieved dynamically through a central SDN interface, it is fair to say that at this stage SDN-based network virtualization does not go beyond sharing or mutualizing a network infrastructure to manage data streams from the application layer.

4.1.2   NFV: changes and benefits

NFV is another emerging paradigm that shares some common traits with SDN, it is called NFV [5]. The goal of NFV is to deploy network functions that are implemented on a dedicated proprietary hardware framework today on a standard software-based platform.

As a result, NFV leads to virtualizing a number of network functions to enable dynamic instantiations of those functions during deployment. The scope within which virtualization can apply is wide and certainly wider than those of SDN which are essentially focused on networking functions such as routing and forwarding/switching; in particular it includes the application layers of the OSI model (layers 4–7) such as firewalls, load balancing elements, CDN solutions, border access controllers and beyond applications even routers and switches. As such, NFV therefore provides virtual networking elements very much like SDN. These network virtualization functions will be installed on virtual machines (VMs) of a virtualized infrastructure offering resources on commodity hardware.

4.1.3   NaaS: the future

The ability of an infrastructure to integrate new technologies is obviously vital for operators. In this respect, the vision in which network virtualization provides a network so flexible that it offers services on demand is for them full of promise. This concept is called "Network as a Service" (NaaS).

We claim that for the operators to take full benefit of the promises of both SDN and NFV, they will have not only to move toward virtualization of network equipment but also address the virtualization another fundamental component of its networks, i.e., its control plane or Network Operating System (NOS). This will allow them to take into account the streams of data relative to a user globally, from the moment it enters the networks to the moment it leaves it and adapt the behavior of networks accordingly and dynamically to the user's usage. This analysis is based on the convergence of user plane and control plane for them to continuously adapt to the user's behavior.

Besides, the underlying principles of virtualization that allow several Operating Systems to run on top of a single physical infrastructure logically lead to the feasibility of having several virtual Network Operating Systems (VNOS) on a single operator network. As a result, on a par with OSes of VMs, the VNOS should allow operators to tailor the functional behavior and the resources of a VN to its needs and those of its users. Further, such flexibility would also open a new way to customize the network response to the user's individual needs in terms of capacity, terminal independence, etc. thereby leading to the notion of "user-controlled services" to integrate on-the-fly mobility constraints and user preferences.

It is both these levels of virtualization, network and control plane, used together and coupled with a high level of automation and finer network monitoring that will enable the on-demand deployment of VNs in this new ecosystem. It is precisely the role of NaaS to manage these personalized elements into an orchestrated service framework that integrate customer demands dynamically. NaaS is therefore a concept that applies to network functions the same approach adopted for Platforms or Infrastructure that lead to PaaS and IaaS, respectively.

The concept of NaaS, however, is not limited to connecting network nodes to the physical network. It takes into account all network services beyond network hardware capacity or platform features to enable VN instantiation aligned with user needs, and doing so thanks to an integrated control layer. It is with these concepts that a NaaS architecture [6] was proposed to integrate SDN (the separation of control and data planes) VNFs (network functions in VMs) and network virtualization on the control plane. The role of NaaS is therefore to:

- Provide APIs that facilitate service selection.
- Offer on demand network services that tenant choose to accommodate the needs of various application flows.
- Orchestrate dynamically network delivery on individual user data flows.
- Control media delivery services in each VN and ensure a continuous and dynamic processing.
- Manage VNs and adapt their deployment on the actual capacity of the underlying infrastructure.
- Instantiate and locate those VNs in the physical infrastructure,

It is worthwhile noting that although the orchestration phase requires foreknowledge of the conditions of deployment but with the new approach, the actual process of deployment can benefit from the virtualized context (user and control planes), and we call it "virtual deployment" to differentiate it from the deployment over the physical infrastructure which answers a problem of placement from our point of view.

In the remainder of this chapter, we will therefore elaborate on this concept and identify the benefits that virtualization and programmability offer with regard to network deployment. Having addressed the general characteristics of SDN, NFV and NaaS environments in this introduction, we then review in Section 4.2 the existing work on the deployment process in virtualized environments, in the Cloud Computing and in Network Cloud. Our analysis leads us to distinguish between a deployment in a virtualized environment which is rather a "placement" from the "virtual deployment" that we introduce in Section 4.3. The latter considers a personalized VN (Section 4.4) which takes into account all network constraints coming as requests from applications and from user SLA. In Section 4.5, we introduce technical tools that can be used to this purpose and illustrate some of their limitations. Finally, Section 4.6 develops our conclusions and elaborates on our perspectives.

4.2   The process of deployment in virtualized environments

In Cloud environments, computing and storage hardware virtualization allows to define VMs where cloud applications are deployed. This allows to effectively manage cloud services. This way, hardware virtualization contributes mainly in the optimization of hardware resources use in the Cloud. Like IT, network operators have also been interested in "cloudifying" the network through NFV to benefit from network functions dynamic deployment in order to achieve a better network service management. Furthermore, virtualization of network links has played a key role in virtualized and Cloud environments. That is why we will also study the deployment process in the Cloud Networking. We have come to the conclusion that in virtualized environments, the notion of "placement" addresses a different problem comparing to the notion of "deployment." In this section, we focus on the deployment process in different virtualized environments. We first consider the deployment process in the Cloud (Section 4.2.1), then the deployment from an NFV perspective in Telco Cloud (Section 4.2.2) that we also call "Cloudified network" and finally the deployment in Cloud Networking (Section 4.2.3).

4.2.1   Deployment process in Cloud computing

Virtualization in the Cloud provides a mechanism to decouple the applications from the hardware support necessary for their operation. This capability enables Data Centers to support diverse applications using a limited number of physical servers. The mechanism allowing this decoupling is the VM that defines an application and its associated functions. A VM defines resources to be associated at the implementation phase. The hosting infrastructure provider offers the requested physical resources in an Infrastructure as-a-Service (IaaS) model [7]. Different virtualization technologies may be used at the IaaS level [8]: Xen [9], KVM [10] and VMware [11] allowing to offer computing resources on demand to respond to requests of VMs creation, migration and scaling [12]. Different cloud infrastructure vendors provide IT resources on-demand like Amazon EC2 [13] and Google Compute Engine [14].

The deployment process in the context of Cloud Computing consists in generating machine images then instantiating them in the form of VMs. These VMs are targeted to ensure the proper execution of a cloud application over the IaaS. However, local aspects of configuration and aspects related to VMs interconnection are to be resolved during the deployment phase. This means that some operations at the placement phase are yet to be addressed. So how to place a maximum of VMs optimally in a minimum number of servers to optimize resource utilization across the physical infrastructure while maintaining the quality required for the application operation? Actually, the placement of VMs is treated in the literature through different techniques of decision making [15,16]: reservations [13], on-demand access [17] and spot markets [18]. As we can see, placement is in all cases guided by physical constraints regarding the utilization of hardware resources in the infrastructure. This is the case for VMware [19] and OpenStack [20] which both follow this model. Computing and storage resources (CPU, disk space and RAM) are the only conditions to satisfy when placing VMs.

Most VMs placement policies focus mainly on the effectiveness and efficiency of computing and storage resources utilization while network aspects are widely ignored and limited to local connectivity aspects. To avoid network problems, solutions tend to place VMs in the same cluster so the communication between VMs would be local to a cluster switch. But, what about the deployment process considering the network environment at a larger scale? Let us study in the next section whether this question is addressed in today's cloudified network through NFV.

4.2.2   Deployment process in Telco Cloud

Like IT, Telcos has worked to "cloudify" the network through NFV in order to be able to implement network functions logics in VMs over a virtualized infrastructure in the same way as in a Cloud Computing environment. This is why NFV is also called Telco or operator Cloud. When applying the Cloud model at a network level (network cloudification) by analogy and in a symmetric manner, different layers can be considered in the delivery of network services.

To go in the direction of this virtualization, ETSI NFV initiative [21] considers cloudification of operator networks through the virtualization of network functions, i.e., the physical network nodes in legacy networks, and defines the following three service models.

4.2.2.1   Virtual network function as a service (VNFaaS)

VNFaaS offers (access to) one or a set of virtual network functions (VNFs) instantiated on Points of Presence (PoPs) of the NFV Infrastructure (NFVI) [22]. This model represents a service in the Network Service as-a-Service layer and provides, similarly to the Cloud SaaS, turnkey services. The difference with the SaaS is that NSaaS services belong to the network service delivery level.

4.2.2.2   Virtual network platform as a service (VNPaaS)

VNPaaS offers a platform to provide VNFs with programming tools and virtual links. This model represents services of the Network Platform-as-a-Service (NPaaS) layer, somehow equivalent to Cloud PaaS, and proposes a service platform with catalogs of exposed services. It thus allows users to choose and compose VNFs such as AAA, NAT, and firewalls, but also functions like routing and addressing.

4.2.2.3   NFV infrastructure as a service (NFVIaaS)

The NFVI, as its name implies, consists in the virtualized physical network infrastructure. In a similar model as Cloud IaaS, this model offers execution environments for VNFs, i.e., computing and storage capabilities but also networking capabilities for functions like routing and switching. It is thus considered as the Network Infrastructure as-a-Service (NIaaS) layer.

VNFs are the central elements within each of these models where VNF Management functions perform VNF life-cycle management based on VNFs' requirements. At instantiation, NFVI resources are assigned to a VNF based on these requirements, but also taking into consideration specific constraints and policies that have been pre-provisioned or accompanying the request for instantiation [23]. During the life-cycle of a VNF, the VNF Management functions may monitor KPIs of a VNF and may use this information for scaling operations. Scaling may include changing the configuration of the virtualized resources: scale up (add CPU), scale down (remove CPU); adding new virtualized resources: scale out (add a new VM), shutting down and removing VM instances (scale in), or releasing some virtualized resources (scale down).

Therefore, VNFs deployment process in NFV does consider only the requirements related to the computing resources needed for the execution of the VMs that contains the VNFs. So the deployment process in Telco cloud focuses as well primarily on the effectiveness and efficiency of computing resources utilization. However, NFV remains a strong component necessary for the definition of a resilient VN deployment process. But is this component enough in the definition of such a deployment?

Let us analyze the deployment process in the Cloud Networking to finalize our response.

4.2.3   Deployment process in cloud networking

It is first important to recall that Cloud Networking has emerged from the need to interconnect not only locally via traditional Virtual Local Area Networks (VLANs), but through the WAN, distant and distributed data centers or elements of a data center such as servers. Cloud Networking started to take shape with the consideration of QoS constraints for applications and cloud services that are hosted in VMs within the Cloud. But from a network point of view, VM placement and operation are not directed only by network links capabilities.

In fact, the links and interconnections between VMs have a strong impact over distributed application performance and over the respect of the required QoS level. These problems have recently been studied in [24,25]. These works are based on distributed Clouds whose problems concern allocation of geographically distributed resources for which interconnection networks are needed. They propose algorithms for resource allocation in distributed data center.

The work in [24] presents a study on the placement of VMs hosting applications and the migration of VM in case of increased latency or congestion over the network. The proposed algorithm is based on minimizing time consumption data transfer. It focuses on VMs placement policy taking into account network requirements. The proposed VM migration policy considered unstable network connections which impacts application performance. In the proposed VMs placement policy, the approach seeks to optimize access to data by placing the VM on the physical machine with the shortest data transfer time. In the same direction, the work in [25] proposes to reduce the distance and latency between selected data centers by minimizing traffic inter and intra data center. It also seeks to minimize data paths and to maintain applications performance. Several steps are involved in the allocation of resources: selection of data centers, VMs partitioning over selected data centers, identification of existing physical resources in each data center and VMs placement.

Moreover, we also have Service Function Chaining mechanisms which are also considered in NFV to connect VNFs and thus the VMs containing them. Chaining mechanisms are based on the notion of Forwarding Graphs [2], equivalent to a set of Service Forwarding Paths (SFPs) that describes the path that packets should follow to be processed through the chained VNFs. The work in [26] has proposed an analytic model for the NFV forwarding graph. However, this model aims at optimizing the execution time of the network services deployment.

But the links defined by such network function chaining mechanisms remain forwarding links defined after VMs placement. Consideration of application requirements at the transport network level before or when deploying and placing VNFs or VMs is not yet considered either in NFV or in Cloud Networking. This is not enough. Indeed, considering links and NFV chaining in the definition of network virtualization does not mean considering each VN as a chaining of network applications or network functions in a customized manner. The two definitions are different.

We can conclude that in all these approaches, the visibility considered at the deployment process is either a placement of VMs or VNFs over commodity hardware with computing resource optimization (the case of Cloud Computing and NFV) or an optimization of physical links (the case of Cloud networking). None of the approaches analyzed above treats the notion of end-to-end network flow which is intended to support applications constraints and needs from the network. This limitation has motivated our reflection approach regarding the deployment which needs to be in line with the "user-centric" behavior of a user in mobility and having preferences linked to his location.

Therefore, as presented in Section 4.1.3, an additional level of abstraction is necessary to define the logical view of the VN to deploy. This additional level is a logical network at the service level. It is called "service network delivery." It is independent from the VN infrastructure level which is called "service media delivery." This latter has requirements on the placement over physical resources.

Network virtualization relies on the virtualization of the control plane and it is to be distinguished from network equipment or hardware virtualization (physical network nodes and links) that deals with placement as introduced in [27].

4.3   Virtual deployment of virtual networks

Of course different technical challenges in terms of instantiation, operation and management in a global network virtualization environment are yet to be raised.

These challenges include mobility integration, configuration of VNs and management of VNs.

In fact, if we consider all the evolutions as we have introduced in this chapter, we should talk about "programmability of personalized VNs." These VNs need to be mapped on physical infrastructure resources. In the underlying heterogeneous environment the demands may be contradictory; management-related paradigms are necessary to allow infrastructure providers to manage global information. The introduction of a common abstraction layer with a unified information model used management applications would be part of an efficient solution [28,29].

In order to provision the different demands in terms of resources coming from different applications or service providers, infrastructure providers (either physical or virtual infrastructures) need to be able to accurately determine the status of the corresponding network elements. For that, a resilient monitoring would also be part of a global efficient solution.

As explained earlier, the issue we are raising is the study of the impact of virtualization (as seen in the new ecosystem) on the deployment process. Indeed, deployment is the installation and configuration of systems over a virtualized infrastructure. But, how to take into account the changes introduced by Cloud, SDN and NFV for the deployment process? We have noticed that network virtualization is not limited to virtualization of network equipment. For that, one of our research directions is to answer the need that comes out from this observation. We need a second intermediary level of abstraction that would allow to build VNs as well as dedicated control plane for each VN. This control plane should be built in accordance with the needs of the different users' and applications' flows and their SLA requests.

In Section 4.2, we have analyzed the impact of cloud model, and we have found that for IT, applications and servers installation process over virtualized hardware is rather a mapping and placement problem that has the objective to benefit from the best allocation of resources (CPU and RAM). It is the same in cloudified network (NFV) for the installation process of VNFs over NFVI.

We propose in this chapter, the definition of "virtual deployment of VNs" as a phase of the network virtual deployment process (Section 4.3.1) in order to fully take advantage of network virtualization beyond the virtualization of network equipment. Then, we show how to consider correct network virtualization in addition to hardware equipment virtualization and the dynamic deployment brought by NFV. We define this new phase in the NaaS life-cycle (Section 4.3.2) through two new distinct operations: the building of VN and association to virtual NOS (building VN + VNOS), and Distribution of VNFs (Section 4.3.3).

4.3.1   Definition of the "virtual deployment of virtual networks"

The NaaS that represents the provider of cloud networking (or a network service for any OTT) must not only provide the most appropriate network service delivery for the application flows and user usage but also supervise the media delivery. For the latter, and according to the concept of SDN, the media delivery is achieved on the data plane that lies on the NFVIaaS level. Service delivery on the other hand must

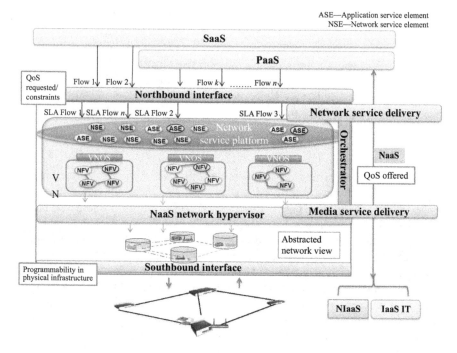

Figure 4.1   NaaS positioning

be designed following SaaS or PaaS demand (which support E2E service delivery) to provide the network service between the access points (that can change in the course of a session because of the user's mobility).

It is the VN that corresponds to the network service delivery, which has to be built and provisioned in a continuous way. This phase must be assumed by an orchestrator to manage the services, to ensure that the QoS requirements are met, to introduce new services as per users' requests, to charge accordingly and to remove the services when they are no longer needed.

It is for building this VN and the associated control (Figure 4.1) that we must plan and evaluate the resources according to the QoS constraints of the demand, which we call "network virtual deployment," the network in this case being a set of virtual components that must be organized according to the NIaaS capabilities. It is this multi-tenant architecture where users and applications share the same infrastructure (IaaS and NIaaS) (Figure 4.1), which can be called "on-demand model." The continuous deployment lies in virtual partitions and slices that enable the model and not hardware- and software-dependent physical stacks to adapt.

Control Metadata define rules that are specific to each domain. These metadata define the behavior, such as the quality of service (QoS) that the orchestrator has to take into account to instantiate the application logic. The shared platform then provides real-time management and configuration of the rules or policies

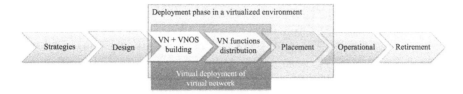

Figure 4.2   Network life-cycle in NaaS

and services. Virtual deployment in this case is positioned as a facilitator of deployment for complex software infrastructure that facilitates the use of a on demand multi-application environment that meets security, reliability or scalability requirements. In other words, we define virtual deployment as a composition of a potential network (VM, NFV) with links that meet the required QoS and whose configuration and optimization are programmable and monitored by a controlling network (VNOS).

In conclusion, we define the virtual deployment as the composition of a potential VN whose virtual nodes (VM or VNF) and virtual links answer a specific requested QoS and for which the dimensioning and optimization are programmed to be performed by a dedicated virtual control plane.

4.3.2   Network life-cycle in NaaS

Virtualization has introduced changes in network services life-cycle. Especially for the Cloud (XaaS), whose solutions are intended to be dynamic and flexible. The traditional processes of designing, deploying and operating the networks must be reviewed for this purpose. Let us describe Figure 4.2 which introduces our proposal for the network life-cycle in NaaS with the new phase of "Virtual deployment of VN."

4.3.2.1   The strategy phase

It guides the implementation of the service provider action plans according to management objectives and business positioning. In the case of current cloud solutions, business strategies need to be as dynamic as possible. Service providers need to be able to adapt their offerings based on market changes. For a NaaS provider particularly, the associated business model allows to consider different technologies, existing and new ones, to program and dynamically deploy their VNs. Thus, they remain flexible in their offerings.

4.3.2.2   The design phase

It allows to design the services offered by the NaaS according to customer demands and business models. At this phase, NaaS describes network services catalog that will be offered through its network platform. It takes position with respect to eventual collaboration with third-party actors. The main task of this phase is to select the appropriate VNFs in accordance with the requirements of the flows and the service logic. This way, a network of functions or a network of services is designed. The nodes of such a network are VNFs, a logical connection topology of VNFs. The links of such a network are the interactions between these VNF components. It defines the nature and the characteristics of exchanged messages and traffic between the access points of the VN to deploy.

4.3.2.3   The deployment phase

Let us first recall that in an uncloudified context, the phase that is supposed to be performed is the deployment phase. In such context, the deployment phase is the process through which the components of an application are distributed to install the application. However, in a Cloud Computing context, the deployment process consists in defining the disk images that will be used to deploy the VMs. As explained, this goes back to a mapping problem on virtualized hardware. The separation of control and data planes and the abstraction of physical resources through SDN, in addition to network virtualization at two levels leads us to decompose the traditional deployment phase into two steps in the NaaS context. The first step is the "virtual deployment of VN" and the second step is "VN Placement."

Virtual deployment step

We position the virtual deployment of VNs in NaaS between the requests expressed through the northbound API and the placement over the physical resources through the southbound API. We define the networks virtual deployment as a two-phase process: the phase of building the VN and the phase of distributing network functions. The first one builds the VN with a dedicated control plane in a form of virtual slice of the network operating system (a vNOS). It is built based on the functional service logic view provided by the design phase. The second phase rings the VN with its VNFs over the abstracted physical network infrastructure.

- **Building "VN + vNOS" phase:** In this phase, we focus on expressing a translation of the needs and requirements coming from applications. We do so to characterize the behavior of the VN to deploy. It concerns an on-demand service composition responding to the request of a service logic to offer. This phase corresponds to the characterization of virtual elements based on the data flow to be processed. It aims at enforcing the needs translation in order to enforce a translation of needs at a network service virtualization level. Thus, a VN is built and its nodes are the VNFs necessary for the implementation and the transfer defined by the network that resulted from the design phase. The virtual control plane (vNOS) is programmed at this phase. It defines, among others, the flow constraints and the routing rules in a Routing Information Base (RIB) (routing information including routes and policy routing rules) for the service of Media Delivery level. The RIB induces the Forwarding Information Base (FIB) that will be placed in the data plane over the NIaaS. The plane where this building of VN associated to a virtual control plane (vNOS) takes place is called the "Controlled Functional Plane."
- **Virtual network functions distribution phase:** This phase takes as an input the VN that resulted from the Building phase. It then distributes the VN component functions over the image that represents the abstraction of physical resources.

The plane that supports this phase is the "Distributed Functional Plane." At this phase, the network obtained is constituted of all nodes that will be traversed. It then defines the network paths. It also includes a proper routing protocol (RIB) and with specific architecture and engineering dedicated to the supported services. The NaaS network hypervisor selects the virtual resources (virtual nodes and virtual links) for all the VNs from the abstract view that he has of the physical infrastructure. This selection is based on negotiated resources from the NIaaS. NaaS also incorporates the description and the translation of the requirements for the mapping and placement of the VN elements allowing to meet flow constraints requests. The application constraints are nothing but the nonfunctional requirements, i.e., the QoS requirements that a NaaS provider has to take into account along with the functional requirements. QoS allows the characterization of non-functional needs and requirements of application flows. Thus, it allows the definition of the set of features that needs to be guaranteed when processing these flows by the VN to deploy. The networks virtual deployment phase is then carried out through the translation at the vNOS level of the constraints to be respected. During the distribution on the NIaaS image, it is necessary to take into account the QoS offered by the physical resources so that the mapping reflects the requested SLA.

Placement step

The second step of the deployment phase is "VN Placement." The placement phase allows to realize mapping of VNs over physical network and computing infrastructure. The infrastructure manager is responsible for the mapping according to available and requested resources.

Figure 4.3 illustrates four planes for NaaS. It also shows the positioning of the design phase and the deployment phases (Building, Distribution and Placement). In the network service plane, we find a logical network that represents the service logic. This a service-level representation of the desired VN. The nodes represent the services that compose the global service of the VN. The phase that correspond to this plane is the Design Phase. The second plane is called "Controlled Functional Plane." Within this plane, the logical representation of the VN is built. This logical representation is associated with dedicated control functions included in a virtual control plane (VNOS). The nodes X1, Y1, A1, and B1 represent the VNFs involved within the composition at this logical level. The phase that correspond to this plane is the Building Phase. The third plane is called "Distributed Functional Plane." It offers an abstracted view of the physical infrastructure. Within this plane, the virtual representation of the VN is built through a distribution. The nodes X, Y, A, B, C and D and the links between them represent virtual nodes and links. The logical nodes of the Controlled Functional Plane are distributed over the virtual nodes of the Distributed Functional Plane. This latter abstracts the physical nodes of the physical infrastructure. The phase that corresponds to this plane is the Distribution Phase. The fourth plane is called "Physical Plane." The nodes X', Y', A', B', C' and D' and their links are physical nodes and links that host one or multiple virtual node and link. The phase that correspond to this plane is the Placement Phase.

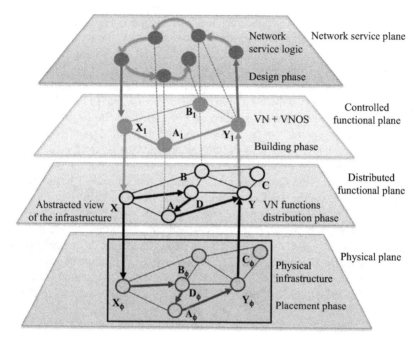

Figure 4.3   Virtual deployment of virtual networks phases and planes for NaaS

4.3.2.4   The operational phase

It corresponds to the actual consumption of the deployed services. In this phase, the network provider guarantees the provisioning of resources to allow consumption of the network service. Also, the implementation of management functions should allow to monitor constantly the behavior of the deployed services. NaaS architecture should guarantee the correct behavior of Media Delivery service.

4.3.2.5   The retirement phase

It will terminate operation of VNFs to remove the network service. Resources are also released at this phase for the services that are not invoked anymore. It is to be mentioned that management and orchestration functions [23] defined in NFV for VNF and network service orchestration and management also play a role in each of these phases. Let us now describe in details the proposed virtual deployment phase.

4.3.3   Virtual network deployment process

The VN deployment process, as shown in Figure 4.4, goes as follows:

1. At first, the PaaS or/and SaaS sends the description of requirements corresponding to the application flows needs. It is an SLA request.
2. Second, during the Design phase, the VNFs that best suits the service logic needs are selected from the exposed VNFs and virtualized network services. The latter

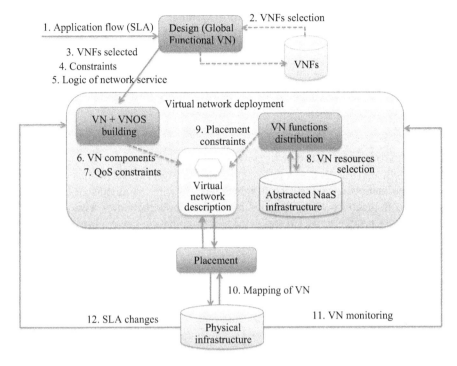

Figure 4.4   Process of virtual deployment of virtual networks

are offered through the NaaS platform (NPaaS). They are exposed with theirs functional and non-functional aspects (QoS). They are thus selected according to the required SLA.

3. The information about the selected VNFs are sent to the VN deployment phase.
4. The application constraints are sent to the VN deployment phase.
5. The service logic are sent to the VN deployment phase.
6. During the phase of building VN + vNOS and programming vNOS, the selected VNFs enable to define the network components to be placed in the next phase.
7. The information associated with the flow description enables to retrieve the QoS constraints to be considered. This ensures that the behavior of the VN is planned according to the SLA needs.
8. During the VN Distribution phase, the VN dimensioning (number of nodes over a network path) is made based on the expressed constraints.
9. Therefore, this phase makes it possible to take into account placement constraints. The selected virtual nodes enable to distribute the components of the VN defined in the VN building phase. The description of the VN is then completed. This description is sent to the next phase of placement.
10. The placement process performs the VN mapping over the physical network infrastructure.

11. Based on monitoring that provides information on the current capabilities of the network, the control plan reviews its routing tables and dynamically updates the forwarding tables.
12. Based on expected changes in the SLA (such as taking into account user preferences, user mobility or the degradation of end-to-end QoS), the association VN + vNOS is reconfigured as the result of a notification concerning one or more of these changes.

4.4   Personalized virtual network

We assume that NaaS builds each personalized VN with the associated control and management functions. The slices of these VNs have their own associated NOS (VN+VNOS). NaaS is expected to provide these elements with the intrinsic properties of virtualization, i.e., isolation, recursivity and the independence of NOS.

For a NaaS architecture to achieve these properties, it must describe the VN that sits on top of the physical infrastructure and how they relate with one another. This includes the specification of the virtual elements (nodes and links) as well as what they demand in terms of external exchanges (SLAs). Each VN is here, defined by three planes: management, control and data. The management plane provides resource management capabilities, the control plane (there is one control plane per VN) takes into account the constraints and establishes routing decisions, and the data plane contains the forwarding tables.

Let us first recall the NaaS components (Section 4.4.1) and we will then move onto the elements to consider in the design of personalized VNs (Section 4.4.2).

4.4.1   NaaS components

As depicted in Figure 4.1, the NaaS architecture consists of the following:

- Two interfaces. The northbound interface collects the demands/requirements/ requests of the flows. The southbound interface instantiates and installs functional update of the various elements in the physical infrastructure. Depending on the specific nature and requirements of the flow, Naas determines the most relevant VNFs to address,
- a service platform that will be described in a catalog,
- one network controller per VN, which we call a Virtual Network Operating System (VNOS). The VNOS must accommodate the specifications and demands of the VN and feedback from the physical infrastructure,
- a network hypervisor, which like the machine hypervisor, will assume the role of NaaS manager for the network and handle resource distribution and sharing on the infrastructure. Through a layer of abstraction that hides the underlying physical infrastructure, it controls and manages the network elements.
- the NaaS orchestrator that automate the continuous deployment of these plans (control, data and management) on the elements of a VN using gathered information on the current state of these elements during operating phase.

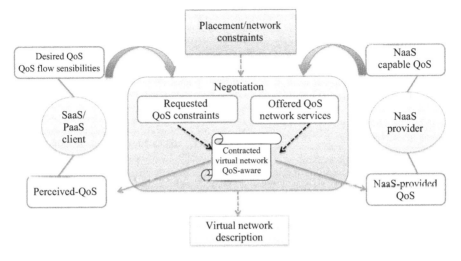

Figure 4.5 A QoS-aware approach for NaaS

4.4.2 Design elements

In our design of the architecture, we should make sure that the actual demands of the application layer properly translate into the instantiation/placement. For this we propose to extend the traditional functional requirements to non-functional requirements, namely QoS. QoS requirements indeed characterize the nonfunctional needs of application flows that the network, and therefore its entities and processes to be deployed, have to meet. This "virtual deployment" then progresses through what can be called a QoS-aware approach, which means that, at all levels of the architecture, the demands in terms of QoS have to be met in the best possible way. The application of the QoS model (see Figure 4.5) reflects the alignment of the needs for QoS at SaaS/Paas level and those available at IaaS for both service and media delivery.

For this, Naas takes into account both:

- the service view where the service delivery characterizes that service offer of network Delivery to address the need of various flows, and
- the placement view with the introduction of QoS constraints to map to the physical infrastructure and conditions of the Media Service Delivery.

The services offered by NaaS depend on the resource capabilities which were negotiated with IaaS. The resources are then represented through an abstraction of the physical infrastructure. Our QoS approach model provides a means of characterizing, controlling and managing the behavior of VNs.

The proposed mechanism therefore provides a way to further consider network virtualization with a second level of abstraction to reflect the nature of application flows. This approach along with the proposed QoS criteria make it possible to express the constraints relative to QoS to be taken into account further down to the

Table 4.1   QoS and placement constraints at node level

High level constraints	
Node affinity (v:set<VNE>)	All virtual nodes (VNEs) that belong to v group should be mapped on the same physical node
Node anti-affinity (v:set<VNE>)	All virtual nodes (VNEs) that belong to v group should be mapped on different physical nodes
Node isolation (v:set<VNE>)	All virtual nodes (VNEs) that belong to v group should not be colocated with existing VNEs
QoS and placement constraints	
Node processing time (vl:VNE, d:time)	Processing time of v1 should be kept below d time
Node availability (vl:VNE, α:rate)	Availability rate of v1 should be at least equal to α
Node integrity (vl:VNE, β:rate)	Integrity rate of v1 should be at least equal to β
Node capacity (vl:VNE, c:string, z:number)	Capacity of type c of node v1 should be at least equal to z

Table 4.2   QoS and placement constraints at link level

High level constraints	
BanPath(vl:VNE, v2:VNE, s:set<INNode>)	Path between nodes v1 and v2 should not include intermediate nodes belonging to s
lsoPath(vl:VNE, v2:VNE, nb:number)	Number of intermediate nodes between nodes v1 and v2 should be kept below nb
QoS and placement constraints	
MaxLinkDelay(vl:VNE, v2:VNE, d:number)	Delivery time between virtual nodes v1 and v2 should be kept below d time
LinkAvailability(vl:VNE, v2:VNE, α:rate)	Availability rate of link between virtual nodes v1 and v2 should be at least equal to α
LinkIntegrity(vl:VNE, v2:VNE, β:rate)	Integrity rate of link between virtual nodes v1 and v2 should be at least equal to β
LinkCapacity(vl:VNE, v2:VNE, t:string, v:numbrer)	Capacity of type t of link between virtual nodes v1 and v2 should be at least equal to v
LinkUtilisationRate(vl:VNE, v2:VNE, E:rate)	Utilization rate of link between virtual nodes v1 and v2 should be kept below E
LinkActivityRate(vl:VNE, v2:VNE, θ:rate)	Activity rate of link between virtual nodes v1 and v2 should be kept below θ

physical layer. In a way, the design phase consists in defining the SLAs both in terms of functional aspects and nonfunctional aspects and in this respect the first phase of the "Network Virtual Deployment" called "Building VN + VNOS" builds up the VN with the control plane (VNOS) with all the applicable constraints. Depending of the level of abstraction of the resources, the second phase called "VN Functions

Distribution" distributes the network functionalities on the nodes and links selected for the QoS they can offer.

Tables 4.1 and 4.2 show how such an approach with a model of network demands including the QoS can be applied to the nodes and links of the VN and then consequently to those of the physical network during placement [30].

4.5   Implementation study

Initial network equipment and networking products related to SDN and NFV and designed for operators' networks are today already available in the market. This first phase of commercialization is accompanied by pilots and tests for cases of applications that are realized by the telcos's first deployments since the second half of 2014. Indeed, the most advanced telcos—including AT&T, Deutsche Telekom, NTT and Telefonica—have announced large-scale migration plan to a new network architecture. AT&T including plans to migrate 90% of its internal applications in the cloud by 2019 while Telefonica announced the migration of 30% of its architecture in 2017.

The global market for SDN and NFV was estimated at EUR 816 million in 2014. This market is essentially deployments in data centers by enterprises and cloud service providers. The weight of the telcos in this market is still marginal and accounts for 5% in 2014, because no major cloud Networking. With an acceleration of adoption in the networks of operators, this share is expected to grow by 2019 to 19%, representing 1.15 billion Euros (out of a total market estimated to be 5.7 billion Euros).

We present in this section the strategies of network equipment vendors (Section 4.5.1) and some tools and existing products allowing us to move toward a NaaS architecture (Section 4.5.2) all by noting the gaps (Section 4.5.3) to achieve the goals and opportunities presented in this chapter.

4.5.1   Strategies of network equipment vendors

The SDN and NFV are at the heart of attention of the industry for new architectures, including of course the equipment that associate directly to the development of the ecosystem. The network equipment providers—routers and switches—are directly involved and are among the most active suppliers: Cisco and Juniper but also Huawei and Alcatel-Lucent. Most of traditional manufacturers are members and participants in the various initiatives of the Open Networking Foundation (ONF) and ETSI but the approaches are quite different. Cisco adopts a broad approach to addressing the issue of SDN and focuses its solutions on a distributed intelligence for programmable networks. Cisco helps telcos in migrating to NFV. Cisco also developed a protocol to keep a share of intelligence in switches and routers, and not only in controllers such as the OpenFlow advocates. As for Juniper, Huawei and Alcatel-Lucent, they have NFV facilities that rely on the SDN to contact operators directly.

Network equipment vendors develop and begin to market SDN and NFV offers to operators [31]. Table 4.3 lists some vendors with the SDN and NFV platforms and

Table 4.3   Some SDN and NFV related vendors' platforms and products

Equipment vendors	SDN controllers or SDN platforms	SDN- and NFV-based products
Alcatel-Lucent [32]	Nuage Virtualized Services Platform (VSP) including an SDN Controller	vEPC, vIMS, radio network virtual controller integrated in Cloudband
Big Switch [33]		SDN switches in collaboration with Extreme Networks
Cisco [34]	ONE Platform	Optimized SDN switches and routers Evolved Services Platform (ESP), different SDN and NFV solutions
Ericsson [35]	Service Provider SDN	vEPC
Extreme Networks [33]		SDN switches integrating Big Switch controller
HP [36]		SDN switches and HP Virtual Application Networks (VAN) SDN controller
Huawei [37]	Softcom Platform SNC Controller for Telcos	Agile SDN switches (for enterprise), NFV products: vEPC, vGGSN, vIMS
IBM [38]		OpenFlow Controller
Juniper [39]	Juniper Contrail Controller	Optimized SDN switches and routers, Fusion Software, vMCG (Mobile Control Gateway)
NEC [40]	ProgrammableFlow	SDN switches

products they propose [32–40]. Switches and routers integrate increasingly SDN and many network functions are now available in a form of VNFs, only as software.

Actors like HP, Dell and IBM are particularly active in SDN participating in standardization and development initiatives. In 2014, IBM has initiated the launch of an SDN controller based on Hydrogen (first version of OpenDayLight [41]). Dell took over the CloudNFV consortium, designed to demonstrate the implementation of an open and cloud-based NFV model in a telecom environment. We should also mention HP [36], Cisco (Insieme) [42], Juniper (contrail) [39] and Brocade (Vyatta) that help move toward a "full NFV/SDN."

4.5.2   Existing tools and products

For an overview of today's ecosystem to wait telcos to provision network services at short notice and optimize network resources to "networks on demand" we report some of the main products in the market:

- Table 4.4 lists and describes some SDN data plane software [43–46].
- Table 4.5 lists and describes some examples of SDN southbound interfaces and protocols [47–55].

Table 4.4   SDN data plane software

Software switch	Description
Indigo [43]	Indigo is an open source project aimed at enabling support for OpenFlow on physical and hypervisor switches. Indigo is the basis of Switch Light by Big Switch Networks.
OpenvSwitch [44]	OvS is an open source vSwitch software stack project that can run as a virtual switch in virtual environments, provide switching to host-based applications, and run as the control stack of hardware switches. OvS plays a vital role in several SDN/NFV open source projects, including OpenStack, OpenNebula, and OpenDaylight. OvS is the vSwitch that is most widely deployed in OpenStack installations, according to the OpenStack.org 2014 survey.1
Pantou [45]	Pantou is an OpenFlow port to the OpenWRT wireless environment. Pantou turns a commercial wireless router/Access Point to an OpenFlow-enabled switch. OpenFlow is implemented as an application on top of OpenWrt. Pantou is based on the BackFire OpenWrt release (Linux 2.6.32). The OpenFlow module is based on the Stanford reference implementation (userspace)
Ofsoftswitch [46]	This is an OpenFlow 1.3 compatible user-space software switch implementation. The code is based on the Ericsson TrafficLab 1.1 softswitch implementation, with changes in the forwarding plane to support OpenFlow 1.3.

- Table 4.6 lists and describes some examples of SDN control plane related software [41,56–59].
- Table 4.7 lists and describes some SDN Network abstraction layers more precisely examples of Network hypervisors [4,60–66].

Figure 4.6 shows the functions of these tools and software onto the NaaS architecture.

4.5.3   Today's shortcomings

In order to reach an automated NaaS, research works need to focus on the evolution of Orchestration where the need for a "full self-service" is the strongest. Orchestrating network services and VNFs would be completely independent of orchestrating infrastructures. This would require compliance of service components with some strongly required properties. Existing VNF managers do not perform automatic provisioning or auto scaling of VNFs. Scalability issues are still presented in SDN and NFV technologies. For more resilient architectures, scalability issued need to be addressed.

Also, SDN [67,68] is meant to ensure the compatibility of network equipment. For that, vendors should adopt the same APIs moving away from control switches and routers which can create overhead signaling. A thorough evaluation is needed to

Table 4.5   SDN southbound interfaces and protocols

Protocol	Description
OpenFlow [47]	OpenFlow® is the first standard communications interface defined between the control and forwarding layers of an SDN architecture. OpenFlow® allows direct access to and manipulation of the forwarding plane of network devices such as switches and routers, both physical and virtual (hypervisor-based).
ForCES [48]	*Forwarding and Control Elements Separation.* The ForCES protocol is a master–slave protocol in which FEs (Forwarding Elements) are slaves and CEs (Control Elements) are masters. Includes both the management of the communication channel and the control messages.
POF [49]	*Protocol-Oblivious Forwarding* as a key enabler for highly flexible and programmable SDN. POF makes the forwarding plane totally protocol-oblivious. The POF FE (Forwarding Element) has no need to understand the packet format. All an FE needs to do, under the instruction of its controller, is to extract and assemble the search keys from the packet header, conduct the table lookups, and then execute the associated instructions (in the form of executable code written in FIS (flow instruction set) or compiled from FIS). As a result, the FE can easily support new protocols and forwarding requirements in the future.
NETCONF [50]	*Network Configuration Protocol* is a protocol defined by the IETF to "install, manipulate, and delete the configuration of network devices" (RFC 6241). NETCONF operations are realized on top of a Remote Procedure Call (RPC) layer using an XML encoding and provides a basic set of operations to edit and query configuration on a network device. A REST-like protocol running over HTTP for accessing data defined in YANG using datastores defined in NETCONF. The Open Networking Foundation (ONF) recently embraced NETCONF and made it mandatory for the configuration of OpenFlow-enabled devices. The specification, called OF-CONFIG, requires that devices supporting it must implement the NETCONF protocol as the transport.
OF-config [51]	*OpenFlow Configuration and Management Protocol.* The OF-Config protocol has been developed since 2011 by the Open Networking Foundation, and it is used to manage physical and virtual switches in an OpenFlow environment.
BGP [52]	*Border Gateway Protocol* is a protocol used for exchanging routing information between gateway hosts in a network of autonomous systems. This protocol is often used between gateway hosts on the Internet and it is also considered a standardized exterior gateway protocol. BGP is often also classified as either a path vector protocol or a distance-vector protocol. Vendors are looking to use BGP in hybrid software-defined networking.
XMPP [53]	*Extensible Messaging and Presence Protocol* is a protocol that is based on Extensible Markup Language. Its intended use is for instant messaging and online presence detection. The protocol functions between or among servers and facilitates near-real-time operation. XMPP has recently emerged as an alternative SDN protocol to OpenFlow in hybrid SDN networks and can be used by the controller to distribute both control plane and management plane information to the server endpoints. It manages information at all levels of abstraction, down to the flow.
OVSDB [54]	*Open vSwitch Database Management Protocol* is an OpenFlow configuration protocol that is meant to manage Open vSwitch implementations. The OVSDB protocol uses JSON/RPC calls to manipulate a physical or virtual switch that has OVSDB attached to it. Nearly all vendors support OVSDB in various hardware platforms.
MPLS-TP [55]	*MPLS Transport Profile* is the transport profile for *Multiprotocol Label Switching*. MPLS-TP is designed to be used as a network layer technology in transport networks. The Open Networking Foundation proposed changes to MPLS that include the use of the standard MPLS data-plane with a simpler control-plane based on SDN and OpenFlow.

Table 4.6   SDN control plane related software

Controller	Description
NOX [56]	NOX is the original OpenFlow controller. It serves as a network control platform that provides a high level programmatic interface for management and the development of network control applications. NOX was initially developed at Nicira Networks side-by-side with OpenFlow— NOX was the first OpenFlow controller. Nicira donated NOX to the research community in 2008, and since then, it has been the basis for many and various research projects in the early exploration of the SDN space.
OpenDayLight [41]	Opendaylight is a collaborative open-source project initiated in April 2013 for which different actors contribute to its development. OpenDaylight controller has multiple network service modules dynamically linked in the Service Abstraction Layer (SAL) which determines a specific network service independently from the underlying protocols. It supports different protocols as southbound interfaces (BGP-LS, OpenFlow 1.0, OpenFlow 1.3).
Beacon [57]	Beacon is a fast, cross-platform, modular, Java-based OpenFlow controller that supports both event-based and threaded operation. Beacon is a Java-based open source OpenFlow controller created in 2010. It has been widely used for teaching, research, and as the basis of Floodlight.
Floodlight [58]	Floodlight Controller, an Apache licensed, Java-based OpenFlow controller, is one of the significant contributions from Big Switch Networks to the open source community. Floodlight's architecture is based on Big Network Controller (BNC), the company's commercial offering.
Ryu [59]	Ryu is a component-based software-defined networking framework. Ryu provides software components with well-defined API that make it easy for developers to create new network management and control applications. Ryu supports various protocols for managing network devices, such as OpenFlow, Netconf, OF-config. About OpenFlow, Ryu supports fully versions 1.0, 1.2, 1.3, 1.4, 1.5 and Nicira Extensions. All of the code is freely available under the Apache 2.0 license.

understand the behavior of COTS switches and SDN controllers. Existing proof-of-concepts limit the maximum number of possible exchanges between virtual switches and SDN controllers. On the other hand, the question concerning the collection and sharing of information, virtualization adds constraints (like latency) that need to be managed. How to collect information about the network status without introducing overhead? which protocols to use? what strategy to follow? In this direction, OpenFlow has increased flexibility but did not completely address the decoupling of the architecture to maintain an overall view showing the state of the network or the dynamic configuration of multiple controllers for better performance avoiding problems related to configuration conflicts, for example, if multiple applications can connect to the same controller.

Regarding NFV, current shortcomings are related to three aspects: Performance (Including networking), Security and Service Assurance. Proposed NFV proof-of-concepts are based on the deployment of VMs to representing VNFs.

Table 4.7   SDN network abstraction layer: Network hypervisor examples

Network hypervisor	Description
FlowVisor [60]	FlowVisor is a tool developed for SDNs to slice the physical switches among its users. This slicing of switches can either be: • On per-port basis of each switch • Or at a higher level (e.g., data link or transport or network layer level). Network admin can define flowspaces (which define the set of packets) for each controller. When a new packet is seen by a switch, it is sent to FlowVisor and then FlowVisor is responsible for directing the packet to its appropriate controller.
ADVisor [61]	*Advanced FlowVisor* is a by-product of OFELIA project from Europe. Although FlowVisor provides slices of the network to its users, the coupling of slices with the underlying topology is stronger than necessary. Particularly, there is no provision in FlowVisor to provide a proper subset of switches in the path that connects two hosts. ADVisor extends FlowVisor to provide users with such an abstraction, i.e., it can provide a proper subset of a path of switches between hosts as a controllable slice to its users.
VeRTIGO [62]	*Virtual Topologies Generalization in OpenFlow* networks. VeRTIGO is a software-defined networking platform designed for network virtualization and based on the OpenFlow original network slicing system FlowVisor. VeRTIGO allows the vSDN controllers to select the desired level of virtual network abstraction. At the "most detailed" end of the abstraction spectrum, VeRTIGO can provide the entire set of assigned virtual resources, with full virtual network control.
AutoSlice [4]	*Automated and Scalable Slicing for Software-Defined Networks.* AutoSlice is a virtualization layer that automates the deployment and operation of software-defined network (SDN) slices on top of shared network infrastructures.
DFVisor [63]	*Distributed FlowVisor* is a hypervisor designed to address the scalability issue of FlowVisor as a centralized SDN virtualization hypervisor. DFVisor realizes the virtualization layer on the SDN physical network itself. This is done by extending the SDN switches with hypervisor capabilities, resulting in so-called "enhanced OpenFlow switches."
HyperFlex [64]	HyperFlex introduces the idea of realizing the hypervisor layer via multiple different virtualization functions. Furthermore, HyperFlex explicitly addresses the control plane virtualization of SDN networks. It can operate in a centralized or distributed manner. The HyperFlex concept has initially been realized and demonstrated for virtualizing the control plane of SDN networks.
OpenSlice [65]	OpenSlice is a hypervisor design for elastic optical networks (EONs). The OpenSlice architecture interfaces the optical layer (EON) with OpenFlow-enabled IP packet routers through multi-flow optical transponders (MOTPs). A MOTP identifies packets and maps them to flows.

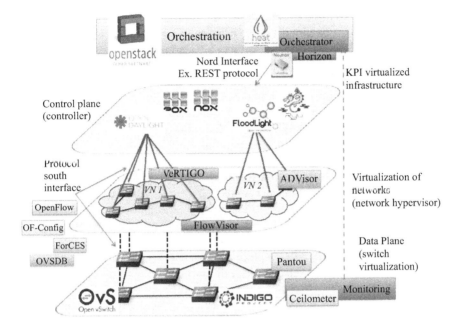

Figure 4.6   Overview of existing tools and products and positioning in NaaS architecture

With regard to the awaited agility, this is a limited solution. To achieve economies of scale expected from NFV, physical resources need to be used effectively in order to determine the potential impact of energy consumption on the move toward NFV? But there are no detailed technical documentation today on the assessment of the consumption of energy. NFV potential lies in its ability to provide a high level of automation and flexibility. In the case of federated or multi-vendor environment, NFV resources and functions will be provided by different entities. To ensure availability of services and resources, an accurate description and of functions, services and resources must be standardized and open to allow large-scale deployment through information models used by the relevant actors. These models must consider both the deployment and management life-cycle.

4.6   Conclusion

The telecommunication networks ecosystem is changing and operators need to be able to offer flexibility, scalability and dynamicity. Indeed, like IT, we are witnessing the cloudification of networks and its softwarization (through SDN) that aim to improve operational efficiency as well as financial and economic benefits. With this transformation, Telcos would be able to offer new services, introduce networking capabilities and network functions on-the-fly without having to install new equipment.

We have analyzed in this chapter the requirements that need to be verified in order to achieve further objectives of network virtualization and to promote the added value of NaaS in offering services through northbound interfaces and to enable building and deploying appropriate network solutions answering OTTs or application users' needs with their new usages. By making the distinction "virtualization of network equipment" and "virtualization of network flow controllers (VNOS)" we advocate a change in the NaaS lifecycle. The network deployment phase refers to VNs associated to their control plane and the implementation phase over the physical infrastructure falls within the problematic of VMs placement. The new phase of virtual deployment of VNs introduces two new processes: the construction of VN + VNOS and the distribution of NaaS components for the construction and programming of VNs prior to their placement. Thus the orchestrator can automate a dynamic and automatic provisioning.

The proposed approach allows to express the constraints to be respected at the nodes and links at each level of abstraction (service constraints, QoS constraints and placement constraints), by declining step-by-step application requirements to physical level. More specifically, the design phase is reflected in the SLA with its functional and non-functional aspects. The first phase of "Virtual Deployment" called "Building VN + VNOS" builds the VN and its control plane (VNOS) with all the constraints to consider. Then, based on the abstraction of resources, the second phase called "VNF distribution," distributes network functionalities in nodes and links selected based on the resources and capacities they offer (CPU, memory capacity, link capacity, bandwidth, etc.).

Following the analysis of the different products we found that the southbound interface meets the promises made by the editors, but for more agility we need to further develop a certain number of features and automation. So we could have a northbound interface that corresponds to the network service delivery level that would enable to satisfy the various applications' requests.

References

[1] ONF, *Software-Defined Networking: The New Norm for Networks. ONF White Paper*, Apr. 2012.
[2] *ETSI NFV ISG, Network Functions Virtualisation (NFV); Architectural Framework. Group Specification*, Dec. 2014.
[3] D. Kreutz, F. M. V. Ramos, P. E. Veríssimo, *et al.*, "Software-defined networking: A comprehensive survey," in *Proceedings of the IEEE*, vol. 103, pp. 14–76, Jan. 2015.
[4] Z. Bozakov and P. Papadimitriou, "Autoslice: Automated and scalable slicing for software-defined networks," in *Proceedings of the 2012 ACM Conference on CoNEXT Student Workshop*, CoNEXT Student'12 (New York, NY, USA), pp. 3–4, ACM, 2012.
[5] B. Han, V. Gopalakrishnan, L. Ji, and S. Lee, "Network function virtualization: Challenges and opportunities for innovations," *IEEE Communications Magazine*, vol. 53, pp. 90–97, Feb. 2015.

[6]   I. Ayadi, G. Diaz, and N. Simoni, "Qos-based network virtualization to future networks: An approach based on network constraints," in *2013 Fourth International Conference on the Network of the Future (NOF)*, pp. 1–5, Oct. 2013.

[7]   R. Montero, "Building IaaS clouds and the art of virtual machine management," in *2012 International Conference on High Performance Computing and Simulation (HPCS)*, p. 573, Jul. 2012.

[8]   S. Varrette, M. Guzek, V. Plugaru, X. Besseron, and P. Bouvry, "HPC performance and energy-efficiency of Xen, KVM and VMware hypervisors," in *2013 25th International Symposium on Computer Architecture and High Performance Computing (SBAC-PAD)*, pp. 89–96, Oct. 2013.

[9]   Z. Jian, L. Xiaoyong, and G. Haibing, "The optimization of Xen network virtualization," in *2008 International Conference on Computer Science and Software Engineering*, vol. 3, pp. 431–436, Dec. 2008.

[10] Z. Guo and Q. Hao, "Optimization of KVM network based on CPU affinity on multi-cores," in *2011 International Conference on Information Technology, Computer Engineering and Management Sciences (ICM)*, vol. 4, pp. 347–351, Sep. 2011.

[11] P. Padala, "Resource management in VMware powered cloud: Concepts and techniques," in *2013 IEEE 27th International Symposium on Parallel Distributed Processing (IPDPS)*, p. 581, May 2013.

[12] D. C. L. Peterson, T. Anderson, and T. Roscoe, "A blueprint for introducing disruptive technology into the internet," in *First ACM Workshop on Hot Topics in Networks (HotNets-I)*, Princeton, NJ.

[13] *Amazon Elastic Compute Cloud (Amazon EC2). Amazon EC2.* https://aws.amazon.com/fr/documentation/ec2/

[14] *Google Compute Engine. High-Performance, Scalable VMs.* https://cloud.google.com/compute/

[15] *Open Virtualization Format (OVF) Specification.* https://www.dmtf.org/standards/ovf

[16] D. Milojii, I. M. Llorente, and R. S. Montero, "Opennebula: A cloud management tool," *IEEE Internet Computing*, vol. 15, no. 2, pp. 11–14, Mar. 2011.

[17] I. Fujiwara, K. Aida, and I. Ono, "Applying double-sided combinational auctions to resource allocation in cloud computing," in *2010 10th IEEE/IPSJ International Symposium on Applications and the Internet (SAINT)*, pp. 7–14, IEEE, 2010.

[18] A. Andrzejak, D. Kondo, and S. Yi, "Decision model for cloud computing under SLA constraints," in *2010 IEEE International Symposium on Modeling, Analysis and Simulation of Computer and Telecommunication Systems (MASCOTS)*, pp. 257–266, IEEE, 2010.

[19] S. Sahasrabudhe and S. Sonawani, "Comparing OpenStack and VMware," in *2014 International Conference on Advances in Electronics, Computers and Communications (ICAECC)*, pp. 1–4, Oct. 2014.

[20] *OpenStack Open Source Project.* http://www.openstack.org/software/.

[21]   *ETSI NFV ISG, Network Functions Virtualisation (NFV); Use Cases. Group Specification ETSI GS NFV 001 V1.1.1*, Oct. 2013.

[22]   E. N. ISG, *Network Functions Virtualisation (NFV); Terminology for Main Concepts in NFV. Group Specification ETSI GS NFV 003 V1.2.1. http://www.etsi.org/technologies-clusters/technologies/nfv*, Dec. 2014.

[23]   E. N. ISG, *Network Functions Virtualisation (NFV); Management and Orchestration. Group Specification ETSI GS NFV-MAN 001 V1.1.1. http://www.etsi.org/technologies-clusters/technologies/nfv*, Dec. 2014.

[24]   J. T. Piao and J. Yan, "A network-aware virtual machine placement and migration approach in cloud computing," in *Grid and Cooperative Computing (GCC), 2010 Ninth International Conference on Grid and Cooperative Computing (GCC)*, pp. 87–92, IEEE, 2010.

[25]   M. Alicherry and T. Lakshman, "Network aware resource allocation in distributed clouds," in *INFOCOM, 2012 Proceedings IEEE*, pp. 963–971, IEEE, 2012.

[26]   J. Ferrer Riera, X. Hesselbach, M. Zotkiewicz, M. Szostak, and J.-F. Botero, "Modelling the NFV forwarding graph for an optimal network service deployment," in *2015 17th International Conference on Transparent Optical Networks (ICTON)*, pp. 1–4, Jul. 2015.

[27]   I. Ayadi, *La Virtualisation de bout-en-bout pour la gestion des services Cloud sous contraintes de QoS*. PhD thesis, Thèse de doctorat. TELECOM ParisTech, 17 mars 2014.

[28]   N. M. K. Chowdhury and R. Boutaba, "A survey of network virtualization," *Computer Networks*, vol. 54, no. 5, pp. 862–876, 2010.

[29]   M. Feridan, M. Moser, and A. Tanner, "Building an abstraction layer for management systems integration," in *Proceedings of the First IEEE/IFIP International Workshop on End-to-End Virtualization and Grid Management (EVGM2007)*, pp. 57–60, Citeseer, 2007.

[30]   *Open Cloudware Project (OCW)*, http://www.opencloudware.org/.

[31]   IDATE, *SDN and NFV in the telco world*, www.idate.org/en/Research-store/Collection/, Jun. 2014.

[32]   J. Van Bemmel, "Topology-aware packet forwarding in a communication network," Apr. 18, 2014. US Patent App. 14/256,040.

[33]   R. Wallner and R. Cannistra, "An SDN approach: Quality of service using big switch's floodlight open-source controller," *Proceedings of the Asia-Pacific Advanced Network*, vol. 35, pp. 14–19, 2013.

[34]   B. Salisbury, "The Cisco one SDN controller," *Network Static*, 2013. http://networkstatic.net/the-cisco-one-sdn-controller/

[35]   J. Kempf, M. Krling, S. Baucke, *et al.*, "Fostering rapid, cross-domain service innovation in operator networks through service provider SDN," in *2014 IEEE International Conference on Communications (ICC)*, Jun. 2014.

[36]   "HP van SDN controller architecture," tech. rep., HP, Sep. 2013.

[37]   D. Fesehaye and J. Wei, "SNC: Scalable NDN-based conferencing architecture," in *2014 IEEE 11th Consumer Communications and Networking Conference (CCNC)*, pp. 872–880, Jan. 2014.

[38] N. McKeown, T. Anderson, H. Balakrishnan *et al.*, "OpenFlow: Enabling innovation in campus networks," vol. 38, no. 2, pp. 69–74, 2008.

[39] A. Singla and B. Rijsman, "Contrail architecture," tech. rep., Juniper Networks, 2013.

[40] NEC, *Award-winning software-defined networking NEC Programmable-Flow networking suite. [Online]. Available: http://www.necam.com/docs/?id=67c33426-0a2b-4b87-9a7ad3cecc14d26a*, Sep. 2013.

[41] *OpenDaylight, A Linux Foundation Collaborative Project. [Online]. Available: http://www.opendaylight.org*, 2013.

[42] C. M. V. Bollapragada, R. White, C. Murphy, "Inside Cisco IOS software architecture," 1st ed., Cisco Press, Jul. 2000.

[43] K.-D. Althoff, U. Becker-Kornstaedt, B. Decker, *et al.*, "The indiGo project: Enhancement of experience management and process learning with moderated discourses," in *Advances in Data Mining*, pp. 53–79, Springer, Berlin, 2002.

[44] B. Pfaff, J. Pettit, T. Koponen *et al.*, "Extending networking into the virtualization layer," in *Workshop Hot Topics Network*, pp. 1–6, 2009.

[45] Y. Yiakoumis, J. Schulz-Zander, and J. Zhu, "Pantou: OpenFlow 1.0 for openwrt@ online," 2012.

[46] K. Shahmir Shourmasti, "Stochastic switching using OpenFlow," 2013.

[47] O. N. F. (ONF), "OpenFlow protocol specification. [online]. available: https://www.opennetworking.org/images/stories/downloads/sdn-resources/onf-specifications/openflow/openflow-spec-v1.4.0.pdf.," tech. rep., Oct. 2013.

[48] R. H. H. K. W. W. L. D. R. G. A. Doria, J. Hadi Salim, R. Haas *et al.*, *IETF. Forwarding and Control Element Separation (ForCES). RFC 5810 (Proposed standard)*, Mar. 2010.

[49] H. Song, "Protocol-oblivious forwarding: Unleash the power of SDN through a future-proof forwarding plane," in *Proceedings of the Second ACM SIG-COMM Workshop on Hot Topics in Software Defined Networking*, HotSDN '13, pp. 127–132, Hong Kong, China: ACM, 2013.

[50] J. S. R. Enns, M. Bjorklund, and A. Bierman, *Network configuration protocol (NETCONF), Internet Engineering Task Force, RFC 6241 (Proposed Standard). [Online]. Available: http://www.ietf.org/rfc/ rfc6241.txt*, Jun. 2011.

[51] O. N. F. (ONF), *OpenFlow management and configuration protocol (OF-Config 1.1.1). [Online]. Available: https://www.opennetworking.org/images/stories / downloads / sdn - resources / onf-specifications / openflow-config/of-config-1-1 1.pdf*, Mar. 2014.

[52] T. L. Y. Rekhter and S. Hares, *A border gateway protocol 4 (BGP-4), "Internet Engineering Task Force", RFC 4271 (Draft Standard). [Online]. Available: http://www.ietf.org/rfc/rfc4271.txt*, Jan. 2006.

[53] L. Li and W. Chou, "Extensible messaging and presence protocol (XMPP) based software-service-defined-network (SSDN)," Feb. 20, 2014. US Patent App. 14/185,661.

[54] B. Pfaff and B. Davie, *The Open vSwitch database management protocol, Internet Engineering Task Force, RFC 7047 (Informational). [Online]. Available: http://www.ietf.org/rfc/ rfc7047.txt*, Dec. 2013.

[55] M. Azizi, R. Benaini, and M. B. Mamoun, "Delay measurement in OpenFlow-enabled MPLS-TP network," *Modern Applied Science*, vol. 9, no. 3, p. 90, 2015.

[56] N. Gude, T. Koponen, J. Pettit *et al.*, "NOX: Towards an operating system for networks," *Computer Communication Review*, vol. 38, no. 3, pp. 105–110, 2008.

[57] D. Erickson, "The beacon OpenFlow controller," in *Proceedings of the Second ACM SIGCOMM Workshop Hot Topics Software-Defined Networks*, pp. 13–18, 2013.

[58] P. Floodlight, *Floodlight Controller. [Online]. Available on: http://floodlight. openflowhub.org/*, 2012.

[59] N. Telegraph and T. Corporation, *RYU network operating system. [Online]. Available: http://osrg.github.com/ryu/*, 2012.

[60] R. Sherwood, G. Gibb, K.-K. Yap, *et al.*, "Can the production network be the testbed?," in *Proceedings of the Ninth USENIX Conference on Operating Systems Design and Implementation*, OSDI'10, (Berkeley, CA, USA), pp. 1–6, USENIX Association, 2010.

[61] E. Salvadori, R. D. Corin, A. Broglio, and M. Gerola, "Generalizing virtual network topologies in OpenFlow-based networks," in *Global Telecommunications Conference (GLOBECOM 2011), 2011 IEEE*, pp. 1–6, Dec. 2011.

[62] R. D. Corin, M. Gerola, R. Riggio, F. D. Pellegrini, and E. Salvadori, "Vertigo: Network virtualization and beyond," in *2012 European Workshop on Software Defined Networking*, pp. 24–29, Oct. 2012.

[63] L. Liao, V. C. M. Leung, and P. Nasiopoulos, "DFVisor: Scalable network virtualization for QoS management in cloud computing," in *Tenth International Conference on Network and Service Management (CNSM) and Workshop*, pp. 328–331, Nov. 2014.

[64] A. Blenk, A. Basta, and W. Kellerer, "Hyperflex: An SDN virtualization architecture with flexible hypervisor function allocation," in *2015 IFIP/IEEE International Symposium on Integrated Network Management (IM)*, pp. 397–405, IEEE, 2015.

[65] L. Liu, R. M. Noz, R. Casellas, T. Tsuritani, R. Martínez, and I. Morita, "Openslice: An OpenFlow-based control plane for spectrum sliced elastic optical path networks," *Optics Express*, vol. 21, pp. 4194–4204, Feb. 2013.

[66] A. Blenk, A. Basta, M. Reisslein, and W. Kellerer, "Survey on network virtualization hypervisors for software defined networking," *IEEE Communications Surveys & Tutorials*, vol. 18, no. 1, pp. 655–685, 2016.

[67] M. Mendonca, B. A. A. Nunes, X.-N. Nguyen, K. Obraczka, and T. Turletti, "A Survey of software-defined networking: Past, present, and future of programmable networks," *hal-00825087*, Jan. 2014.

[68] W. Xia, Y. Wen, C. H. Foh, D. Niyato, and H. Xie, "A survey on software-defined networking," *IEEE Communications Surveys & Tutorials*, vol. 17, no. 1, pp. 27–51, 2015.

Chapter 5

Network service discovery, selection, and brokerage

Guanhong Tao[1] and Zibin Zheng[1]

The next generation Internet is expected to cope with new challenges to support a wide spectrum of network applications with highly diverse requirements due to the coexisting heterogeneous network environment. One of the challenges to achieve this objective lies in enabling network domain collaboration and network application interaction without exposing the internal structure and implementation details of each domain, where network virtualization will play a pivotal role in allowing a large number of service providers to offer various network services upon shared network infrastructure. Service-Oriented Architecture (SOA) [39] offers an effective architectural principle for heterogeneous system integration and provides a promising approach to support network virtualization, which can be applied in network service discovery, selection, and brokerage for the special requirements of future Internet. Due to the heterogeneity of network systems in ubiquitous and pervasive computing environments, network service discovery, selection, and brokerage face one of the main challenges to specify network demands of various applications. A key to solve this problem lies in flexible and effective interactions among the heterogeneous networks, various implementations and ubiquitous architectures with scalable information update, network-platform-independent methods, multi-attribute decision-making techniques, etc.

5.1 Introduction

With the rapid development of network architectures and technologies, the Internet is facing great challenges to support the wide spectrum of network applications based on such diversity of heterogeneous network systems. As the various network requirements are proposed by numerous network applications, it has motivated the appearance of new network technologies providing all kinds of functionalities. Existing network architectures, however, do not have the capability of satisfying the requirements of all the network applications completely. The next generation Internet

[1] School of Data and Computer Science, Sun Yat-sen University, China

is facing the fact that there will be various heterogeneous network systems coexisting on the Internet, where the collaboration among these network systems becomes a urgent and significant problem to address.

As to facilitating the collaboration among heterogenous network systems, a great deal of research work has been conducted in both academia and industry. Turner *et al.* [36] provided an exposition of the diversified Internet concept to address the problem of heterogeneous networks. Feamster *et al.* [16] presented an architecture, Cabo, to separate infrastructure providers and service providers in order to support the next generation Internet. DRAGON [25] proposed by Lehman *et al.* allowed dynamic provisioning of network resources by using a distributed control plane across heterogeneous network technologies. These approaches aiming to address the problem of heterogeneous network architectures and technologies were all based on the idea of network virtualization. Network virtualization can decouple and abstract network resources into independent functional components, which can subsequently be discovered, selected, and composed to support various network applications. As an approach first proposed to evaluate new network architectures, network virtualization has shown its capability to collaborate heterogeneous network systems for the next generation Internet.

Service discovery, selection, and brokerage in network virtualization environments face the challenge to effectively and efficiently interact with different network applications. The Service-Oriented Architecture (SOA) [39] is a system architecture designed to coordinate computational resources among heterogeneous systems. These resources are virtualized into services that can subsequently be discovered, selected, and composed to support various computing applications. Such characteristic of SOA can be utilized in network virtualization to address the problem of heterogeneous network systems. Based on the principles of SOA, heterogeneous network systems can be virtualized into network services. The network services are the reusable network functional components decoupled from various heterogeneous network systems. Therefore, the requirements of network applications can be met by discovering, selecting, and composing the existing network services.

Performance-based discovery and selection are the main topics widely studied. Commonly, the requirements of network applications include network performance parameters, such as network bandwidth, delay time, and load balance. The network service broker needs to find the optimal network service that meets the requirement of network applications. Therefore, the capabilities of network services should be evaluated and predicted as to compare them with the requirements of network applications.

In this chapter, the network virtualization concept and its importance for the next generation Internet are discussed. Based on network virtualization, the principle of SOA is introduced to cope with the challenge of various heterogeneous network systems. By adopting the idea of SOA, network resources can be decoupled into independent reusable network components, and then virtualized into network services, which can subsequently meet various network requirements from numerous network applications through service discovery, selection, and brokerage. Network-as-a-Service (NaaS) can considerably facilitate the development of the next generation Internet.

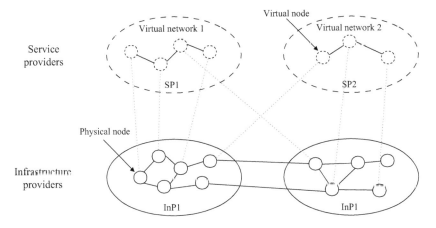

Figure 5.1 Network virtualization environment

5.2 Network virtualization for next generation Internet

With the explosion of various heterogeneous network systems, the next generation Internet is facing the fact that these network systems will coexist, and support various network requirements from numerous network applications. The diversity of network systems, however, is challenging the capability of the current Internet architecture as to support various network applications. Therefore, in order to collaborate diverse network systems, it is urgent and significant to develop new network architectures to flexibly interact with various heterogeneous networks. Network virtualization decouples the roles of the traditional Internet Service Providers (ISPs) into two entities: Infrastructure Providers (InPs) who manage the physical infrastructure, and Service Providers (SPs) who create virtual networks to offer end-to-end services by aggregating resources from various InPs [6]. It can effectively and efficiently address the problem of cooperating with heterogeneous network systems.

Infrastructure providers manage and maintain the physical network infrastructure, and provide physical resources [11]. Differing from the traditional Internet service providers, resources provided by infrastructure providers are employed by service providers instead of end users. Service providers, subsequently, lease the physical resources from various infrastructure providers and construct different kinds of virtual networks for end users. Each service provider manages and maintains one virtual network, which utilizes the resources allocated in the underlying infrastructure [13]. Separating infrastructure and service providers effectively decouples the correlation between network functions inside the same infrastructure provider, and virtualize the physical resources into standard services for end users. As the physical resources decoupled and virtualized into services, virtual nodes based on these resources can be connected by virtual links to form a virtual topology, which is finally composed into the virtual network. Each virtual node can be one particular function provided by the network system or one whole network system. The virtual links present the paths connected between different virtual nodes, where network resources are transferred along these virtual links.

As shown in Figure 5.1, network virtualization separates the traditional Internet service providers into infrastructure providers and service providers. The infrastructure providers provide physical network infrastructures, and InP1 and InP2 are both infrastructure providers. The physical network infrastructures provided by InP1 and InP2 are decoupled into several independent components, which can be utilized by service providers to form virtual networks. SP1 and SP2 are service providers who manage and maintain virtual network 1 and virtual network 2, respectively. The dotted line between virtual nodes in virtual network and physical nodes in InPs denotes the physical resources that service providers lease from infrastructure providers. The solid line between virtual nodes are the resource transfer paths between different virtual nodes. These virtual nodes connected with each other form network services, which then can be employed by end users, i.e., network applications.

A great deal of research work has been conducted on network virtualization in both academia and industry. An exposition of the diversified Internet concept was proposed by Turner *et al.* [36] to address the problem of heterogeneous networks. Similarly, the architecture, Cabo *et al.* [16], separated infrastructure providers and service providers to support various network requirements based on shared physical infrastructures. To facilitate the coexisting approaches and architectures, 4WARD [30] developed a systematic and general approach to network virtualization, which was also investigated by Cabo. Dynamic composition of network resources is also a significant problem that researchers aim to address. DRAGON [25] proposed by Lehman *et al.* allowed dynamic provisioning of network resources by using a distributed control plane across heterogeneous network technologies. Belqasmi *et al.* [2] introduced a novel networking paradigm, Ambient Networks, to enable on-demand and transparent cooperation between heterogenous networks. In the article, they illustrated that three degrees of composition were possible: network interworking, control sharing, and network integration, where network interworking was the most common degree in real situations. Douville *et al.* [8] proposed a service plane architecture that can automatically composite multi-domain network services. More research work about network virtualization is summarized by Chowdhury *et al.* [6].

Diverse and purpose-built proprietary appliances and various network devices make it increasingly difficult to add new services into current networks and raise the network ossification problem. Network function virtualization (NFV) was recently proposed to address these problems and to reduce capital expenditures (CapEx) and operating expenditures (OpEx). According to ETSI white paper [5], NFV decouples the functionality from hardware equipment in specific locations to provide software that can run on industry standard servers, switches, and storage anywhere. By consolidating equipment and workload, NFV can reduce equipment costs, power consumption, as well as energy consumption. Due to the virtualization of hardware in different locations, targeted and tailored services can be rapidly introduced and scaled based on customer needs and geography. Although Han *et al.* [21] pointed out that NFV may bring several challenges to network operators, the application of NFV can greatly benefit the development of NaaS for next generation Internet.

Network virtualization hides the diversify of heterogeneous network infrastructures from end users by providing on-demand network services to meet various requirements of network applications. Differing from the traditional network environments, the next generation Internet is facing the fact that a great number of heterogeneous network systems coexist and the collaboration between these network systems cannot be avoided. Hence, virtualizing the physical resources and abstracting them into services become urgent and inevitable, where network virtualization plays a crucial role in.

Due to the migration from traditional Internet environments to the next generation Internet, one of the most significant technologies, i.e., service discovery and selection techniques, cannot be applicable in network virtualization scenarios. The major reason is that traditional Internet is based on the Internet Service Providers (ISPs) who provide physical resources directly. In network virtualization, however, ISPs is separated into two entities: Infrastructure Providers (InPs) and Service Providers (SPs). The physical resources are provided by InPs, and they are virtualized and abstracted into services managed and provided by SPs. As end users interact directly with SPs instead of ISPs, the discovery and selection techniques that cope with physical resources can be inapplicable in dealing with virtual networks. Besides, traditional service discovery and selection techniques are based on the access performance generated by local access network infrastructures [13]. As network services are virtualized from the physical infrastructures, the network access is no longer limited to the local infrastructures. That is, the performance criteria adopted for discovery and selection should be adapted and improved as to meet the wide perspective of network services.

The great challenge of network service discovery and selection in network virtualization is to interact with numerous heterogeneous network systems and to meet various requirements of network applications. In order to address this problem, researchers have made a lot of efforts and the SOA is widely employed by them. The SOA is an effective architecture for heterogenous system integration, which provides a promising mechanism to facilitate service discovery and selection in network virtualization. The SOA principles and its application in network virtualization is discussed in the next section.

5.3 Service-oriented architecture in network virtualization

SOA is a system architecture initially employed to coordinate computational resources among diverse heterogenous systems to meet various computing requirements of applications in distributed computing environment. As described in article [3], an SOA might be an application architecture within which all functions are defined as independent services with well-defined invokable interfaces. In the SOA, heterogeneous resources are virtualized into services, which can subsequently satisfy the requirements of all kinds of applications through service discovery and selection. The virtualized services are self-constrained and modular, and contain entities that can be used to cooperate with other services. The loose-coupling mechanism is the pivotal feature of SOA that allows services and applications to flexibly interact with each other. Employing this feature of SOA in network virtualization can effectively and efficiently address the problem of the coexistence of numerous heterogeneous network systems.

SOA is widely employed in the Web service research [14,29,41]. As illustrated in the article [22], SOA in Web service includes two features: interfaces with Internet protocols and messages in machine-readable documents. According to Figure 5.2, there are three parties in the SOA system: service provider, service broker, and service customer. The service provider provides service for customers. As there are a plenty of service providers and various requirements from service customers, a service provider needs to publish its service description at the service registry. Service descriptions are organized in standard format that contain functions and access methods. In Web service, the service description is standardized by Web Service Description Language (WSDL) [40]. The service registry is maintained by the service broker that provides various services including service discovery, service selection, service composition, etc. If a service customer (e.g., a Web application) requests a service, it sends its criteria to the service broker. After the service broker receive the query, it applies service discovery and selection algorithms (e.g., Universal Description Discovery and Integration (UDDI) [31]) to search for the optimal service that matches the criteria of the service customer in the service registry. If there exists a matching service, the service broker returns the location and interface information to the service customer. With the service information, the service customer can invoke the service based on the service description. In contrast to the traditional Web service techniques, Simple Object Access Protocol (SOAP) [7], Representative State Transfer (REST) defined by Fielding [17] is much more lightweight. The REST architectural style includes four principles: resource identification through Universal Resource Identifier (URI) [28], uniform interface, self-descriptive messages, and stateful interactions through hyperlinks [32]. These principles guarantee the simpleness of RESTful Web services as REST utilizes existing well-known W3C/IETF standards (HTTP, XML, URI, MIME), which makes the construction of services low-cost and easy-to-build. Due to its stateless feature that provides possibility of serving a very large number of clients, RESTful techniques is promising for realizing the SOA principle and the development of NaaS.

The pivotal feature of SOA, the loose-coupling mechanism, makes SOA applicable in the environment of numerous coexisting heterogenous network systems. A great deal of research work has been carried out in applying SOA. PlanetLab [23] is a global research network that supports the development of new network services, which firstly adopted the virtualization concepts into network services. A slice is a network of virtual machines that run on the computer notes of PlanetLab. Each slice is isolated from each other and allowed to use by different members independently. GENI [15] addressed the shortcomings of PlanetLab and allowed the virtualized network to maintain its own management protocols. Similar to GENI, FEDERICA [34] leveraged virtualization with appropriate hardware and software capabilities of the infrastructure. FEDERICA provided virtual networks via a centralized decision

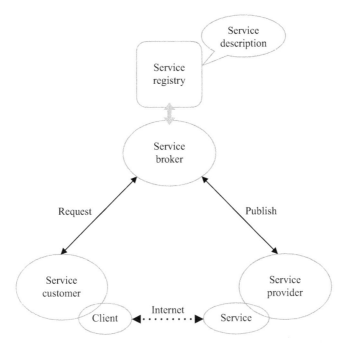

Figure 5.2   The service-oriented architecture in Web service

making procedure, which allowed users to fully configure and manage the resources without affecting the physical infrastructure. UCLPv2 [19] was designed as an SOA that provided virtual networks for users to build their own services or applications. Based on UCLPv2, there was no need for users to deal with the complexities of the underlying network infrastructure. Additionally, articulated private networks (APNs) was presented upon UCLPv2, which can be considered as a next generation virtual private network (VPN). SOA has been considered state-of-the-art techniques for service-delivery platforms [26], which is widely employed in the telecommunications industry. The research work conducted on SOA in telecommunications is summarized by Griffin *et al.* [20].

Applying SOA in network virtualization is to utilize the advantage of the loose-coupling feature of SOA. SOA virtualizes the resources into services, and the paradigm of SOA is illustrated in Figure 5.2. Similar to Web services, network resources can also be encapsulated into network services, and managed and provided by network service providers. The descriptions of various network systems are published in the service registry, which can be maintained by network service brokers. The request from network service customers (e.g., network applications) can be sent to the network service brokers, and then brokers can discover and select the optimal network services that meet the requirements of customers. A layered structure of service-oriented network virtualization proposed by Duan *et al.* [13] is shown in Figure 5.3. This structure separates the traditional structure of Internet

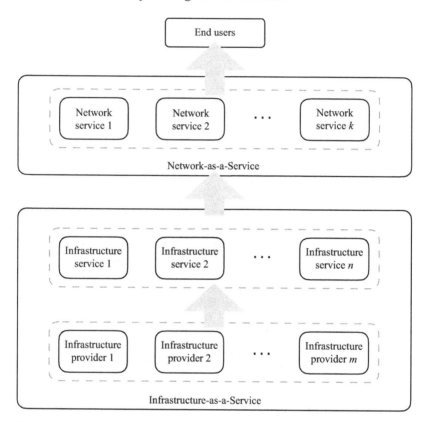

Figure 5.3   The structure of service-oriented network virtualization

service providers into two layers: a Network-as-a-Service layer and an Infrastructure-as-a-Service layer. The two layers are composed by service providers discussed in the previous section of network virtualization. Infrastructure providers virtualize the physical network resources into infrastructure services, where these services are the base of Infrastructure-as-a-Service. Based on the infrastructure services, network service providers construct various network services according to the requirements of end users (e.g., network applications). Therefore, end users can utilize the underlying network resources by accessing the network services provided by service providers. This constructs the Network-as-a-Service paradigm.

By adopting the principle of SOA in network virtualization, the physical network resources are decoupled and virtualized into independent network services. It addresses the problem of the diversity of numerous network systems, which may lead to the complexity of the interaction between service providers and end users. Based on SOA paradigm, the requirements of end users can be satisfied by discovering, selecting, and composing the network services provided by service providers. Hence, in order to assist network users to find the optimal network services that meet their requirements, the techniques of network service discovery, selection, and brokerage need to be proposed.

5.4 Network service discovery and selection

The research problem of network service discovery and selection has attracted a great attention from academia and industry [4,24,38]. As to assisting the network service broker to discover and select optimal network services that meet the requirements of network applications, performance-based discovery and selection techniques have be carried out by researchers. Based on the requirements of network applications, the discovery and selection procedure can be conducted via performance prediction, which is the major concern of network applications. After predicted the performance of network services, sufficient amount of network resources needs to be allocated as to guarantee network provisioning.

5.4.1 Performance prediction for network service discovery and selection

For the network service broker, the network service discovery and selection is based on the performance requirements of network applications, the characteristic of network traffic generated by network applications, and the provisioning capabilities of available network services [12]. The performance requirements and the characteristic of network traffic are provided by network applications during the request process. Therefore, the capabilities of network services should be evaluated and predicted by the network broker as to seek for the optimal network services that meet the requirements of network applications.

Due to the rapid development of Internet, the next generation Internet is facing the burst of various network applications. As there exists a vast number of network applications that demand various network services, proposing a common approach that describes the requirements is of urgency and significance. In order to cope with this important problem, Duan *et al.* [12] defined a *Demand Profile* $\mathbf{P}(\mathbf{d}, \mathbf{L}, \mathbf{a})$ that presented the general specification of network requirements, where $\mathbf{d}$ denoted the address set, $\mathbf{L}$ was a traffic load descriptor, and $\mathbf{a}$ indicated the performance requirement set. Let $T^{in}(t)$ be the accumulated amount of network traffic generated by a network application by the time t. The network application that has a nondecreasing, nonnegative function $A(\cdot)$ for any nonnegative time t and s satisfies the following:

$$T^{in}(t) - T^{in}(s) \leq A(t - s). \tag{5.1}$$

This equation gives the upper bound of the amount of network traffic that a network application may load on the network service. Commonly, the arrival traffic generated by network applications is shaped by the traffic regulation mechanism [27,37] using leaky buckets [1,35]. Hence, the arrival curve is defined as follows:

$$A(t) = \min\{pt, \sigma + \rho t\}, \tag{5.2}$$

where p is the peak rate, σ is the maximal burst size of the traffic flow, and ρ is the sustained rate.

As to discovering and selecting network services for network applications that meet their requirements, the performance of network services needs to be predicted based on the *Demand Profile* $\mathbf{P(d, L, a)}$. The performance prediction methods were proposed by previous research work [9,10]. Let $S(t)$ denote the service curve of the network route R provided by a network service. Then the minimum bandwidth guaranteed by the network service on the network route R can be defined as:

$$b_{min} = \lim_{t \to \infty} \left\lfloor \frac{S(t)}{t} \right\rfloor. \tag{5.3}$$

Assume arrival curve $A(\cdot)$ denotes the traffic load descriptor $\mathbf{L}$ in *Demand Profile*, then the maximum delay d_{max} for data transportation can be calculated by the following equation:

$$d_{max} = \max_{t \geq 0}\{\min\{\delta : \delta \geq 0 \; \& \; A(t) \leq S(t + \delta)\}\}. \tag{5.4}$$

The article [9] also proposed the performance prediction on Latency-Rate (LR) server [33]. Assume the arrival curve of the traffic load generated by a network application is $A(t)$ defined in (5.2), and the service curve of the network route R is $S(t) = r(t - \theta)$, where r and θ are the service rate and latency of the network flow, respectively. Therefore, the minimum bandwidth guaranteed by the network service on the network route is:

$$b_{min} = \lim_{t \to \infty} \frac{r(t - \theta)}{t} = \lim_{t \to \infty} \left(r - \frac{r\theta}{t}\right) = r. \tag{5.5}$$

Based on (5.4), the maximum delay for data transportation guaranteed by the network route can be calculated as:

$$d_{max} = \theta + \left(\frac{p - r}{p - \rho}\right)\frac{\sigma}{r}, \quad r \geq \rho. \tag{5.6}$$

With the predicted performance of a network service on a network route R defined in (5.3) and (5.4), network brokers can discover and select network services that meet the requirements of network applications. As the requirements of network applications are presented in the performance requirement set $\mathbf{a}$ of the *Demand Profile* $\mathbf{P}$, the discovery and selection procedure is based on the comparison between the predicted performance and the required one. The article [12] conducted the following analysis and proposed a service selection approach. The request of network services generated by a network application can be classified into three categories: (1) With only bandwidth requirement, i.e., $\mathbf{a} = b_{req}$; (2) with only delay requirement, i.e., $\mathbf{a} = d_{req}$; (3) with both bandwidth and delay requirements, i.e., $\mathbf{a} = \{b_{req}, d_{req}\}$. Therefore, based on the request of network applications, the selection process is decided as follows: (1) If $\mathbf{a} = b_{req}$, select network services with $b_{min} \geq b_{req}$; (2) if $\mathbf{a} = d_{req}$, select network services with $d_{max} \leq b_{req}$; (3) if $\mathbf{a} = \{b_{req}, d_{req}\}$, select network services with $b_{min} \geq b_{req}$ and $d_{max} \leq b_{req}$. Network services that meet the performance requirements of network applications can be selected as the candidate services. If there are more than one network services selected as candidate services, then other criteria such as service cost or load balance will be adopted to select the optimal network service [10].

5.4.2 *Resource allocation for network service provisioning*

As to making sure that allocated network resources satisfy the network provisioning for the network application, the selected network service ought to allocate adequate amount of resources, e.g., the bandwidth of underlying network systems. In order to address this problem, research work in articles [10,12] discussed the bandwidth allocation for Quality of Service (QoS) provisioning in network services. Detailed discussion is presented in the following paragraphs.

Analyzed in the article [10], the delay performance is in the scope of $[\theta, \theta + \frac{\sigma}{\mu}]$, and the service rate is in the range of $[\rho, p]$. Therefore, the optimal allocation of bandwidth is to find the most reasonable service rate r_a that satisfies the delay requirement d_{req} given by the network application, which is defined as:

$$r_a = \min\{r : \rho \le r \le p \ \& \ d_{max}(r) \le d_{req}\}. \tag{5.7}$$

In order to meet the requirement of network applications, the delay performance guaranteed by the network service should not be greater than the required delay performance d_{req}, which is to satisfy the following function:

$$d_{max} = \theta + \left(\frac{p-r}{p-\rho}\right)\frac{\sigma}{r} \le d_{req}. \tag{5.8}$$

Therefore, the minimum bandwidth that ought to be allocated to satisfy the requirement of the network application can be obtained as:

$$r_a = \frac{p\sigma}{(p-\rho)(d_{req}-\theta)+\sigma}. \tag{5.9}$$

From (5.9), it can be seen that the service rate r_a is upper bounded by the peak rate p, and it approaches the peak rate when the sustained rate ρ gets close to the peak rate. The sustained rate denotes the current condition of the network, and if the sustained rate is near the peak rate, it means that the network is under the best condition with the full utilization of resources.

To summarize the allocation of bandwidth for guaranteeing a delay requirement d_{req}, here gives a viable function:

$$r_a = \begin{cases} \rho & d_{req} \ge D_{max} \\ \frac{p\sigma}{(p-\rho)(d_{req}-\theta)+\sigma} & D_{min} \le d_{req} < D_{max} \\ \text{no valid value} & d_{req} < D_{min} \end{cases} \tag{5.10}$$

where $D_{max} = \theta + \frac{\sigma}{\rho}$ and $D_{min} = \theta$. If the requirement of the network application includes both bandwidth and delay performances, i.e., $\mathbf{a} = \{b_{req}, d_{req}\}$, the minimum bandwidth allocation can be obtained by the following function:

$$b_{min} = \max\{r_a, b_{req}\}. \tag{5.11}$$

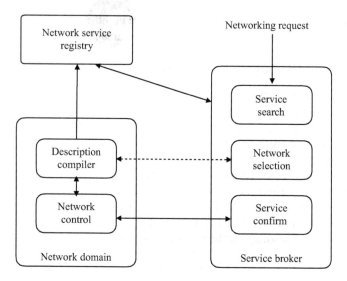

Figure 5.4   The structure of network service broker system

5.5   Network service brokerage

The SOA changes the relation between network end users and network service providers by introducing the network service broker. The purpose of network service broker is to register network services based on the service description provided by network service providers, and to discover and to select appropriate network services according to the requirements of network end users. With the network service broker, network end users can submit their various requirements to the broker and do not have to consider the underlying infrastructures provided by network service providers. Similarly, the network service providers virtualize the network resources into different types of network services in the same format, and publish them on the service registry with unified service descriptions. Therefore, network end users and network service providers do not have to be compatible and seek for each other as the network service brokers mediate between them. To be more specific, the article [9] gave a detailed analysis of networks service brokerage, and a network service broker system is shown in Figure 5.4.

The service broker provides the functionalities of network service discovery and selection. To illustrate the whole structure of network service broker system, network domain and network service registry are also illustrated in Figure 5.4, which are significant components in providing and supporting network services. Network domain is the part where the network service provider interacts with the network to publish service descriptions and build network connections. As the network fluctuates all the time, the network information provided by network service providers is not stable and invariant. Therefore, the description compiler is employed to collect the current network status and generate the network service description. Based on the current network status, the network service description can be published on the network service registry. The network control component copies with the connection between network service providers and end users, which is at the last stage of the whole process. The service broker is the main and significant part in the interaction between network providers and network end users. When a networking request sent from a network end user, the network service discovery and selection process is conducted by the service broker. Firstly, based on the requirement submitted by the end user, the service broker searches the registered network services in the network service registry to find appropriate network services. As analyzed in Section 5.4.1, the request **P(d, L, a)** includes the destination address **d**, the traffic load descriptor **L**, and the performance requirement **a**. Those network services that satisfy the *Demand Profile* **P** are selected as candidates for further selection. The dotted line between description compiler and network selection denotes the selection process based on the current network status. That is, by checking the current network status, the service broker selects the optimal network service under the current network condition from the network service candidates. After selecting the satisfied network service, the service broker check with the network control to confirm resource allocation for network service provisioning. Once the connection between the network service provider and the end user is established, then the network service can be utilized by the network end user.

5.6   Information update for network service discovery and selection

The next generation Internet can be a large scale dynamic network environment with various network services provided different network service providers. With the highly dynamic update of service information, it is urgent and significant to keep the latest service information available at the network service registry, which is the foundation of network service discovery and selection. But the update of service description may occupy a large part of network resources between the network service provider and the network service broker/registry if the entire service description is republished at the registry. In order to address this problem, an efficient and scalable protocol for network service information update was proposed in the article [11].

The protocol proposed in the article [11] was based on subscription-notification mechanism, where one-way notification messages are sent by providers to multiple information consumers. A standard approach for notification was provided by the Origination for the Advancement of Structured Information Standards (OASIS) [18], which employed a topic-based publish/subscribe mechanism. By adopting this approach, the network broker/registry can subscribe a certain set of network states as topics to network services. Then the update of network service information can be completed by only changing the subscribed network states, which can greatly enhance the update process of network service information and reduce the frequent communication between the network service provider and the network service broker/registry. Figure 5.5 proposed by Duan *et al.* [11] presents the detailed updating process.

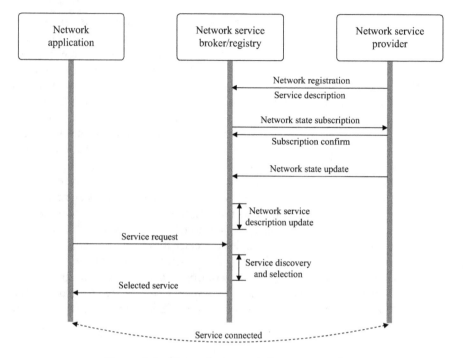

Figure 5.5 Network service information update

The network service provider publishes a service at the network service registry with a service description document. According to the service description, the network service registry can subscript interested network states to network service providers by sending a subscription notification. The network service provider confirms the subscription of the network service registry and updates the scripted network states. After the network service broker/registry subscribes the specified network states of network services, the notification can be pushed to the network service broker/registry whenever there is change occurred on network services. The network service broker/registry updates the network service description based on the notification message including network state update. Therefore, if a network application sends a service request to the network service broker/registry, the network service discovery and selection process can be carried out according to the latest network information. The most appropriate network service that meets the requirement of the network application is selected and sent to the network application (ID of the selected service) by the broker. The dotted line represents the connection between the network application and the network service provider, and the network application starts accessing the selected network service.

Dynamic update of network information is significant for the discovery and selection of network services due to the instability of Internet and network services.

In order to provide the optimal network service that meets the requirement of network applications, the broker has the responsibility to choose the most appropriate one from thousands of network services, which shows the best performance at that specific time. The update procedure implements the description compiler discussed in Section 5.5.

5.7   Conclusions

The next generation Internet has to face the fact of the coexistence of various heterogenous network systems. Due to the rapid increase of network requirements from all types of network applications, it is crucial to provide network functionalities without considering the underlying infrastructures. In order to address this critical problem, network virtualization provides the capability to decouple and abstract network resources into independent functional components. The underlying infrastructures are provided and managed by the infrastructure providers, and the service providers provide network services for network end users by utilizing the virtualized underlying network resources. The technique of network virtualization is widely adopted by researchers in solving the problem of coexisting heterogenous network systems and show its capability to collaborate these network systems. As to cope with the diversity of requirements of network applications, SOA is introduced and applied in the network virtualization environment. The key feature of SOA is the loose-coupling mechanism, which makes SOA applicable in the environment of numerous coexisting heterogenous network systems. The network services provided by service providers can be published and managed by the service brokers, which makes it convenient and efficient for network applications to acquire the network services that meet their requirements. The communication between network applications and network service providers is surrogated by network service brokers who discover and select the most appropriate network service for network applications. Therefore, the network service discovery and selection can be conducted in the service brokers. The request from network applications include the requirements of network performance. Due to the instability of Internet, the performance of network services needs to be predicted based on the service description under the corresponding circumstance. Additionally, sufficient network resources should be allocated to satisfy the network provisioning for the network applications. Network service brokerage optimizes the communication between network applications and network service providers. The architecture and technologies proposed and discussed by previous researchers are network-independent, which are applicable to heterogenous network environments of the next generation Internet.

References

[1]   Leaky bucket. https://en.wikipedia.org/wiki/Leaky_bucket.

[2]   Fatna Belqasmi, Roch Glitho, and Rachida Dssouli. Ambient network composition. *IEEE Network*, 22(4):6–12, 2008.

[3]   Kishore Channabasavaiah, Kerrie Holley, and Edward Tuggle. Migrating to a service-oriented architecture. *IBM DeveloperWorks*, 16, 2003.

[4]   Liang Chen, Yipeng Feng, Jian Wu, and Zibin Zheng. An enhanced QoS prediction approach for service selection. In *Proceedings of the Eighth IEEE International Conference on Services Computing (SCC'11)*, pages 727–728, 2011.

[5]   Margaret Chiosi, Don Clarke, Peter Willis, *et al.* Network functions virtualisation: An introduction, benefits, enablers, challenges and call for action. In *SDN and OpenFlow World Congress*, pages 22–24, 2012.

[6]   NM Chowdhury and Raouf Boutaba. Network virtualization: State of the art and research challenges. *IEEE Communications Magazine*, 47(7):20–26, 2009.

[7]   Francisco Curbera, Matthew Duftler, Rania Khalaf, William Nagy, Nirmal Mukhi, and Sanjiva Weerawarana. Unraveling the Web services web: An introduction to SOAP, WSDL, and UDDI. *IEEE Internet Computing*, 6(2):86, 2002.

[8]   Richard Douville, J Le Roux, J Rougier, and Stefano Secci. A service plane over the PCE architecture for automatic multidomain connection-oriented services. *IEEE Communications Magazine*, 46(6):94–102, 2008.

[9]   Qiang Duan. Service-oriented network abstraction and discovery for the diversified Internet. In *Proceedings of the IEEE Sarnoff Symposium*, pages 1–5, 2007.

[10] Qiang Duan. Network service description and discovery for high-performance ubiquitous and pervasive grids. *ACM Transactions on Autonomous and Adaptive Systems (TAAS)*, 6(1):3, 2011.

[11] Qiang Duan. Automatic network service discovery and selection in virtualization-based future Internet. *Network Protocols and Algorithms*, 4(2):30–48, 2012.

[12] Qiang Duan and Enyue Lu. Network service description and discovery for the next generation Internet. *International Journal of Computer Networks (IJCN)*, 1(1):46–65, 2009.

[13] Qiang Duan and Yong Lu. Service-oriented network discovery and selection in virtualization-based mobile Internet. *Journal of Computer Information Systems*, 53(3):38–46, 2013.

[14] Thomas Erl. *Service-oriented architecture: A field guide to integrating XML and web services*. Upper Saddle River, NJ, USA: Prentice-Hall PTR, 2004.

[15] D. Clark, S. Shenker, A. Falk. GENI Research Plan. Technical Report, GENI Research Coordination Working Group, April 2007.

[16] Nick Feamster, Lixin Gao, and Jennifer Rexford. How to lease the Internet in your spare time. *ACM SIGCOMM Computer Communication Review*, 37(1):61–64, 2007.

[17] Roy Thomas Fielding. *Architectural styles and the design of network-based software architectures*. PhD thesis, University of California, Irvine, 2000.

[18] Steve Graham, David Hull, and Bryan Murray. Web services base notification 1.3. *Organization for the Advancement of Structured Information Standards (OASIS)*, 2006.

[19] Eduard Grasa, Gabriel Junyent, Sergi Figuerola, Albert Lopez, and Michel Savoie. UCLPv2: A network virtualization framework built on web services [web services in telecommunications, part II]. *IEEE Communications Magazine*, 46(3):126–134, 2008.

[20] Dayton Griffin and Dirk Pesch. A survey on web services in telecommunications. *IEEE Communications Magazine*, 45(7):28–35, 2007.

[21] Bo Han, Vijay Gopalakrishnan, Lusheng Ji, and Seungjoon Lee. Network function virtualization: Challenges and opportunities for innovations. *IEEE Communications Magazine*, 53(2):90–97, 2015.

[22] Hao He. What is service-oriented architecture. *Publicação eletrônica em*, 30:50–55, 2003.

[23] Scott Karlin and L Paterson. PlanetLab: A blueprint for introducing disruptive technology into the Internet. In *joint Princeton ACM/IEEE Computer Society meeting*, 2003. https://www.planet-lab.org/.

[24] George Lee, Peyman Faratin, Steven Bauer, and John Wroclawski. A user-guided cognitive agent for network service selection in pervasive computing environments. In *Proceedings of the Second IEEE Annual Conference on Pervasive Computing and Communications (PerCom'04)*, pages 219–228, 2004.

[25] Thomas Lehman, Jerry Sobieski, and Bijan Jabbari. DRAGON: A framework for service provisioning in heterogeneous grid networks. *IEEE Communications Magazine*, 44(3):84–90, 2006.

[26] Thomas Magedanz, Niklas Blum, and Simon Dutkowski. Evolution of SOA concepts in telecommunications. *Computer*, 40(11):46–50, 2007.

[27] Domingo Marrero, Elsa M Macías, and Alvaro Suárez. An admission control and traffic regulation mechanism for infrastructure WiFi networks. *IAENG International Journal of Computer Science*, 35(1):154–160, 2008.

[28] Larry Masinter, Tim Berners-Lee, and Roy T Fielding. Uniform resource identifier (URI): Generic syntax. 2005.

[29] Eric Newcomer and Greg Lomow. *Understanding SOA with web services (independent technology guides)*. Boston, MA, USA: Addison-Wesley Professional, 2004.

[30] Norbert Niebert, Stephan Baucke, Ibtissam El-Khayat, *et al.* The way 4WARD to the creation of a future Internet. In *Proceedings of the IEEE 19th International Symposium on Personal, Indoor and Mobile Radio Communications (PIMRC'08)*, pages 1–5, 2008.

[31] Organization for the Advancement of Structured Information Standards (OASIS). Universal Description, Discovery and Integration (UDDI) v3.0.2. https://www.oasis-open.org/standards.

[32] Cesare Pautasso, Olaf Zimmermann, and Frank Leymann. Restful web services vs. 'big' web services: Making the right architectural decision. In *Proceedings of the 17th international conference on World Wide Web*, pages 805–814, 2008.

[33] Dimitrios Stiliadis and Anujan Varma. Latency-rate servers: A general model for analysis of traffic scheduling algorithms. *IEEE/ACM Transactions on Networking (ToN)*, 6(5):611–624, 1998.

[34]   Peter Szegedi, Sergi Figuerola, Mauro Campanella, Vasilis Maglaris, and Cristina Cervelló-Pastor. With evolution for revolution: Managing FEDERICA for future Internet research. *IEEE Communications Magazine*, 47(7):34–39, 2009.

[35]   Jonathan Turner. New directions in communications (or which way to the information age?). *IEEE communications Magazine*, 24(10):8–15, 1986.

[36]   Jonathan S Turner and David E Taylor. Diversifying the Internet. In *Proceedings of the Global Telecommunications Conference (GLOBECOM'05)*, volume 2, pages 1–6, 2005.

[37]   Jing Wang and Suparerk Premvuti. Distributed traffic regulation and control for multiple autonomous mobile robots operating in discrete space. In *Proceedings of the IEEE International Conference on Robotics and Automation (ICRA'95)*, volume 2, pages 1619–1624, 1995.

[38]   Shangguang Wang, Zibin Zheng, Qibo Sun, Hua Zou, and Fangchun Yang. Cloud model for service selection. In *Proceedings of the 30th IEEE Conference on Computer Communications Workshops (INFOCOM WKSHPS)*, pages 666–671, 2011.

[39]   Sanjiva Weerawarana, Francisco Curbera, Frank Leymann, Tony Storey, and Donald F Ferguson. *Web services platform architecture: SOAP, WSDL, WS-policy, WS-addressing, WS-BPEL, WS-reliable messaging and more.* Upper Saddle River, NJ, USA: Prentice-Hall PTR, 2005.

[40]   World Wide Web Consortium (W3C). Web Services Description Language (WSDL) Version 2.0. https://www.w3.org/TR/wsdl20/.

[41]   Zibin Zheng and Michael R Lyu. Collaborative reliability prediction of service-oriented systems. In *Proceedings of the 32nd ACM/IEEE International Conference on Software Engineering (ICSE'10)*, pages 35–44, 2010.

Chapter 6

Service selection and recommendation in integrated network environment

Lingyan Zhang[1], Mingzhe Yang[1], Yan Guo[1], and Shangguang Wang[1]

6.1 Introduction

With the rapid development of information technology and network technology, the Internet is becoming more intelligent, personalized, and social, influencing and changing people's way of life. Many heterogeneous networks (Mobile network, Internet, Television broadcasting network, etc.) and technology are integrated into an open communication network, that is, the integrated network.

In the context of network integration, a large number of services, provided by service providers, have sprung to meet various needs of service consumers or users. Service is becoming a "collective term" that is also known as "everything as a service." Various services, including online shopping, music download, live streaming video, social networking, and various mobile apps, improve the efficiency of people's work and facilitate people's life. The ubiquitous and openness characters of the integrated network provide users with personalized services better.

However, along with the explosion of services in the integrated network, there have emerged new problems. On the one hand, to the user, with the rapid growth of the type and number of services, the users run into the trouble of information overload, and usually users need to spend a lot of time finding themselves services they need; on the other hand, to the service provider, the process that users browse a large number of irrelevant services will no doubt make consumers submerged in the problem of information overload in the continuous loss. How to achieve fast and reliable selection and recommendation of optimal services for users in integrated network environment has become one of the most challenging issues in the field of service computing [1].

One of the most typical examples is social networking services. Because social networking service always has a huge user groups and users frequently update status, causing social networking services would produce lots of users and redundant

[1]State Key Laboratory of Networking and Switching Technology, Beijing University of Posts and Telecommunications, China

data every day. How to find useful information from these data, and how to provide users with personalized recommendation services, such as friends recommendation and advertising targeting, becomes the focus and keystone of social networking services [2].

In order to let users get information they want accurately, service selection and recommendation systems emerge as the times require. Because it can bring tremendous commercial value and interest, whether in academia area or industry area, service selection and recommendation systems have attracted great attention. In academic area, there have appeared many efficient methods and algorithms of service selection and recommendation, and in the industrial area, service selection and recommendation systems have been widely used in various occasions. With continuous growth in the number of services, selecting and recommending optimal services for users will become more and more important and acute in the field of service computing [3,4].

6.2　Integrated network

The integrated network is an open communication network which integrates many technologies and heterogeneous networks (mobile network, Internet, Television broadcasting network, etc.). As network convergence evolves, its characters bring special requirements in service selection and recommendation [5–7].

- Different networks have different user groups, and the integrated network integrates many heterogeneous networks, so the number of users in integrated network is far more than any previous network.
- The ubiquitous of integrated network makes the provision of personalized services in a much convenient and popular way, and more and more user requirements are raised on the integrated network.
- An integrated network needs to satisfy all sorts of service requests of a large number of users, and the service requests of these users generally have the same or similar function and different preferences, it is often the case that too many user requests lead to overloading of the service, which further leads to the decline of the quality of service (QoS).
- The dynamic feature of integrated network will lead to the changing of service execution environment. To ensure the continuity of service, appropriate services should be selected and recommended for their adaptability. And this change process should be automatic, cannot be aware by the users.

6.3　Service selection

6.3.1　Selection problem definition

With the rapidly growing number of available services, users are presented with a choice of functionally similar services. This choice allows user to select services that match other criteria which is often referred to as QoS criteria. This line of work opens

Table 6.1   Major QoS attributes

QoS attributes	Description	Type
Availability	The probability that a service can respond to requests	Positive
Capacity	The limit on the number of requests a service can handle	Positive
Economic	The economic conditions of using a service, such as price	Negative
Throughput	The rate of successful service request completion	Positive
Response time	The delay from the request to getting a response from the service	Negative
Reliability	The likelihood of successfully using a service	Positive
Scalability	Whether the service capacity can increase as needed	Positive
Security	The level or kind of security a service provides	Positive
Stability	The rate of change of service attributes, such as its service interface	Negative

up two fundamental questions: how can these extra attributes be described and how can one select the most appropriate service [8].

These questions should be addressed on both the selection of isolated services and the selection of composite services. In some cases, an individual service cannot satisfy users' requirements, and we need to select and integrate several individual services from multiple service classes to create new value-added composite services [9].

6.3.1.1   Quality of service attributes

In general, nonfunctional extra attributes are defined as QoS attributes. A QoS attribute can be static or dynamic. A static QoS property value has been defined at the time it is described, whereas the dynamic QoS property value requires measuring and updating its value periodically. The QoS value from the service user's perspective can be positive and negative. For example, users expect to buy a service at a low price and expect to call the service in a low response time.

Positive attributes refer to the QoS attributes that the higher the attribute value is, the better the quality is (e.g., reliability and availability);

Negative attributes refer to the QoS attributes that the higher the attribute value is, the worse the quality is (e.g., response time, delay time, and price).

Table 6.1 lists the major QoS attributes of services.

6.3.1.2   Service selection definition

In order to better understand the service selection problem, the related concepts [10] are given below.

Service candidate is a service which can satisfy a user's particular functional demand;

Service class is a set composed by multiple service candidates which have the same functions but different nonfunctional attribute values;

QoS attribute is nonfunctional attribute that defines the performance of a service, such as error rates, bit rate, and throughput;

Utility of a service indicates what degree user-defined constraints will be satisfied;

Service selection is to select the most suitable service candidate(s) to meet the functional and QoS requirements of user from one or more service class.

More formally, the selection problem can be formulated as follows: let CS be the set of the required service classes according to the functional requirements of the user, i.e., $CS = \{S_1, \ldots, S_n\}$; and the composite service cs is integrated by component service $s_1, s_2, \ldots, s_n$; and let $Q(s)$ be the vector of the QoS value of service s, i.e., $Q(s) = \{q_1(s), \ldots, q_m(s)\}$; let u be a utility function that measures the fitness of the composite service cs to user constraints; then, for each individual service s, we want to choose such service that maximizes the utility of the composite service. More formally:

$$\max \sum_{i=1}^{m} c_i \overline{Q_i(s_1, s_2, \ldots, s_n)}, \tag{6.1}$$

where $\overline{Q_1(s_1, s_2, \ldots, s_n)}$ is the normalized value of $Q_i(s_1, s_2, \ldots, s_n)$ which represents the ith QoS attribute value of the composite service integrated by component service $s_1, s_2, \ldots, s_n$, c_i is the weight coefficient corresponding to the ith attribute, and there are $c_i \in [0, 1]$ and $\sum_{i=1}^{m} c_i = 1$.

The constraints are defined by (6.2), which is used to describe the users' personalized requirements.

$$\text{constraint } s \begin{cases} Q_1(s_1, s_2, \ldots, s_n) \leq b_1 \\ \vdots \\ Q_m(s_1, s_2, \ldots, s_n) \leq b_m \end{cases}, \tag{6.2}$$

where b_i is the ith QoS constraint of user.

6.3.2   Problem induction

According to the different needs of users under different scenarios, service selection can be converted into different math problems and then designs different models to choose appropriate selection strategies and algorithms.

6.3.2.1   Multiple attribute decision making problem

In the service selection problem, it requires consideration of specific preferences and constraints by a service requester as well as evaluation of combinations of different QoS constraints. Hence, QoS-based service selection is basically the problem of multiple attribute decision making.

Although regular QoS attributes are listed in Table 6.1, it remains some issues on selection of services. First, the perception on QoS of services is distinct between the user and provider. Second, service requestors may have varying preferences for the QoS attributes depending on the situation.

In a real service selection scene, because of the complexity of QoS attributes, it is difficult for users to provide the weights or the precise value of QoS attributes to express their preferences; besides, it is difficult to standardize the QoS attribute values to the unified metric space and accurately assess a comprehensive QoS value of service. Therefore, the service selection problem can be converted to a multiple attribute decision-making problem, a problem with the multiple and conflicting attributes.

6.3.2.2   Integer linear programming problem

Service selection problem for composite services becomes a decision problem to maximize the fitness of the constraints of service requesters. Actually, any decision problem in which the decision variables must assume nonfractional or discrete values can be classified into an integer optimization problem. And one way of solving this optimization problem is to model it as an integer linear programming (ILP) problem. In the ILP approach, decision variables are integers representing whether a particular service is selected for composite services. To define an ILP problem, three inputs should be provided: a set of decision variables, an objective function, and a set of constraints.

Decision variables refer to service candidates from service classes.

Objective function is to reflect some benefit or utility of service requesters. It is defined as weighted sum of QoS attributes that are favorable to the requesters.

Constrains refer to the functional and QoS requirements of the service requester.

However, the computational complexity and cost of the ILP solution will increase exponentially with the growth of the size of services. And it requires the objective function and constraint conditions are linear, which limits the usefulness of the algorithm in a certain extent.

6.3.2.3   Knapsack problem

In order to deal with the computational complexity and cost in large-scale service selection, service selection for a composite service can be modeled as a multidimension multichoice knapsack problem (MMKP), which is known to be an NP-hard problem in the strong sense. This means that for large problems, it is unlikely that an optimal solution can be found given a reasonable amount of computational effort.

In certain cases, in the process of service selection, there is no need to search all candidate services to find global optimal solution. Looking for near-optimal service solution can ensure the needs of users, and can greatly reduce the time cost as well.

MMKP is about taking items into a knapsack with a limited capability. Thus, a selection must be performed to identify the optimal subset. Service selection problem can be mapped into a knapsack problem, when each item represents a candidate service, the profit is on behalf of the utility of service, the weight represents the service quality attribute, and the capacity of backpack is on behalf of the QoS constraint. The objective of MMKP is to select exactly one service candidate from each service class to be included in the knapsack within the functional and QoS constraint while maximizing the total profit.

6.3.2.4   Combinatorial optimization problem

At first, researchers only focus on a certain indicator in the process of service selection, such as to minimize response time or maximize service utility, so this kind of problem is often depicted as single objective optimization problem. With the diverse needs of users and service providers, the concerned indexes and attributes gradually increase, and the optimize target is transformed from a single index into multi indexes, so the service selection turned into a multiobjective optimization problem, i.e., combinatorial optimization problem.

The combinatorial optimization is often difficult to have an optimal solution meeting all the objective functions. For example, it may be impractical if you want to minimize the response time while minimizing the system cost, because the low response time means the high service level, and it is likely to pay more fee for a high level of service. So the methods adopted to solve the single-objective optimization problems are difficult to directly be applied to combinatorial problems.

The most common thinking is transforming the multiobjective optimization problem into the single-objective optimization problem. The linear weighted sum method and the ε-constraint method are the most widely used methods. In addition, the Pareto model is also used to the multiobjective optimization problem of service selection.

6.3.3   Service selection algorithm

Traditional service selection algorithms mainly contain exhaustive algorithm and greedy algorithm. However, the complexity of exhaustive algorithm seeking the optimal solution is too high, and greedy algorithm cannot necessarily obtain global optimal solution. Therefore, a effectively approximate solution has been put forward to attain near-optimal solution. In general, service selection algorithms include optimization algorithms, heuristic algorithms, and fusion algorithms.

Optimization algorithms: The optimization algorithms can guarantee the optimality of the solution, and the specific optimal algorithms applied to the service selection problem mainly include the exhaustive algorithm and graph algorithm.

Heuristic algorithms: Relative to the optimization algorithm, a heuristic algorithm can be defined as a constructed algorithm based on intuition or experience. The specific heuristic algorithms applied to the service selection problem mainly include genetic algorithm (GA), particle swarm optimization algorithm (PSO), and ant colony optimization algorithm (ACO).

Fusion algorithms: Although heuristic algorithm can solve the feasible solution of the optimization problem under the acceptable costs (such as computing time), the deviation between the feasible solution and the optimal solution cannot be expected necessarily in advance. Fusion algorithms can combine the advantages of various algorithms and improve the performance and efficiency of the optimization.

6.3.3.1   Optimization algorithm

Optimization algorithms, which try to find the maximum values of utility functions, have been widely applied in service selection.

Exhaustive algorithm is a straightforward one by computing the QoS value for each possible execution plan and selecting the best one from them. It lists all possible execution plans by exhaustive searching and compares their QoS values. It can always produce the optimal solution but consume time and memory, and the complexity depends on the size of the solution space. Assuming that there are n service classes and m service candidates for each class, the total number of execution plans is m^n, making this approach impractical. So it is only suitable when the number of classes and the candidates of each class are all small.

Greedy algorithm is an algorithm that follows the problem solving heuristic of making the locally optimal candidate services at each class with the hope of finding a global optimal service. In service selection, a greedy strategy does not in general produce a global optimal solution, but a greedy algorithm may yield locally optimal solutions that approximate a global optimal solution in a reasonable time.

Graph algorithm [11] is used to solve the abstracted graph theory problems of service selection. For example, the service selection problem can be transformed into the shortest path problem with functional and QoS constraints. First, build the directed graph corresponding to the composite service, in which the graph nodes represent the candidate services, and the QoS parameters are represented by edges. The optimal path obtained by traversing the graph edge satisfies all the constraint conditions and has the best utility, and it corresponds to the optimal solution of the original service selection problem for the composite service. Using the graph algorithm often needs to access the topology information of service deployment, so the access to information before optimization may cost a lot of time.

6.3.3.2 Genetic algorithm

GA [12,13] solves the formulated optimization problem by using the idea of Darwinian evolution. As a powerful search algorithm based on natural selection and population genetics, GA represents the problem solving process by the survival process of the fittest chromosome, through the iterative evolution of population, including selection, crossover, and mutation operation, combines populations to produce the next generation individuals, gradually evolves toward the optimized population, and ultimately obtains the chromosome most adapted to the environment, which is the optimal solution or near-optimal solution of problem.

The implementation of GA in application of the service selection problem incorporates three basic steps so that the algorithm is formulated for the specific application: the presentation of individual according to the specific service selection, the formulation of the fitness function, crossover and mutation operators. Each chromosome represents a possible service combination which consists of candidate services selected from different service classes. Each gene corresponds to a candidate service, and the value of the ith gene is the serial number of selected candidate service in the ith service class. The determination of fitness function is according to the QoS value of each candidate service. Crossover operation is to exchange the value of some genes in two different service combinations by a crossover method, and to produce two new chromosomes. Mutation operation is to replace the candidate service by another corresponding available candidate service.

6.3.3.3 Particle swam optimization algorithm

PSO is a computational method that optimizes a problem by iteratively trying to improve a candidate solution with regard to a given measure of quality. It solves a problem by having a population of candidate solutions (i.e., particles) and moving these particles around in the search space according to simple mathematical formulae over the particle's position and velocity. Each particle's movement is influenced by its local best known position, but is also guided toward the best known positions in the search space, and the positions are updated as better positions found by other particles. This is expected to move the swarm toward the best solutions.

Using PSO algorithm to solve service selection problem [14,15], the key is to establish a mapping relationship between the particle position vector and combination plan, the expression of particle position, the determination of velocity formula, and the evaluation of fitness function. Each particle represents a candidate solution, which is a complex service composed of several candidate services in a certain order. The position of each particle represents the serial number of candidate services selected from each service class. The QoS attribute values of composition service represented by a particle is the particle's fitness, the fitness function in accordance with the specific composition service example. The update formula of speed and position depends on the coding scheme of service selection which is the presentation of particle position.

However, PSO exhibits some disadvantages: it sometimes is easy to be trapped in local optima (it is often called premature convergence), and the convergence rate decreased considerably in the later period of evolution.

6.3.3.4 Ant colony optimization algorithm

ACO is a simulated evolutionary algorithm which is aiming to search for an optimal path in a graph, based on the behavior of ants seeking a path between their colony and a source of food. The basic principle of ACO comes from the observation of entomologists is as follows: the ants can release a kind of gas (pheromone) on the path they have walked when they find food. The pheromone can be known by other ants and influences their behavior. As more and more ants walk the same path, the density of pheromone will be bigger and bigger, which improves the selected probability of this path. And the attractiveness of this path will be bigger and bigger too. The behavior of releasing the pheromone is the foundation of positive feedback mechanism of ant colony algorithm. The basic thinking of ant colony algorithm is as follows: at the certain point, an ant chooses the next path, then, the path many ants have walked before has the bigger probability to be selected again, and the more pheromone means the shorter path length, which means a better answer.

The selection problem for composite services can be transformed into a problem seeking the QoS optimal path from a starting point to the target point. Suppose that there is a composite service which is combined with n services with different functions, the QoS-aware problem of the complex service selection is to select n services from n services classes, and each class has n_i independent services which have the similar or the same function. So the number of the possible selection plan is $n_1 \times n_2 \times \cdots \times n_n$.

The essence of process of service selection based on the ACO algorithm [16] is to find a shortest path from the initiative point to the end point, and the ant must travel each class only once. The key is to solve how to determine state transition rule according to the QoS parameters in a composite service, update methods of pheromone and assessment of the fitness function, etc.

6.3.3.5 Fusion algorithm

Each algorithm has its advantages and disadvantages, and a single algorithm will not be able to solve the more complicated problem with larger scale. Fusion algorithms which can combine the advantages of various algorithms are used to improve the performance and efficiency of service selection in more complex scenarios.

Many fusion algorithms [17] are proposed to solve service selection problem. For example, on the basis of redefining the location, speed, add/subtraction and multiplication of PSO, combining the crossover and mutation operations in the GA, a QoS scheduling method based on hybrid particle swarm is designed to select the best candidate services satisfying users' QoS requirements. Scalable QoS computation method [1] based on heuristic algorithm is designed, which decomposes a services selection problem into several sub optimization problems. Compared with the ILP method, it can find nearly optimal solution more quickly. A reliable random QoS-aware service selection algorithm is designed based on the Markov Decision Process. The methods based on skyline reduce a large number of redundant candidate services, and they are more effectively for service selection.

6.3.4 Service selection summary

There has been much research done on service selection over the past several years that have inducted service selection problem into Multiple Attribute Decision Problem, Integer Programming Problem, Knapsack Problem, Combinatorial Optimization Problem, and so on. And many optimization algorithms, heuristic algorithms, and fusion algorithms have been proposed to solve the above problems. A classification and summary of the service selection research is presented in Table 6.2.

It can be concluded that most approaches contribute specific aspects to the overall picture of service selection, which requires methods for expressing user requirements, expressing service offerings, and also the actual service selection method. With the increasing availability of services as a solution to application integration, the QoS parameters offered by services are becoming the chief priority for service providers and their service users. Due to the agile and dynamic nature of the Internet, providing the suitable QoS for various services is really a challenging task. In addition to this, important aspects that need addressing are powerful mechanisms to capture user requirements that are both user friendly and also expressive enough to capture large numbers of preferences and the logical relationships between preferences. One aspect that falls into this area is the measuring of weights, that is, users' fuzzy view on QoS parameters has to be modeled and weighted in universal manner. Also, in the process of capturing the needs of users, their preference of data, research has to show interest and capability to automatically capture this, to reduce the burden on the user part, and to react to changes in circumstances automatically [18,19].

6.4 Service recommendation

Nowadays, lots of applications are available for recommending books and products in online shopping websites, friends and hot topics in social networks, music and movies

Table 6.2   Classification and summary of service selection research

Selection approach		Techniques	Papers
Optimization algorithm	Exhaustive algorithm	Weighted multistage graph	Gao *et al.* (2006) [20]
	Greedy algorithm	Relational algebra	Ding *et al.* (2009) [21]
	Graph algorithm	Multigrain clustering	Schröpfer *et al.* (2007) [11]
	ILP algorithm	Fuzzy synthetic evaluation	Ardagna and Pernici (2007) [22]
		Mixed integer programming cloud computing	Wang *et al.* (2011) [23] Wang *et al.* (2015) [24]
Heuristic algorithm	GA	Global optimization	Mardukhi *et al.* (2013) [12]
		Local optimization	Ma and Zhang (2008) [13]
		Hybrid GA	Tang and Ai (2010) [25]
	PSO	Fuzzy logic	Wang *et al.* (2013) [15]
		Skyline operator	
	ACO	Swarm intelligence	Wang *et al.* (2010) [16]
	Fusion algorithms	Service clustering	Guidara *et al.* (2015) [1]
		Constraints decomposition	
		Neural network	

in video websites. Moreover, some of the vendors have incorporated recommendation capabilities into their commerce servers. With the increasing amounts of applications that help users deal with information overload, service recommendation is playing an increasingly significant role [26,27].

At the same time, users have more independence and options when choosing services including movie, music, and video. And they can choose services depending on their own personal preferences. With movies as an example, they can be tagged with fiction, action, war, comedy, etc., or classified as director and main actor. When watching movies, users always make decisions according to these tags. Based on user's history records and film scores, service recommendation system can dig up active users' preferences and find out users with similar interests, then, we can recommend similar users' movies which have never been watched by the active user.

6.4.1   Recommendation scenarios

Service recommendation systems are applied to various occasions, especially in commercial vocation. In our daily life, there are lots of recommendation examples and scenarios. Examples of such applications include recommending products at Amazon.com [28], movies by MovieLens [29], and news at VERSIFI Technologies. Following is a list of several common service recommendation scenarios.

6.4.1.1   Commodity recommendation in e-commerce site

Commodity recommendation in e-commerce site is one of the most widely used applications. Most of these websites recommend relevant items to users depending on users' browsing history. As an international e-commerce company, Amazon has always been an active participant and a user of recommendation system. The most important applications are personalized and relevant items' recommendation lists in Amazon's recommendation system.

6.4.1.2 Advertisement targeting and recommendation

When browsing webpages, we always see all kinds of advertisements which are elaborately designed to draw users' attention. In addition to advertisement content, manifestation pattern, the most important factor is increasing the accuracy of advertising, which means precisely carrying advertisements to users with purchasing demand. In fact, the precisely advertisement targeting is the personalized product recommending to user and the advertisement is the output product of recommendation system. YouTube has a variety of targeting options, such as age, gender, location, interest, and others, that help advertisers reach the right customer for their business.

6.4.1.3 Location based service recommendation

Through Global Position System (GPS), Wi-Fi, and base station signal, users' location information can be acquired in real time. Location-based service recommendation is widely used in modern life. For example, when a user stays in a strange place, as long as opening the map of mobile phone, recommendation system can immediately provide restaurants suitable for user's taste and hotels meeting user's consumption level by mobile phone location.

6.4.2 Recommendation problem definition

Service recommendation problem can be reduced to a problem of estimating ratings for the services that have not been seen by a user. Intuitively, this estimation is usually based on the ratings given by the user to other services and on some other information. Once we can estimate ratings for the yet unrated services, we can recommend to the user the service with the highest estimated rating.

More formally, the recommendation problem can be formulated as follows: let C be the set of all users and let S be the set of service candidates that can be recommended; let u be a utility function that measures the usefulness of service s to user c, i.e., $u : C \times S \to R$, where R is a totally ordered set; then, for each user c, we want to choose such service that maximizes the user's utility. More formally:

$$\forall c \in C, s'_c - \arg \max_{s \in S} u(c, s) \tag{6.3}$$

In recommendation systems, the utility of a service is usually represented by a rating, which indicates how a particular user liked a particular service, for example, John Doe gave the movie "Harry Potter" the rating of 7 (out of 10). Each element of the user space C can be defined with a profile that includes various user characteristics, such as age, gender, income, and marital status. Similarly, each element of the service space S is defined with a profile that includes various service characteristics, such as title, genre, director, year of release, and leading actors.

6.4.3   Recommendation systems and techniques

Much research over the past decade has focused on developing service recommendation systems and techniques from disciplines such as human-computer interaction, statistics, machine learning, and information retrieval. These methods are often classified into broad categories according to recommendation approach, as well as to algorithmic technique.

Recommendation systems [30] are usually classified based on recommendation approach as:

Content-based recommendation [31]: Content-based methods analyze the common features among the services a user has already rated highly. Only the services similar to user's past preferences are then recommended;

Collaborative recommendation [32]: Collaborative methods recommend services to the user that people with similar tastes and preferences have liked in the past;

Hybrid recommendation: It is a combination of collaborative and content-based methods to avoid certain limitations of content-based and collaborative systems.

Recommendation systems methods can be classified based on the algorithmic technique as:

Heuristic-based technique: Heuristic-based techniques make rating predictions based on the entire collection of previously rated services by the users;

Model-based technique: Model-based techniques use previous transactions to learn a model (usually using some machine-learning [33] or statistical technique) that is then used for making recommendations.

6.4.3.1   Content-based recommendation with heuristic-based technique

In content-based recommendation methods, the utility $u(c, s)$ of service s for user c is estimated based on the utilities $u(c, s_i)$ assigned by user c to services $s_i \in S$ that are "similar" to service s. For example, in a movie recommendation application, in order to recommend movies to the target user, the content-based recommendation system tries to understand the commonalities among the movies user c has rated highly in the past (specific actors, directors, subject matter, etc.). Then, only the movies that have a high degree of similarity to whatever user's preferences would get recommended.

More formally, let *ServiceP*(s) be a service profile, that is, a set of attributes characterizing service s. It is usually computed by extracting a set of features from service s and is used to determine appropriateness of the service for recommendation purposes. Let *UserP*(c) be the profile of user c containing preferences of the user. These profiles are obtained by analyzing the content of the services previously seen and rated by the user. Thus, the utility function $u(c, s)$ is usually defined as:

$$u(c, s) = score(UserP(c), ServiceP(s)). \tag{6.4}$$

Note, both service profile *ServiceP*(s) and user profile *UserP*(c) are constructed using keyword analysis techniques from information retrieval. And one of the best-known measures for specifying keyword weights in Information Retrieval is the term frequency/inverse document frequency (TF-IDF) measure. By using the TF-IDF measure method, both *UserP(c)* and *ServiceP(s)* can be represented as vectors w_c and w_s of keyword weights. Thus, utility function $u(c, s)$ is usually represented in information retrieval literature by some scoring heuristic defined in terms of vectors w_c and w_s, such as cosine similarity measure:

$$u(c,s) = \cos(w_c, w_s) = \frac{w_c \cdot w_s}{\|w_c\|_2 \times \|w_s\|_2} = \frac{\sum_{i=1}^{K} W_{i,c} W_{i,s}}{\sqrt{\sum_{i=1}^{K} w_{i,c}^2} \sqrt{\sum_{i=1}^{K} w_{i,x}^2}} \qquad (6.5)$$

where K is the total number of keywords in the system.

6.4.3.2   Content-based recommendation with model-based technique

Model-based techniques differ from information retrieval-based approaches in that they calculate utility predictions based not on a heuristic formula, but rather based on a model learned from the underlying data using statistical learning and machine learning techniques [34].

For example, based on a set of services that are rated as "relevant" or "irrelevant" by the user, we can use the naïve Bayesian classifier to classify unrated services. More specifically, the naïve Bayesian classifier is used to estimate the following probability that service s_j belongs to a certain group G_i (e.g., relevant or irrelevant) given the set of keywords, $k_{1,j}, \ldots, k_{n,j}$, on that service:

$$P(G_i | k_{1,j} \& \ldots \& k_{n,j}). \qquad (6.6)$$

Moreover, we can assume that keywords are independent and, therefore, the above probability is proportional to:

$$P(G_i) \prod_x P(k_{x,j} | G_i). \qquad (6.7)$$

While the keyword independence assumption does not necessarily apply in many applications, naïve Bayesian classifiers still produce high classification accuracy. Furthermore, both $P(k_{x,j} | G_i)$ and $P(G_i)$ can be estimated from the underlying training data. Therefore, for each service s_j, the probability $P(G_i | k_{1,j} \& \ldots \& k_{n,j})$ is computed for each group G_i, and service s_j is assigned to the group G_i having the highest probability.

As in the case of content-based approaches, the main difference between model-based techniques and heuristic-based techniques is that the model-based techniques calculate utility (rating) predictions based not on some ad hoc heuristic rules, but rather based on a *model* learned from the underlying data using statistical and machine learning techniques.

6.4.3.3   Collaborative recommendation with heuristic technique

Collaborative recommendation systems try to predict the utility of services for a particular user based on the services previously rated by other users. More formally, the utility $u(c, s)$ of services s for user c is estimated based on the utilities $u(c_j, s)$ assigned to services s by those users $c_j \in C$ who are "similar" to user c. For example, in a movie recommendation application, in order to recommend movies to user c, the collaborative recommendation system tries to find the "peers" of user c, that is, other users that have similar tastes in movies (rate the same movies similarly). Then, only the movies that are most favored by the "peers" of user c would get recommended.

Collaborative recommendation with heuristic-based techniques [35] makes rating predictions based on the entire collection of previously rated services by users. That is, the value of the unknown rating $r_{c,s}$ for user c and service s is usually computed as an aggregate of the ratings of the N most similar users for the same services:

$$r_{c,s} = \underset{c'=\hat{C}}{\text{aggr}} \, r_{c',s}, \tag{6.8}$$

where $\hat{C}$ denotes the set of N users that are the most similar to user c and have rated service s. The most common aggregation approach is to use the weighted sum as:

$$r_{c,s} = \bar{r}_c + k \sum_{c' \in \hat{C}} \text{sim}(c, c') \times (r_{c',s} - \bar{r}_{c'}), \tag{6.9}$$

where k serves as a normalizing factor and is usually selected as $k = 1/\sum_{c' \in \hat{C}} |\text{sim}(c, c')|$, $\bar{r}_c$ is the average rating of user c, and $\text{sim}(c, c')$ is the similarity between user c and user c'.

In most of these approaches, the similarity between two users is based on their ratings of services that both users have rated. The two most popular approaches are correlation-based and cosine-based approaches. The Pearson correlation coefficient is shown as follows:

$$\text{sim}(x, y) = \frac{\sum\limits_{s \in S_{xy}} (r_{x,s} - \bar{r}_x)(r_{y,s} - \bar{r}_y)}{\sqrt{\sum\limits_{s \in S_{xy}} (r_{x,s} - \bar{r}_x)^2 \sum\limits_{s \in S_{xy}} (r_{y,s} - \bar{r}_y)^2}}, \tag{6.10}$$

where S_{xy} is the set of all services correlated by both user x and user y.

6.4.3.4   Collaborative recommendation with model-based technique

Collaborative recommendation with model-based techniques makes rating predictions based on a model learn from the collection of ratings. For example, we can use a simple probabilistic model to implement collaborative filtering, where the unknown ratings of services are calculated as:

$$r_{c,s} = E(r_{c,s}) = \sum_{i=0}^{n} i \times \Pr(r_{c,s} = i | r_{c,s'}, s' \in S_c), \tag{6.11}$$

where $r_{c,s}$ is the rating value for user c and service s, and it is assumed that rating values are integers between 0 and n, and the probability expression is the probability that user c will give a particular rating to service s given that user's ratings of the previously rated services.

To estimate this probability, there are two alternative probabilistic models: cluster models and Bayesian networks. In the first model, like-minded users are clustered into classes. Given the user's class membership, the user ratings are assumed to be independent, that is, the model structure is that of a naïve Bayesian model. The number of classes and the parameters of the model are learned from the data. The second model represents each service in the domain as a node in a Bayesian network, where the states of each node correspond to the possible rating values for each service. Both the structure of the network and the conditional probabilities are learned from the data. One limitation of this model is that each user can be clustered into a single cluster, whereas some recommendation applications may benefit from the ability to cluster users into several categories at once. For example, in a book recommendation application, a user may be interested in one topic for work purposes and a completely different topic for leisure.

Several other model-based collaborative recommendation approaches [36] have been proposed in the literature. Statistical model, Bayesian model, probabilistic relational models, linear regression models, and maximum entropy models have been developed to gain more ideal performance for collaborative recommendation.

6.4.3.5 Hybrid recommendation

Several service recommendation systems use a hybrid approach by combining collaborative and content-based methods, which helps to avoid certain limitations of content-based and collaborative systems. Different ways to combine collaborative and content-based methods into a hybrid recommendation system can be classified as follows:

Combining separate recommendations [37]

One way to build hybrid recommendation systems is to implement separate collaborative and content-based systems. Then we can have two different scenarios. First, we can combine the ratings obtained from individual recommendation systems into one final recommendation using either a linear combination of ratings or a voting scheme [38]. Alternatively, we can use one of the individual recommendations, at any given moment choosing to use the one that is "better" than others based on some recommendation "quality" metric.

Adding content-based characteristics to collaborative models

Several hybrid recommendation systems, including Fab and the "collaboration via content" approach, are based on traditional collaborative techniques but also maintain the content-based profiles for each user. These content-based profiles, and not the commonly rated services, are then used to calculate the similarity between two users. This type of methods can overcome some sparsity-related problems of a purely collaborative approach, and users can be recommended with the service not only when this service is rated highly by users with similar profiles, but also when this service scores highly against the user's profile.

Adding collaborative characteristics to content-based models

The most popular approach in adding collaborative characteristics to content-based models is using some dimensionality reduction technique on a group of content-based profiles. For example, we can use latent semantic indexing to create a collaborative view of a collection of user profiles, where user profiles are represented by term vectors, resulting in a performance improvement compared to the pure content-based approach.

Developing a unifying recommendation model [39]

Many researchers have followed this approach in recent years, such as single rule-based classifiers based on content-based and collaborative characteristics, unified probabilistic methods based on the probabilistic latent semantic analysis, Bayesian mixed-effects regression models based on Markov chain Monte Carlo methods, and so on.

Hybrid recommendation systems can also be augmented by knowledge-based techniques in order to improve recommendation accuracy and to address some of the limitations (e.g., new user, new service problems) of traditional recommendation systems. For example, we can use some domain knowledge about restaurants, cuisines, and foods to recommend restaurants to its users. The main drawback of knowledge-based systems is the need for knowledge acquisition—a well-known bottleneck for many artificial intelligence applications. However, knowledge-based recommendation systems have been developed for application domains where domain knowledge is readily available in some structured machine-readable form, for example, as an ontology. For example, Quickstep and Foxtrot systems use research paper topic ontology to recommend online research articles to the users.

6.4.3.6   Mobile recommendation systems

Mobile recommendation system is the hot spot of the current research. With the increasing ubiquity of Internet-accessing smart phones, it is now possible to offer personalized, context-sensitive service recommendations. This is particularly a difficult area of research as mobile data are complex, heterogeneous, and noisy, and these require spatial and temporal autocorrelation. Mobile recommendation systems have to deal with its validation and generality problems.

One example of a mobile recommendation system is one that offers potentially profitable driving routes for taxi drivers in a city. This system [40] takes as input data in the form of GPS traces of the routes that taxi drivers took while working, which include location (latitude and longitude), time stamps, and operational status (with or without passengers). It then recommends a list of pickup points along a route that will lead to optimal occupancy times and profits. This type of system is obviously location-dependent, and as it must operate on a handheld or embedded device, the computation and energy requirements must remain low.

Another example of mobile recommendation is the system developed for professional users. This system takes as input data the GPS traces of the user and his agenda to suggest him suitable information depending on his situation and interests. The system uses machine learning techniques and reasoning process in order to adapt dynamically the mobile recommendation system to the evolution of the user's interest.

Mobile recommendation systems have also been successfully built using the Web of Data as a source for structured information. A good example of such system is SMARTMUSEUM [41]. The system uses semantic modeling, information retrieval, and machine learning techniques in order to recommend services matching user's interest, even when the evidence of user's interests is initially vague and based on heterogeneous information.

6.4.4 Service recommendation summary

There has been much research done on service recommendation over the past several years that have used a broad range of statistical, machine learning, information retrieval, and other techniques and that significantly advanced the state of the art in comparison to early recommendation systems that utilized collaborative- and content-based heuristics. However, both the content-based recommendation and collaborative-based recommendation have their own limitations.

6.4.4.1 Limitation of content-based recommendation

New user problem
A user has to rate a sufficient number of services before the system can really understand the user's tastes and preferences and present the user with reliable recommendations. Therefore, a new user, having very few ratings of services, would not be able to get accurate recommendations.

Limited content analysis
Content-based techniques are limited by the features that are explicitly associated with the objects these systems recommend. Therefore, in order to have a sufficient set of features, the content must either be in a form that can be parsed automatically by a computer, or the features should be assigned to services manually. While information retrieval techniques work well in extracting features from text documents, some other domains have an inherent problem with automatic feature extraction. Moreover, it is often not practical to assign attributes by hand. Another problem is that, if two different services are represented by the same set of features, they are indistinguishable in content-based recommendation systems.

Overspecialization
When the system can only recommend services that score highly against a user's profile, the user is limited to being recommended services similar to those already rated, and it cannot recommend services that are different from anything the user has seen before. Therefore, some content-based recommendation systems filter out services not only if they are too different from user's preferences, but also if they are too similar to something the user has seen before. In summary, the *diversity* of recommendations is often a desirable feature in recommendation systems. Ideally, the user should be presented with a range of options and not with a homogeneous set of alternatives.

6.4.4.2 Limitation of collaborative recommendation

New user problem
It is the same problem as with content-based systems. In order to make accurate recommendations, the system must first learn the user's preferences from the ratings that the user makes. Several techniques have been proposed to address this problem.

New service problem
New services are added regularly to recommendation systems. Collaborative systems rely solely on users' preferences to make recommendations. Therefore, until the new

service is rated by a substantial number of users, the recommendation system would not be able to recommend it.

Sparsity

In any recommendation system, the number of ratings already obtained is usually very small compared to the number of ratings that need to be predicted. Effective prediction of ratings from a small number of examples is important [42,43]. Also, the success of the collaborative recommendation system depends on the availability of a critical mass of users.

6.4.4.3   Summary of the recommendation systems research

As discussed above, recommendation systems can be categorized as being (a) content-based, collaborative-based or hybrid-based recommendation approach, and (b) heuristic-based or model-based recommendation techniques used for the rating estimation. We use these two orthogonal dimensions to classify the recommendation systems research as shown in Table 6.3.

6.5   Evaluation index

There are several major indexes that are commonly considered when comparing different selection or recommendation approaches. As different applications have different needs, the designer of the system must decide on the important index to measure the concrete application at hand, and some of the indexes can be traded off. Therefore, when suggesting a method that improves one of the indexes, it should evaluate the influence of user experience affected by changes in this index either through a user study or through online experimentation.

6.5.1   User preference

When we need to choose one out of a set of candidate services, an obvious option is to run a user study (within subjects) and ask the participants to choose one of the candidates. The service with the largest number of votes is the ideal one to recommend to users. And we need to further weight the vote by the importance of the user and provide nonbinary answers for the preference question in the user study.

6.5.2   Accuracy

A basic assumption in a selection/recommendation system is that a system providing more accurate selections/predictions will be preferred by the user. The root-mean-square deviation (RMSD) or root-mean-square error (RMSE) is the most popular metric used in evaluating accuracy of predicted ratings. The RMSD represents the sample standard deviation of the differences between predicted values and actual values. The RMSD serves to aggregate the magnitudes of the errors in predictions for various times into a single measure of predictive power. At the same time, we could measure the diversity of a selection/recommendation list based on the sum, average, min, or max distance between service pairs, or measure the value of adding each

Table 6.3   Classification of service recommendation system research

Recommendation approach		Techniques	Papers
Content-based	Heuristic-based method	TF-IDF KNN algorithm Clustering	Zhu *et al.* (2013) [31] Deldjoo *et al.* (2016) [44]
	Model-based method	Bayesian classifiers clustering Artificial neural networks	Liu *et al.* (2013) [33] Meehan *et al.* (2013) [34]
Collaborative	Memory-based method	KNN algorithm Nearest neighbor clustering Graph theory	Kim *et al.* (2010) [32] Chen *et al.* (2015) [35] Park *et al.* (2015) [45]
	Model-based method	Bayesian networks Artificial neural networks Linear regression Markov process	Breese *et al.* (1998) [36] Wei *et al.* (2016) [46] Nilashi *et al.* (2015) [47]
Hybrid	Feature combining	Bayesian clustering	Sattari *et al.* (2015) [37]
	Recommendation result combining	Linear combination	Dooms *et al.* (2015) [48]
	Unique models	Probabilistic model Maximum entropy	He *et al.* (2016) [39]

service to the selection/recommendation list as the new service's diversity from the services already in the list.

6.5.3   Coverage

The term coverage can refer to several distinct properties of the system as follows:

Service space coverage, which refers to the proportion of services that the recommendation system can recommend; this is often referred to as catalog coverage; the measure of catalog coverage may be the percentage of all services that can ever be recommended or the sales diversity which measures how unequally different services are chosen by users when a particular recommendation system is used;

User space coverage, which refers to the proportion of users or user interactions for which the system can recommend services; coverage can be measured by the richness of the user profile required to make a recommendation. For example, in the collaborative filtering case, this could be measured as the number of services that a user must rate before receiving recommendations;

Cold start, which concerns the issue that the system cannot make any recommendation for users when the system has not yet gathered sufficient information; a more generic way is to consider the "coldness" of a service using either the amount of time it exists in the system or the amount of data gathered for it.

6.5.4   Confidence and trust

Confidence refers to the system's trust in its selection or recommendations. The most common measurement of confidence is the probability that the predicted value is indeed true, or the interval around the predicted value [49,50]. For example, a recommendation may accurately rate a movie as a four-star movie with probability 0.85. The most general method of confidence is to provide a complete distribution over possible outcomes. Trust refers to the user's trust in the selection/recommendation system. The most common method for evaluating user trust on the recommendation system is by asking users whether the system recommendations are reasonable in a user study. In an online test, one could associate the number of recommendations followed with the trust in the recommendation, assuming that higher trust in the recommendation would lead to more recommendations being used. Alternatively, we could also assume that trust in the system is correlated with repeated users.

6.5.5   Robustness and privacy

Robustness is the stability of the recommendation in the presence of fake information, typically inserted on purpose in order to influence the recommendations. Such attempts to influence the recommendation are typically called attacks. It is more useful to estimate the cost of influencing a recommendation, which is typically measured by the amount of injected information. We can simulate a set of attacks by introducing fake information into the system data set, empirically measuring average cost of a successful attack. Privacy refers to the right that the private messages of users are not infringed. It is important for most users that their preferences stay private, that is, no third party can use the recommendation system to learn something about the preferences of a specific user. It is generally considered inappropriate for a service recommendation system to disclose private information even for a single user.

6.6   Conclusion

This chapter introduces the background and characteristics of service selection and recommendation in integrated network. Then it details the scenarios and definitions of service selection and recommendation, analyzes various algorithms, techniques, and strategies of service selection and recommendation, and finally summarizes limitations and evaluation indexes for service selection and recommendation approaches.

References

[1]   Guidara I., Guermouche N., Chaari T., Tazi S. 'Heuristic based time-aware service selection approach'. *Proceedings of International Conference on Web Services*; New York, USA, Jun 2015 (New York, IEEE, 2015), pp. 65–72.

[2]   Ma Y., Wang S.G., Yang F.C., Chang R.N. 'Predicting QoS values via multi-dimensional QoS data for web service recommendations'. *Proceedings of*

IEEE International Conference on Web Services; New York, USA, Jun 2015 (New York, IEEE, 2015), pp. 249–256.

[3] Guo Y., Wang S.G. 'Skyline service selection based on QoS prediction'. *Proceedings of IEEE International Conference on Cluster Computing*; Taipei, Taiwan, Sep 2016 (New York, IEEE, 2016), pp. 150–151.

[4] Wang S.G., Hsu C-H., Liang Z.J., Sun Q.B., Yang F.C. 'Multi-user web service selection based on multi-QoS prediction'. *Information Systems Frontiers*. 2014;16(1): 143–152.

[5] Wang S.G., Zhou A., Hsu C.-H., Xiao X.Y., Yang F.C. 'Provision of data-intensive services through energy- and QoS-aware virtual machine placement in national cloud data centers'. *IEEE Transactions on Emerging Topics in Computing*. 2016;4(2): 290–300.

[6] Wang S.G., Lei T., Zhang L.Y., Hsu C.-H., Yang F.C. 'Offloading mobile data traffic for QoS-aware service provision in vehicular cyber-physical systems'. *Future Generation Computer Systems*. 2016;61: 118–127.

[7] Wang S.G., Sun Q.B., Zou H, Yang F.C. 'Web service selection based on adaptive decomposition of global QoS constraints in ubiquitous environment'. *Journal of Internet Technology*. 2011;12(5): 757–768.

[8] Yu H.Q., Reiff-Marganiec S. 'Non-functional property based service selection: a survey and classification of approaches'. *Proceedings of Non-Functional Properties and Service Level Agreements in SOC Workshop*; Dublin, Ireland, Nov 2008 (New York, IEEE, 2008), pp. 12–14.

[9] Sun Q.B., Wang S.G., Zou H, Yang F.C. 'QSSA: a QoS-aware service selection approach'. *International Journal of Web and Grid Services*. 2011;7(2): 147–169.

[10] Wang S.G., Zheng Z.B., Sun Q.B., Zou H, Yang F.C. 'Reliable web service selection via QoS uncertainty computing'. *International Journal of Web and Grid Services*. 2011;7(4): 410–426.

[11] Schröpfer C., Binshtok M., Shimony S.E., *et al*. 'Introducing preferences over NFPs into service selection in SOA'. *Proceedings of International Conference on Service-Oriented Computing*. 2007; Vienna, Austria, Sep 2007 (Springer, Berlin, 2009), pp. 68–79.

[12] Mardukhi F., Nemat Bakhsh N., Zamanifar K., Barati A. 'QoS decomposition for service composition using genetic algorithm'. *Applied Soft Computing*. 2013;13(7): 3409–3421.

[13] Ma Y., Zhang C. 'Quick convergence of genetic algorithm for QoS-driven web service selection'. *Computer Networks*. 2008;52(5): 1093–1104.

[14] Li F., Huang Y. 'An web service selection optimization method based on particle swarm optimization'. *Proceedings of International Conference on Computer Design and Applications*; Qinhuangdao, China, Jun 2010 (New York, IEEE, 2010), pp. 477–481.

[15] Wang S.G., Sun Q.B, Zou H., Yang F.C. 'Particle swarm optimization with skyline operator for fast cloud-based web service composition'. *Mobile Networks and Applications*. 2013;18(1): 116–121.

[16]   Wang R., Ma L., Chen Y. 'The research of Web service selection based on the ant colony algorithm'. *Proceedings of International Conference on Artificial Intelligence and Computational Intelligence*; Sanya, China, Oct 2010 (New York, IEEE, 2010), pp. 551–555.

[17]   Yang Z., Shang C., Liu Q. Zhao C. 'A dynamic web services composition algorithm based on the combination of ant colony algorithm and genetic algorithm'. *Journal of Computational Information Systems*. 2010;6(8): 2617–2622.

[18]   Wang S.G., Sun Q.B., Yang F.C. 'Towards web service selection based on QoS estimation'. *International Journal of Web and Grid Services*. 2010;6(4): 424–443.

[19]   Sun L., Wang S.G., Li J.L., Sun Q.B., Yang F.C. 'QoS uncertainty filtering for fast and reliable web service selection'. *Proceedings of IEEE International Conference on Web Services*; Alaska, USA, Jun 2014 (New York, IEEE, 2014), pp. 550–557.

[20]   Gao Y., Na J., Zhang B., Yang L., Gong Q. 'Optimal web services selection using dynamic programming'. *Proceedings of IEEE Symposium on Computers and Communications*; Sardinia, Italy, Jun 2006 (Washington, IEEE, 2006), pp. 365–370.

[21]   Ding K., Deng B., Zhang X., Ge L. 'Optimization of service selection algorithm for complex event processing in enterprise service bus platform'. *Proceedings of International Conference on Computer Science & Education*; Nanning, China, Jul 2009 (New York, IEEE, 2009), pp. 582–586.

[22]   Ardagna D., Pernici B. 'Adaptive service composition in flexible processes'. *IEEE Transactions on Software Engineering*. 2007;33(6): 369–384.

[23]   Wang S., Zheng Z., Sun Q., Zou H., Yang F. 'Cloud model for service selection'. *Proceedings of IEEE International Conference on Computer Communications Workshops*; Shanghai, China, Apr 2011 (New York, IEEE, 2011), pp. 666–671.

[24]   Wang S., Sun L., Sun Q., Li X., Yang F. 'Efficient service selection in mobile information systems'. *Mobile Information Systems*. 2015;2015(1): 1–10.

[25]   Tang M., Ai L. 'A hybrid genetic algorithm for the optimal constrained web service selection problem in web service composition'. *Proceedings of IEEE Congress on Evolutionary Computation*; Barcelona, Spain, Jul 2010 (New York, IEEE, 2010), pp. 1–8.

[26]   Ma Y., Xin X, Wang S.G., Li J.L., Sun Q.B., Yang F.C. 'QoS evaluation for web service recommendation'. *China Communications*. 2015;12(4): 151–160.

[27]   Wang S.G., Ma Y., Cheng B., Yang F.C., Chang R.N. 'Multi-dimensional QoS prediction for service recommendations'. *IEEE Transaction on Services Computing*. 2016;99(1): 1–12.

[28]   Su X.Y., Khoshgoftaar T.M. 'A survey of collaborative filtering techniques'. *Advances in Artificial Intelligence*. 2009;2009(12): 1–19.

[29]   SG12 I. 'Definition of quality of experience'. *TD 109rev2 (PLEN/12)*; Geneva, Switzerland, 2007. pp. 16–25.

[30]   Ricci F. 'Introduction to recommender systems handbook', in Rokach L. (ed.). *Recommender System Handbook*. New York: Springer; 2011. pp. 1–35.

[31] Zhu Q., Shyu M.L., Wang H. 'Videotopic: content-based video recommendation using a topic model'. *Proceedings of International Conference on Multimedia*; Anaheim, CA, USA, Dec 2013 (New York, IEEE, 2013), pp. 219–222.

[32] Kim H.N., Ji A.T., Ha I., Jo G.S. 'Collaborative filtering based on collaborative tagging for enhancing the quality of recommendation'. *Electronic Commerce Research and Applications*. 2010;9(1): 73–83.

[33] Liu J., Wu C., Liu W. 'Bayesian probabilistic matrix factorization with social relations and item contents for recommendation'. *Decision Support Systems*. 2013;55(3): 838–850.

[34] Meehan K., Lunney T., Curran K., McCaughey A. 'Context-aware intelligent recommendation system for tourism'. *Proceedings of International Conference on Pervasive Computing and Communications Workshops*; San Diego, CA, USA, Mar 2013 (New York, IEEE, 2013), pp. 328–331.

[35] Chen M.H., Teng C.H., Chang P.C. 'Applying artificial immune systems to collaborative filtering for movie recommendation'. *Advanced Engineering Informatics*. 2015;29(4): 830–839.

[36] Breese J.S., Heckerman D., Kadie C. 'Empirical analysis of predictive algorithms for collaborative filtering'. *Proceedings of Conference on Uncertainty in Artificial Intelligence*; Madison, WI, USA, Jul 1998 (San Francisco, Morgan Kaufmann Publishers, 1998), pp. 43–52.

[37] Sattari M., Toroslu I.H., Karagoz P., Symeonidis P., Manolopoulos Y. 'Extended feature combination model for recommendations in location-based mobile services'. *Knowledge and Information Systems*. 2015;44(3): 629–661.

[38] Claypool M., Gokhale A., Miranda T., Murnikov P., Netes D., Sartin M. 'Combining content-based and collaborative filters in an online newspaper'. *Proceedings of ACM SIGIR Workshop on Recommender Systems*; Berkeley, CA, USA, Aug 1999 (New York, ACM, 1999), pp. 1–8.

[39] He P., Yuan H., Chen J., Zhao C. 'An effective and scalable algorithm for hybrid recommendation based on learning to rank'. *Proceedings of International Congress on Signal and Information Processing, Networking and Computers*; Beijing, China, Oct 2015 (Boca Raton, CRC Press, 2016), p. 59–67.

[40] Ge Y., Xiong H., Tuzhilin A., Xiao K., Gruteser M., Pazzani M. 'An energy-efficient mobile recommender system'. *Proceedings of ACM SIGKDD International Conference on Knowledge Discovery and Data Mining*; Washington, DC, USA, Jul 2010 (New York, ACM, 2010), pp. 899–908.

[41] Ruotsalo T., Haav K., Stoyanov A., *et al.* 'SMARTMUSEUM: a mobile recommender system for the Web of data'. *Web Semantics*. 2013;20(2): 50–67.

[42] Ma Y., Wang S.G., Hung P.C.K., Hsu C.-H., Sun Q.B., Yang F.C. 'A highly accurate prediction algorithm for unknown web service QoS value'. *IEEE Transactions on Services Computing*. 2016;9(4): 511–523.

[43] Sun Q.B., Wang L.B., Wang S.G., Ma Y., Hsu C.-H. 'QoS prediction for web service in mobile internet environment'. *New Review of Hypermedia and Multimedia*. 2016;22(3): 207–222.

[44]   Deldjoo Y., Elahi M., Cremonesi P., Garzotto F., Piazzolla P., Quadrana M. 'Content-based video recommendation system based on stylistic visual features'. *Journal on Data Semantics.* 2016;5(2): 99–113.

[45]   Park Y., Park S., Jung W., Lee S.G. 'Reversed CF: a fast collaborative filtering algorithm using a k-nearest neighbor graph'. *Expert Systems with Applications.* 2015;42(8): 4022–4028.

[46]   Wei J., He J., Chen K., Zhou Y., Tang Z. 'Collaborative filtering and deep learning based recommendation system for cold start items'. *Expert Systems with Applications.* 2016;69(1): 29–39.

[47]   Nilashi M., Jannach D., Bin Ibrahim O., Ithnin N. 'Clustering-and regression-based multi-criteria collaborative filtering with incremental updates'. *Information Sciences.* 2015;293(293): 235–250.

[48]   Dooms S., De Pessemier T., Martens L. 'Online optimization for user-specific hybrid recommender systems'. *Multimedia Tools and Applications.* 2015;74(24): 11297–11329.

[49]   Wang L.B., Sun Q.B., Wang S.G., Ma Y., Xu J.L., Li J.L. 'Web service QoS prediction approach in mobile internet environments'. *Proceedings of IEEE International Conference on Data Mining*; Shenzhen, China, Dec 2014 (New York, IEEE, 2015), pp. 1239–1241.

[50]   Ma Y., Wang S.G., Sun Q.B., Zou H, Yang F.C. 'Predicting unknown QoS value with QoS-prophet'. *Proceedings Demo & Poster Track of ACM/IFIP/USENIX International Middleware Conference*; Beijing, China, Dec 2013 (New York, ACM, 2013), pp. 1–2.

Chapter 7
Cloud networking evolution

Gladys Diaz[1] and Noëmien Simoni[2]

On-demand service is the paradigm that reflects the way that users want to access their applications today. In the last decade, the technologies were adapted to this 'User Centric' approach, but they still do not reach a large number of users. A necessary transformation and innovation must continue to progress! The questions that we address in this chapter are: What evolution should allow this objective? What are the fundamental novelties that would enable us to improve the agility of the network in dynamic and competitive environments? Thus, we analyze the models of: Cloud, Virtualization, Programmability and the 'As a Service' model. This analysis will enable us to consider a convergent global view for the near future, with more pragmatism of the digital world to offer the hyper-connectivity desired by users.

7.1 Introduction

To meet the market demands today, we can note that the technology landscape responds either to IT requirements—Application Centric—or network requirements—Network Centric—while today it is the user who is at the center of this ecosystem that is completely transformed by the digitization. These are the requirements of users that have to be addressed. The user wants a personalized and simplified access; he/she wants access to all resources, i.e., the application resources, those of networks and those of storage. In face of the completely and anytime connected the 'User Centric' still needs an overall and dynamic overview in order to do self-services and 'on-demand' services.

7.1.1 The new landscape: changes

Most Cloud services are delivered remotely via the Internet. From the provisioning perspective, the Cloud services involve not only IT functions provided by the infrastructure, but also communication functions provided by the network.

[1]L2TI Department, Paris 13 University, Sorbone Paris Cité, Villanelle, France
[2]Networking and Computer Science Department, Télécom ParisTech, Paris, France

We can say that the 'Cloud model' impacts the entire ecosystem and in particular the network model. These changes are seen at different layers.

- At the infrastructure layer: The Cloud computing is running on generic hardware, which can be moved, or instantiated in various nodes as needed. The infrastructure becomes virtualized and programmable and servers and applications become mobile.
- At the application layer: The Cloud enables the integration of various types of applications, for example, intensive data analysis, parallel processing and clustering, telemedicine and other community Cloud services. These different applications have very different needs relative to the network.
- At the network layer: The access and traffic levels have been impacted by the virtualization.
    1. At the access level: Today, access to virtualized spaces and services in the Cloud is also possible through mobile devices. Thus, access must be guaranteed regardless of the location and the equipment used.
    2. At the traffic level: The 'Cloud Networking' integrates primarily server-to-server models with an independent location at both ends of a service or transaction.

All these changes and these new uses require new architectures, with new properties in order to get the desired target benefits.

7.1.2   The new landscape: needs

Several network changes are necessary to take account of these new usages and to combine the advances of both computers and networks. The network architecture must become flexible and agile, able to support and respond to changing demands for services, for the associated QoS (Quality of Service) and for the location of Cloud application services. Network resources must be allocated in a differentiated manner to respond to the request of services. Ensuring network capacity and the associated QoS of data flows becomes an indispensable ingredient for maintaining a high performance of Cloud services. But it is still difficult to characterize the behavior and QoS of network services. The description of the QoS for network services still remains an open issue.

We can also see that the 'Cloud paradigm' introduces the properties of flexibility and dynamicity. Flexibility means to be adaptable to different demands. At the network layer, dynamicity involves being able to deploy and adapt network services according to the needs expressed at the application layer—the needs of data flows. But these properties are still a challenge today for the implementation of the new network solutions.

All these considerations enforce to the network infrastructure underlying to be open and expose its services to application at upper layers. Like the Cloud, this upper view of the system must be supplemented by a downward view, based on the abstraction of the network's physical resources. This abstraction of network resources is based on virtualization technologies. Applied at the network layer, the virtualization allows the decoupling of the services from the network infrastructure and the network equipment.

Based on these observations, we examine the challenges that have changed the 'Cloud Networking' to meet the needs of resource efficiency, the flexibility of solutions and the agility for the service creation, adaptation and dynamicity in networking.

7.1.3 The new landscape: a ranking of evolution

We will present the research that has helped to advance the Cloud Networking. We will propose a classification according to the needs covered and the technologies used. The analysis of this new landscape will allow us understand how the convergence of these new paradigms leads to a flexible targeted architecture for on-demand network services.

The first solution concerns the network links to establish the connection and access for data centers, by using the virtualization of network links. Then there is the virtualization of network nodes. In the latter case, the virtual routers load several protocol stacks to be used under suitable networks. That is why we call this phase of Cloud Networking the 'virtualization of network equipment' (Section 7.2).

Considering that Cloud Networking must meet the needs of cloud applications, differentiated offers based on QoS have been provided by the academics. We call this phase 'QoS-based network offers' (Section 7.3). But this solution is based on the current network infrastructure (e.g., IP, MPLS), whereas the network must undergo a more significant change for the dynamicity that has been long awaited by our new ecosystem.

Next we find what we can call the 'Network Cloudification' (Section 7.4). In fact, based on the principle of NFV (Network Function Virtualization), i.e., the virtualization of network functions, it is possible to deploy the VMs—virtual machines containing the NFV—in commoditized equipment. Thus, it is possible to present a full menu of *composable* and on-demand network functions.

Meanwhile, the need for agility—in the deployment—privilege the programmability capabilities offered by SDN (Software Defined Networking). Quickly an SDN/NFV complementarity is required (Section 7.5). We call this phase the 'Network Softwarization'.

For maximum flexibility in terms of organization, i.e., distribute the service components as needed, it is necessary that the functional components have weak linkages, i.e., they have at least the properties of the SOA (Service-Oriented Architecture). The introduction of the 'as a Service' approach (Section 7.6) and its associated APIs are still an open issue to be treated to automate the life cycle process.

Finally, we realize that it would be necessary to bring together all of these advances because each has an important role for the convergence and creation of an overall solution, that we call 'NaaS' (*Network as a Service*) (Section 7.7).

We go in this chapter, from this introduction, look at the evolution of Cloud Networking. For this purpose, we analyze the solutions to understand the issues of the convergence of services and its implementation through target architecture

NaaS—virtualized and programmable—allowing flexibility and openness of the network. This vision promotes a platform of services for the purpose of functional customization and dynamical organization. Through a use case, we will show how the convergence allows each operator to have a customized and scalable strategy.

7.2   Virtualization of network equipment

Network virtualization began through the virtualization of links. Link virtualization was made very early through protocols to enable the sharing of resources. Thus, Virtual Local Area Networks (VLANs), Virtual Private Networks (VPNs) and tunneling mechanisms are the virtualization techniques of the network links, and have long been used to connect remote sites and secure data exchange. What is it in the world of Cloud?

Cloud Networking was originally intended to set up and ensure connectivity and access to services and resources placed in Data Centers. Thus the virtualization applied to the IaaS layer (Infrastructure as a Service), which is the capacity to share physical resources in several equivalent logical isolated resources, must be analyzed at the links (Section 7.2.1) and the nodes (Section 7.2.2) to form virtual networks (Section 7.2.3).

7.2.1   Virtualization of links

The virtualization of network links is usually associated with the provision, by the owners of the network infrastructure, of a virtual network such as VPN or VLAN, where the routers—network nodes—are not necessarily virtualized. Cloud providers are focusing on server virtualization for more flexibility and scalability in the data center with 'Customer-managed VPN solution' because the VPN tunnels created are insulators at the CE (Customer Edge). This 'Data-path virtualization' is either at layer 2 in single-hop or layer 3 in multiple-hop with IPSec tunnels (RFC-2401) or GRE (Generic Routing Encapsulation) (RFC-2784).

We find other solutions proposed by various actors.

Those of the IT world, with the VPC (Virtual Private Cloud) provided by Amazon for secure cloud connectivity, Amazon EC2 [1]. Similarly, Microsoft Windows Azure [2] plans to construct virtual private networks in the Cloud by creating connectivity between their sites. These examples constitute a first step to add flexibility in the offers of network services, but they do not yet offer an on-demand virtual network in the Cloud, nor network traffic performance guarantees.

Those of the research community, with CloudNaas, SecondNet and LV2 placed at layer 2, and also VICTOR, Diverter and NetLord at layer 3, are examples of these architectures. These architectures are compared and analyzed in [3]. Most of these architectures provide solutions with limitations, for example, the allocation of bandwidth is not dynamic but static and predictive, if not non-existent; others are limited in terms of their scalability, performance and deployment, which depend on the physical topology of the network infrastructure and physical facilities and services installed.

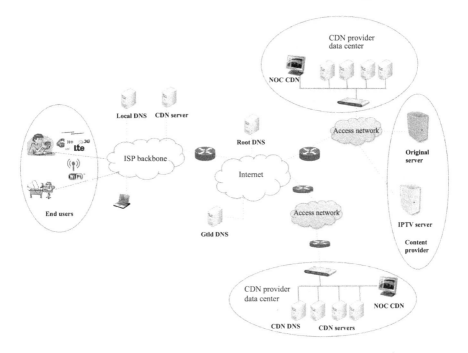

Figure 7.1   Cloud content delivery network (CCDN)

One of the most common applications is the CDN (Content Delivery Network). A CDN consists of networked computers across the Internet that cooperates to make the content or data—usually large media content—available to users. CDNs provide tools and strategies of service placements—and distribution of servers—in order to streamline and accelerate the access and consultation of content. The advantages and principles of the Cloud applied to CDN (called CCDN (see Figure 7.1) will provide a viable and cost effective solution for the creation of networks and content delivery services. In this context Akamai [4] and Amazon CloudFront [5] are examples of CDNs.

7.2.2   Virtualization of nodes

In configurations mentioned previously the virtualized nodes are not ones of the network, but those application servers. They are based on hypervisor technologies, which became ubiquitous with the result of a better server utilization and greater flexibility by partitioning physical servers into multiple virtual machines. Gartner calls this phenomenon *physicalization*.

For the aspects of suppleness at the level of the development, we can note the additional technology, which is one of the containers. The container allows encapsulating an application, and all the resources and dependencies which it needs.

As in a VM, the container supplies an abstraction of all the underlying layers until the equipment itself, but by adding a total abstraction of the operating system.

Like these server nodes, the network equipment (e.g., the routers, the switches, the gateways) will adopt the virtualization, some on a hypervisor technology; others will be based on the complementarity of containers and hypervisors.

The first virtual routers have emerged with solutions 'Provider-provisioned VPN solutions' (RFC-2917 and RFC-2764). The virtual router can use any combination of routing protocols (e.g., OSPF, RIP, BGP-4). A physical virtual router supports multiple routing areas isolated from each other. Each logical virtual router handles the routing tables and a specific switching by VPN.

7.2.3   Virtualization of the network

With the nodes virtualization the companies can have—on the same virtualized router—all access types and different virtual networks to reach different applications, instead of having a specific router by solution. It is also a solution for MVNOs (Mobile Virtual Network Operator), so each 'slice' of the network can have an adequate and separate management and security.

These approaches aim to provide flexibility in the support of the evolution of network services, and in the context of networking for the cloud. But these virtual networks for Cloud Networking are still defined with static configurations, and offered without QoS guarantees based on transported flows. Virtualization is still applied to resources from the infrastructure layer—without integrating the consideration of the service layer.

7.3   QoS-based network offers

It is in the context of research from the last decade, that many efforts have been made to analyze application traffic in order to provide the most appropriate network. The proposed solutions are based on overlay networks [6] that enable several networks to coexist on the same existing substrate infrastructure. For example, Hass *et al.* [7] introduce the concept of VPSN (Virtual Private Service Network) and VPCN (Virtual Private Connectivity Network) that define overlay networks as a set of autonomous, shareable and self-manageable components—nodes and links. We also find the GENI project [8], 4WARD [9], PlanetLab [10], VINI, Trellis, etc. These projects use the virtualization technologies to create and test virtual networks across multiple network technologies (e.g., wireless, sensors, cellular networks) enabling the assessment of multiple network protocols under realistic environments. A recent survey [11] presents a complete analysis on all aspects of network virtualization.

Most of the approaches mentioned above define a virtual network from an equipment point of view—because they focus in the south interface of the network architecture. This is the conventional case when the deployment focuses on the search of the path to connect servers and network equipment, which are nowadays virtualized. These approaches are designed to provide flexibility in support of the development of network services, and in the context of networking for the cloud. Even with this vision, there is a lack of flexibility providing a customized networking in accordance with the application requirements. Thus, the composition of a 'network as a service' based on demands and properties of data flows still remains a challenge.

A first response has been to create logical virtual networks in accordance with the requirements of the flows and according to a QoS model. The following sections provide: the QoS model (Section 7.3.1) that defines the process called 'QoS-aware' introduced in [12], and the application of the 'QoS model'—the VN-CoS approach—for modeling proposals for Cloud Networking (Section 7.3.2).

7.3.1   QoS-aware model

The QoS model, which we describe in this section, is intended to be a generic solution, which can be applied to any element of architecture regardless of the level considered—equipment, network or service. Each level consists of a set of service elements having functional and non-functional requirements—those representing the QoS. That is why we define the QoS as the ability to provide service in accordance with the non-functional requirements.

Non-functional requirements describe the properties of the system and the effects of these characteristics on the service provided to the user. These features are in addition to functional requirements. To be qualified like 'requirements' they must be measurable—otherwise it represents just the goals or design intent. They are therefore associated with a number of QoS criteria, as well as various metrics or measurable parameters.

In the design phase, we propose the 'conception values' characterizing the behavior of the service to some use, and 'threshold values' from which there is dysfunction of the device. In the operational phase, we need to monitor—through measurements—with the purpose to detect a change in behavior and to notify that the service violates its contract.

In other words, to offer a QoS management it is necessary to define in advance an agreement about the intended quality objectives of the services. That is, there is not management without measurement.

How to define behavioral objectives?
Non-functional requirements include demands, related to behavioral aspects of the service, affected by the execution of the processing or transferring information function.

A user who asks for a service indeed expects that the support is 'the most transparent' possible, in other words that the service of support has least possible impact on the treatment or the transfer of the contents. This notion of *transparency* can be formulated in several ways. Four of them seem generic:

* First, the *temporal and spatial transparency* expresses the fact that it is possible to treat or transfer content; each time the user produces, and for as long as its generation, where it is located. This behavior corresponds to the concept of availability.

- Second, the *semantic transparency* expresses the fact that the processing or transfer function must preserve the accuracy and completeness of the content; it should not alter the information. This behavior corresponds to the concept of integrity.
- Third, the *distance transparency* means being able to operate any treatment or linking without changing the intrinsic temporal relationship to the generated information. This behavior corresponds to the notion of delay.
- Finally, the *source behavior transparency* expresses the fact of treating or transfer of the content generated at every moment. This behavior corresponds to the concept of capacity.

The QoS can be associated with these four generic criteria, necessary and sufficient, to describe the behavior of any service [13]:

- **Availability**: This criterion reflects the accessibility of a resource taking into account the temporal and spatial terms and conditions. The availability rate (e.g., 99.5%) is focused on the supplier's point of view, while accessibility (e.g., 990 successful requests in 1,000) represents the user's point of view.
- **Integrity**: This criterion reflects the ability of a service to be performed without deterioration of the processed information, while respecting the demands and contractual conditions. Integrity can indicate, for example loss or modification of information in a node or a link.
- **Delay**: This criterion reflects the ability of a service to be executed respecting the time specified in the request and the contractual conditions. It expresses the duration of treatment in a node, and duration of waiting and crossing on the network links.
- **Capacity**: This criterion corresponds to the means to be implemented, enable to perform a service according to the requirements and contract terms. This may be the load that the nodes can treat or the amount of information capable of being transmitted over the network links.

Note: We can obviously consider more criteria, such as the reliability, which is the ability of an entity to perform a required function under given conditions for a given period of time. But we can say that reliability is a composite feature providing an additional notion of compliance of the above criteria for a given period of time.

The generic QoS criteria—availability, integrity, delay, capacity—as described above, must be linked to measurable parameters in order to give quantitative guarantees to the user.

The information model describes the QoS criteria for each service element.

Figure 7.2 shows the QoS information model. The first level represents the four QoS criteria—availability, integrity, delay, and capacity—for each item. The second shows the values for each criterion measured. The third level represents parameters necessary for measurement.

The service is essential unit exposable in the architectures of today.

A QoS agent may be placed in each service element. This agent will manage the access and use of resources of each element in a contract. The idea is to have an integrated mechanism to manage dynamic allocation of resources in accordance

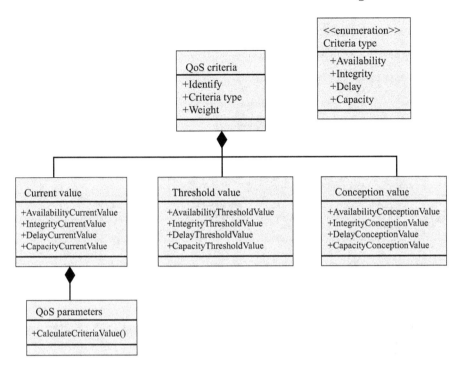

Figure 7.2   QoS informational model

with the QoS elements. QoS criteria are assessed through three types of measurable values: design, current and threshold values [14] (see Figure 7.3).

The proposed QoS information model is associated with several management profiles, which are solicited during the different phases of life cycle—design, deployment, operation, etc. [15].

Applying this model at the network level makes it possible to describe the non-functional aspects of network services—transmission, routing, security, etc. The network service parameters to be measured are based on application level requirements. For example, the delay is a more critical parameter in real-time applications than in file transfer services.

In the context of Cloud computing, the QoS model can characterize the behavior of network services of Cloud Networking providers by defining the QoS criterion associated with the virtual network to deploy. That is the main idea of the VN-CoS (Virtual Network Class of Services) proposal presented in Section 7.3.2 to adapt the network to the application request.

7.3.2   VN-CoS for cloud networking

The idea of this proposal [12,16] is to allow the implementation of cloud network services with a QoS-aware approach to address the needs of data flow of cloud

Life cycle phases		
Design	*Deployment*	*Operational*

	Design	*Deployment*	*Operational*
Profiles	Resource profile	Resource usage profile	Real-time profile
QoS criteria values	Conception values	Threshold values	Current values

Figure 7.3   QoS Criteria and measurable values

applications—coming from SaaS (Software as a Service) and PaaS (Platform as a Service).

The main contribution here is the classification of *Cloud Networking Services* (called VN-CoS) to represent the different flow types. This classification changes the current offerings of Cloud Networking by integrating the QoS into the offer of the virtual network service.

Two important issues about the Cloud Networking suppliers have been addressed by this proposal: (i) determining the QoS provided by each networking service for the cloud; (ii) providing service differentiation to ensure the inclusion of application performance requirements. First, the 'Cloud Networking model' is developed by using the 'QoS-aware cloud services' in [16]. This model aims to define the Cloud Networking Services (CNS) with a self-control behavior at the moment of execution. And second, a new classification of CoS (Class of Services) is made for the cloud application flows, based on QoS sensitivities in [12]. The customization of the composition and the networking of a virtual cloud are addressed by choosing the most appropriate networking services and determining the most appropriate class of service. The adaptation is possible and can be done in both cases: changes in the application flow requirements or network degradations.

To integrate these two aspects, an architecture is proposed in [17], to select, compose and configure the virtual network services that meet the requirements of the flow, regardless of network devices and underlying network technologies. To do this the proposed architecture has two solvers, the south interface and the north interface as well as a hypervisor managing the network resources. Figure 7.4 shows the proposed architecture.

The 'solver 1'—the north interface in Figure 7.4—addresses the classification of flow by determining its sensitivity to QoS criteria—availability, delay, capacity

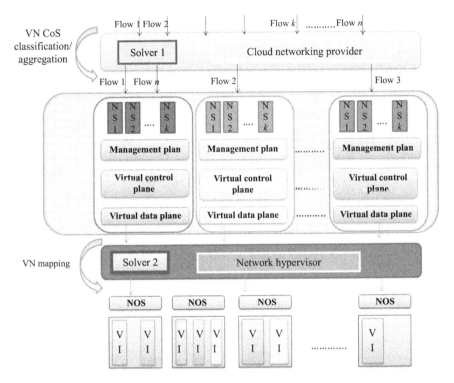

Figure 7.4   Cloud networking provider—architectural model

and integrity. Depending on the degree of sensitivity—low, medium, high—to each QoS criterion, the possible service classes are defined. The Cloud Networking provider associates the data stream to a class of service (VN-CoS). By using traffic engineering mechanisms, an aggregation on selecting CoS can be performed by the Cloud Networking provider. The logical solution, i.e., the virtual network, is built.

To describe the stream classes for Cloud Networking a classification of services—the Virtual Network Class of Service (VN-CoS)—has been proposed in [12]. Figure 7.5 shows the examples of VN-CoS.

The 'solver 2'—the south interface in Figure 7.4—handles the mapping between the virtual solution and its placement in the physical infrastructure. Once the description of the data plane for each virtual network is created, a mapping between virtualized elements—nodes and links—and the underlying network infrastructure elements is necessary. The mapping of the virtual network includes: the graph of the logic of the network service—its nodes and links—its topology and constraints that must be guaranteed during the mapping. The 'solver 2' makes it possible to integrate the various placement constraints to describe the virtual network to be mapped.

Unlike the differentiation of classical CoS with similar treatments for the data streams belonging to the same priority level, the proposed VN-CoS classes have been

	Flow sensitivity				Existent CoS					
	Availability	Delay	Capacity	Reliability	Y.1541	Diffserv PHB	MPLS (DS-TE)	UMTS	802.1d	WIMAX
VN-CoS1	x	x	x	x	Class 5	BE	0	Background	1	BE
VN-CoS2	x	xx	x	x	Class 1	AF3.3	3	Streaming	5	ErtPs
VN-CoS3	x	xxx	x	x	Class 0	EF	5	Conversational	7	ErtPs
VN-CoS4	x	x	xxx	x	Class 2	AF4.1	5	Streaming	5	RtPs
VN-CoS5	xxx	x	xx	xxx	Class 4	AF2.1	2	Interactive	2	NrtPs
VN-CoS6	x	xxx	xxx	x	Class	AF4.1	5	Streaming	4	ErtPs
...	...	...	...	...	...	...	...	...	...	...
VN-CoSk	xx	xx	xx	x	Class 1	AF4.2	4	Streaming	5	RtPs
...	...	...	...	...	...	...	...	...	...	...
VN-CoS81	xxx	xxx	xxx	xxx	Class 0	EF	5	Conversational	6	UGS

Figure 7.5 VN-CoS classification

designed to customize the treatment for each stream by appropriately determining its sensitivities to QoS criteria. For example, the class 'VN-CoS1' represents the 'best effort' treatment for data streams that do not need to fulfill all the requirements in terms of availability, delay, integrity and capacity. The class 'VN-CoS6' is designed for flows with a sensitivity to the criteria of time and capacity. The class 'VN-CoS81' represents the most demanding flows, which are very sensitive to all four QoS criteria. The differentiation of services proposed helps the Cloud Networking provider control the network performance and determine the protocol stack suitable for the construction of the virtual network. In fact, once the correct VN-CoS class is selected, a one-to-one relationship map between the class and existing traffic classes is then performed (e.g., Y.1541, Diffserv PHB, 802.1d).

But the network landscape is moving with the virtualization of network functions (NFV). The paradigms evolve, from a predefined classification of services to the composition of virtualized services on demand.

7.4   Network Cloudification

As mentioned in Section 7.2, like the Cloud computing, most network operators use virtualization of networks—nodes and links—for the design and deployment of network solutions. Virtualization makes it possible to be close to applications without having to worry about the physical aspects of implementation or the nature of the network equipment. This allows operators to offer more adapted and dynamic networks without having to handle their physical infrastructure. The similarity was concretized with the formalization of NFV [18] which aims to virtualize the functions of the network and make it possible to support all types of telecommunication services performed today in a Cloud IT standard.

NFV thus proposes to implement by software a standard platform of network services operating today on a proprietary hardware.

Figure 7.6 illustrates the reference architecture adopted by ETSI NFV [19]. This architecture allows the deployment and implementation of virtualized network functions (VNF) on cloud infrastructure called 'NFV Infrastructure' (NFVI). The NFVI consists of physical and software resources of computing, storage and networking. Thus, the network becomes a platform that can be open to third party software developers. In the context of NFV, some functions such as routing and the session controller have been studied in [20,21], other network functions are still not virtualized.

It is this approach that we call *cloudification*.

With the virtualization of network functions, it is also necessary to manage the deployment of network services—the virtual networks appliances—over a virtualized infrastructure. By automating the virtualization and by applying it to the different network layers, operators will be able to offer a menu of on-demand and composable network services.

This *network cloudification* is a network modeling according to the cloud model such as SaaS, PaaS and IaaS. It is the same for the NaaS modeling, which consists

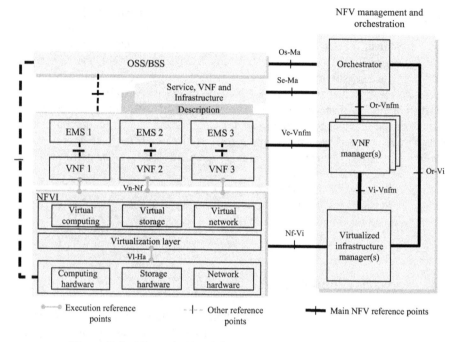

Figure 7.6   Network Cloudification—ETSI NFV architecture

of different several layers: NSaaS (Network Software as a Service), NPaaS (Network Platform as a Service) and NIaaS (Network Infrastructure as a Service).

- The NSaaS, like the SaaS, offers the *Network Service Delivery* for a specific application from the SaaS or PaaS.
- The NPaaS (or VNPaaS) offers the composition of network services (e.g., routing, addressing, NAT, firewall) and the dedicated application (e.g., IT services, location, billing, authorization) to a function network necessary to provide the *Media Service Delivery* of NSaaS. NPaaS programs the usage plane—data plane—of the virtual network to deploy.
- The NIaaS (or NFVIaaS) involves the virtualization of network infrastructure (IaaS) where the NaaS programs the control plane and deploys the virtual network providing the requested service of media delivery.

7.5   Network *Softwarization*

The next paradigm that promotes another development is the SDN (Software Defined Networking) [22]. SDN introduces the programmability of the data plane by separating it from the control plane. Thus, network intelligence is separated from forwarding mechanisms, which are implemented in devices that become programmable.

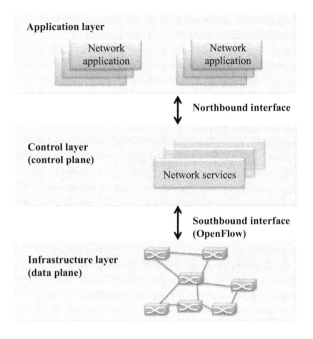

Figure 7.7   The generic SDN framework

This approach allows the network to be more agile and respond quickly to configurations changes to handle different data streams.

It is this approach that we call *Network Softwarization*. Indeed, by disregarding physically the network equipment, the SDN provides a central interface to dynamically control the network, while isolating the constraints of interfaces and other specificities related to that material. This programmatic capacity allows to develop the network architectures to make them simpler and more dynamic.

Such as that shown in Figure 7.7, the generic SDN framework defines three layers: the infrastructure layer (called also data plane), the control layer or control plane and the application layer. The data plane corresponds with the forwarding network elements, and it ensures different functionalities such as: the data forwarding and the gathering and monitoring of local information. The control plane is responsible for programming and managing the data plane and defines network operation and routing. The application layer aids the network configuration made through control layer. This layer provides the means to deal with the introduction of new network features through network applications (APIs).

SDN framework defines also two interfaces.

The southbound interface enabling the support of the communication needed between the controllers—located at control plane—and the programmable network elements—located at data plane. The most common protocol for the SDN southbound

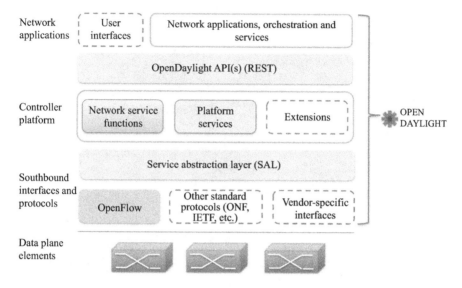

Figure 7.8   SDN: OpenDaylight project

interface is the OpenFlow protocol [23], which is standardized by the ONF (Open Networking Foundation).

The northbound interface enables to avoid to applications deal with the details of implementation of the south interface (e.g., the network topology) and provides a fine-grained control of the network elements. This interface is still not standardized. In current solutions, the northbound interface is provided with own API attached at the control software. Recent and extensive surveys about SDN have been presented in [24–26].

An implementation of SDN was launched through the OpenDaylight project [27] in April 2013 by the 'Linux Foundation'. The OpenDaylight Controller is able to deploy in a variety of production network environments. It can support a modular controller framework, but can provide support for other SDN standards and upcoming protocols.

As shown in Figure 7.8, the architecture of the OpenDaylight controller includes an abstraction layer called SAL (Service Abstraction Layer) that is dynamically updated in function of data models of the network elements associated with different resources. OpenDaylight offers several network southbound interfaces integrating different protocols such as OpenFlow and Netconf, and Restfull as the northbound interface.

The current evolution of the 'SDN control plane' leads it to become a Network Hypervisor supporting multiple virtual control planes. This is the case of Flowvisor [28], one major representative of network hypervisors for realizing SDN control plane virtualization. Further study on all SDN controllers was carried out in [25]. A complete survey about network virtualization hypervisors for SDN is presented in [29].

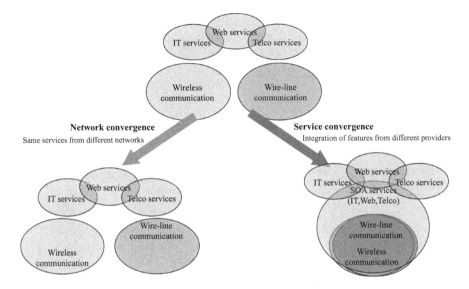

Figure 7.9   Service convergence

7.6   Network *ITification*: the convergence of service

The third paradigm in evolution is the 'as a service' resulting from the SOA approach (Section 7.6.1). It is again necessary that the set of components of new service architectures (Section 7.6.2) have all properties advocated by SOA (Section 7.6.3).

7.6.1   Why the SOA?

SOA (Service Oriented Architecture) enables the desired agility in the development of services: '*develop once, reuse many*'. Indeed, the tightly coupled applications of today, whether technically coupled at protocol level or functionally at semantic level, do not allow reuse and reconfiguration of services. We need in particular 'stateless' components and 'loosed' links (i.e., with a functional decoupling) allowing the customization of the service requested.

Indeed, the introduction, replacement or removal of a component is more easily to do with the loosed binding and whether the component's functional scope is well defined. Services encapsulate treatments (i.e., operations) and exposable data through a standardized and referenced interface. Thus, a library or a service repository can be accessible to all. Operations are actions that the service can achieve.

Duan *et al.* [30] presents a comprehensive study on how the principles of SOA can be applied at the Cloud Computing and Cloud Networking in the future Internet.

It is in the project 'OpenCloudware' [31] where we find the elements that lead us to name this phase as the network *ITification* because it characterizes the service convergence (see Figure 7.9), that is 'a service integration'—the composition of services (whether Telco, Web or IT) provided by multiple vendors. In the SOA properties have been added to other features including those having a QoS contract conformed to the model of QoS presented above. In this fashion are modeled the service components of the Cloud and Cloud Networking.

7.6.2   QoS-based SCC components

These components are described by their functional and nonfunctional (QoS) aspects, and are referred to as SCC (Self-Controlled Components) [32]. This modeling lets service suppliers know precisely the capabilities of their offered services.

The network functions are seen as a set of services that can be composed.

As shown in Figure 7.10, a SCC component contains:

- A functional content (business).
- Two external interfaces (client and server) used to communicate with the environment.
- A membrane to specify the non-functional aspects.

The membrane concept comes from GCM (Grid Component Model) that defines a high-level framework for the development of management capabilities. Thus, the monitoring components should be placed in the membrane of each service component. The logic of the membrane structure makes the reusability and genericity of our components possible.

Three types of interfaces are needed to fulfill the function of self-control of the SCC: the functional interface, the management interface and the control interface.

- There are two functional interfaces: 'JeeRequest' and 'JeeAnswer'. A server interface that includes processing functions (service methods) that can be executed by the service component. A client interface that performs the invocation to the next service in the chain, and the transmitting of its current operations for further processing or for the operation of the final result.
- Management interfaces that are non-functional server interfaces (ConfigInM, ConfigOutM and ActivateQoS). They contain the necessary mechanisms to manage the configuration of non-functional components in the membrane.
- The control interface is a non-functional client interface (OutOfQoS). It contains the transmission mechanisms of self-control information, allowing the manager to process the QoS events. It sends 'InContract' notifications when the behavior conforms to the contract; otherwise it triggers the 'Out-Contract' notification. The absence of the 'In-Contract' signal may be used by the Manager to detect serious defects of the SCC component. Management and control interfaces, associated with each service component make it possible to introduce a homogeneous management of the service component.

SCC service components are interconnected by networks, through the definition of a VPSN. A VPSN represents the logic of the user's application. The integrated QoS control for each SCC service component is used to manage contract violations (Out-contract) when they occur: it makes appropriate operational decisions.

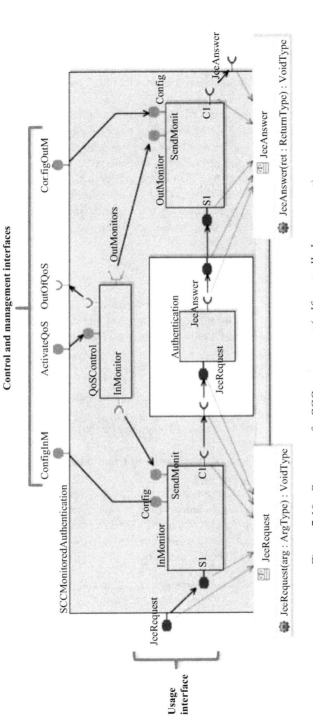

Figure 7.10   Structure of a SCC component (self-controlled component)

The monitoring components make it possible to measure the data needed to calculate the criteria for the control of SCC behavior—the nonfunctional capabilities—in order to conform to the model described above:

- Availability (A): which is the part of the time that SCC can handle the requested service without failure.
- Integrity (I): which is the rate of offered service in relation to compliance service required. It can be assessed by the loss rate, error rate, etc.
- Capacity (C): which is the ability of SCC to provide the necessary service. This criterion, depending on the service level considered, can be calculated by the bandwidth, the number of PDUs, the number of transactions/requests, etc.
- Delay (D): which is the total time taken by SCC to achieve its functions. It is measurable from several parameters such as latency and jitter to the network level, propagation period at the physical level and the processing time at the service level.

7.6.3   The properties 'as a Service'

The *ITification*, of Cloud Networking or of networks in general, allows introducing the necessary agility to provide the new features and respond more quickly to demand. This contribution of IT is based on properties that service components must meet in order to be 'as a Service'.

They are first SOA:

- **Descriptive**: by using metadata (e.g., WSDL, XML) the description of the service is made by decoupling it of its implementation.
- **Recordable**: the services must be possible to use by third parties across domains 'registry'.
- **Accessible**: Services must be discovered and interpreted effectively through directories, repositories and portals.
- **Reusable**: A service expresses an agnostic logic—stateless. This is an important property for agility avoiding redundancy and encourages the reuse of existing components in order to promote the Time To Market, main element of ROI.
- **Autonomous**: the service has to exercise a strong control over its underlying environment of execution to be the most predictive possible.

But for the service to be as autonomous as possible and be an 'as a service' and *composable*, they must be:

- **Invocable**: by the definition of open and standardized APIs. Three types of contracts are necessary for successful interconnection:
  - *Syntactic contract* that represents the use contract of the service (i.e., its interface). It shows the name of the treatment, its input and output parameters and structural constraints.
  - *Semantic contract* that provides an informal description of the treatment. It specifies the rules and constraints of the use of the service.

- *Service level agreement* (QoS and SLA) which specifies the commitments of service, for example, response time, accessibility time slots, recovery time after interruption and the procedures implemented in case of failure.
- **Programmable**: according to the three planes for one successful interoperability:
  - *Use plane*: stateless to facilitate the composition and be interoperable.
  - *Control plane*: resource reservation according to the QoS requirements.
  - *Management plane*: services must be aware of their states, and they must be able to share them to facilitate the management.
- **Mutualisable**: to ensure that links are really loose and the interworking of service components is successful.

The current model of the Internet must evolve to provide the integration of all mentioned approaches in order to create the network as a service. The following section provides the consideration of all these 'evolutions' allowing go toward an integrated and convergent solution with a flexible offer by considering both points of view: functional and organizational.

7.7   NaaS: toward an on-demand customization

As we have presented in the previous sections, the convergence between the Network and Cloud has gone through several phases. The Cloud imposes a new model for the design, deployment and consumption of IT resources that lead network architectures to take into account the new constraints and take advantage of new technologies.

In view of the above analysis, we can say that virtualization (i), programmability (ii) and SOA (iii) are the three main ingredients of the future NaaS. The regrouping of the three approaches can provide both conceptual and operational benefits, affecting not only the resource-oriented processes (such as provisioning, management and network monitoring) but also the service-oriented processes (such as sales, configuration and activation of new services). Thus:

(i)   Virtualization is applied at all levels, i.e., the equipment layer (NaaS infrastructure), at the network layer (virtual networks decoupled from the infrastructure) and at the component/service layer (virtual platform offering NFV).

(ii) Programmability is applied at the equipment layer, which becomes customizable to process the needs of the data streams; at the control plane level (SDN), which exposes the underlying network features through the abstraction of their resources, allowing their composition and achievement.

(iii) The SOA approach, with the 'As a Service' vision, applied to the network takes the automation of the supply network into consideration, and thus facilitates the maintenance and creation of new on-demand services.

Thus, by applying of SDN and NFV, it is possible to define customizable virtual networks (Section 7.7.1) through a composition of network functions. But for this to be a reality, it is necessary that mechanisms of design, deployment and management become automatic and dynamic, and allow the composition of the network services on the fly as required (Section 7.7.2).

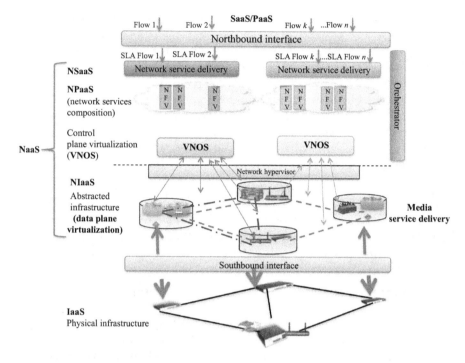

Figure 7.11   Virtualized and programmable NaaS QoS-based architecture

7.7.1   Customizable virtual networks

The evolution of cloud-network convergence, according to the analysis made above, therefore requires the introduction of programmable and virtualized NaaS QoS-based.

But what is this virtualized NaaS?
In other words, the function of virtualization should not be limited to network equipment levels, that are nodes (e.g., routers, switches, access points) or circuits (e.g., VC, WLAN, VPN), but must also include the network control plane associated with each data stream. The latter includes the network operating system (NOS) makes decisions management and control of flow, which must be associated with the requirements of the application flows. Thus each of the 'slices' must be checked according to the QoS and according to the SLA appropriate to every flow. It is necessary that all the changes, which will occur during the exploitation, be taken into account dynamically, as the mobility of the user, the user preferences according to his location, etc. The virtualization of NOS makes it possible to create customized virtual networks offering appropriate network services according to the types and requirements of the flow to be transferred.

In this context, the target architecture offers a distributed SDN architecture; where the SDN controller is personalized for each virtual network (Figure 7.11) [27]. Thus, for each virtual network, a SDN network controller is built—called VNOS for 'Virtual Network Operating System'—with customizable control features to manage the QoS requirements of the transferred flow. The definition of monitoring functions and QoS agents come in complement of our vision and ensuring the architecture management.

The orchestrator of NaaS automates the continuous deployment of different planes (control, data and management) within the virtual network elements, and a Network Hypervisor—NaaS Controller—manages the sharing and distribution of the virtualized infrastructure of NaaS.

Applying the QoS model (described in Section 7.3.1), NaaS provides virtualization of networks and equipment based on the evaluation of the QoS defined by the SLA. By combining SDN and NFV, the future NaaS offers both: flexibility through the customization of network services and the programming of equipment, and dynamicity through the virtualization of the network and programming of the control plane.

The future QoS-based NaaS architecture aims to provide a customized virtual network solution as a service. Through the northbound interface, NaaS captures the service requirements, i.e., the QoS constraints associated to data flow; through the south interface, NaaS communicates with IaaS to enable the programming of the equipment. So, NaaS allows providing composable network services in its NPaaS layer. NaaS must also configure the most appropriate *Media Service Delivery*—located at NIaaS—to support the *Network Service Delivery*, makes by the layer NSaaS (Figure 7.11).

7.7.2 The customized session

To implement dynamic designs and deployments, and to enable a true customization of networks, we must also consider the organizational aspects. The organizational aspects refer to the distribution of services in the components of the architecture to build a customizable end-service. Such services and theirs needs vary depending on the application context, and the organization must allow to deploy services using different strategies.

The orchestration mechanisms must be integrated into the new architecture of services, to enable both: service composition, and also the programing of control and management process—enabling to service providers an integration (horizontal and vertical) of their solutions.

This is the 'network session' (Figure 7.12) through orchestration mechanisms that combine horizontal and vertical composition efforts. The network session includes all components ensuring a customized network service for the user. The NaaS therefore supports two types of services—located at several different levels: the *Network Service Delivery* and *Media Service Delivery*. Figure 7.13 shows both levels.

The network service delivery solicits the media service delivery to transport the content of flow in accordance with the QoS constraints expressed by the application.

The logic of the network service is described by the virtual network corresponds to the network service delivery—placed at the NSaaS/NPaaS. The control plane and the data plane are programmed accordingly, to provide together the media service delivery.

By decoupling these issues, this approach makes it possible to create isolated and controlled virtual networks. We define a controller for each virtual network.

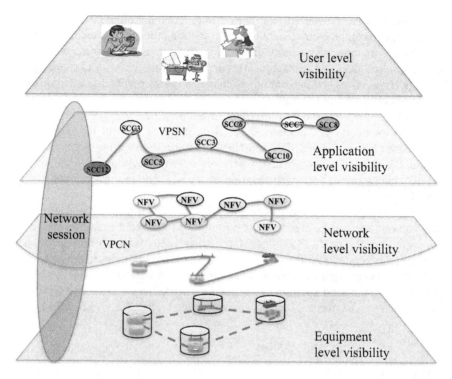

Figure 7.12   Customized network session

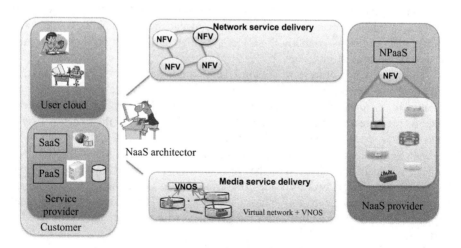

Figure 7.13   Network and media service delivery

In this way heterogeneous networks—from a service media delivery point of view— can coexist in the same virtualized infrastructure.

In the virtualized environment, the equipment layer includes the virtualized equipment—enabling the isolation and independence of the OS. It is on this virtualized infrastructure (NIaaS) that will be placed the entities of the virtual network (NFV). The network layer will have two views: the north interface that represents the application flow and that of the south interface that represents connectivity components through an abstract view of the physical infrastructure.

The management plane, by using monitoring components, will be able to monitor the behavior of the different service compositions, as much vertically as horizontally.

7.8   Use case: provisioning of on-demand cloud services

We will illustrate in the following paragraphs a use case on Cloud Gaming to develop our approach for provisioning 'on-demand Cloud services' (see Figure 7.14).

How to represent, store, connect and organize information on service components are the questions to be addressed to consolidate the implementation of an on-demand service deployment. For this purpose it is necessary to review the existing service infrastructures and integrate the SOA approach to break down complex and monolithic systems into applications and ecosystems with simple and well-defined components.

Our model can define the components of services with a functional and nonfunctional (i.e., QoS) point of view, and integrate the resource profiles associated with the components during the various phases of the life cycle—design, deployment, execution. Our model is not bound to a particular language; its implementation can be achieved using existing service description technology, such as USDL or OWL-S, in the context of web services. Thus, once the dynamic deployment was realized, at the time of the effective consumption of the service by users the allocation of the resources is made under constraint of QoS.

The example in Figure 7.14 shows a use case in the context of cloud gaming. Initially, the game session is open according to the home user configuration. The gamer connects to the Internet through wireless access to their 3D session, which is stored in the cloud gaming server.

The cloud gaming provider is responsible for composing and provisioning the service components—at application level—required providing support for the session according to the usage of user. The Cloud NaaS provider must provide the network services required to support the application flows of the game session, defined by cloud gaming provider. Thus, NaaS is responsible for the deployment of virtual network service that will support the execution of application services.

One or more suppliers of cloud gaming can have a hand in providing access to specific codes of the games. The games offered to the user depend on their configuration and resources to view the video stream—audio sent. To this purpose, the cloud service provider—Cloud Gaming provider, in our example—analyzes the needs of the user based on the services and resources (i.e., the user profile) and composes—at usage level—the VPSN (i.e., the services associated with the video game according to the required QoS of the user).

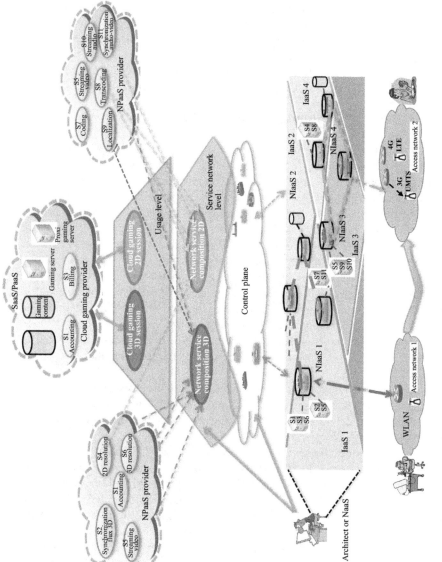

Figure 7.14 Cloudified, virtualized and programmable NaaS for personalized service

The overall composition of services includes the application services at the PaaS layer, and at the NaaS layer, the network services that will support the service delivery to the application flows. To support the gaming sessions, multiple network services can compose the VPCN, among them:

- Connectivity services: DHCP server, NAT router, APL (Application Level Gateway), etc.
- Media streaming services: coding, video streaming, audio streaming, synchronization of 3D streams, audio-video synchronization, multi-screen, media cache, etc.
- Security services: firewall, IPS, Parental control, etc.

Different suppliers could provide these network services. At network level, the NaaS supplier must analyze the QoS requirements of flows and dimensioning the network in accordance with them. NaaS must decide what network services select and distribute to define the virtual network that will support the game session.

Finally the NaaS architect orchestrates the composition of services and performs the necessary deployment at the NIaaS layer to support the session. In this example we find the red network that supports the user's home session. Secondly, the user changes equipment and the location. These changes are reflected in the cloud gaming service composition, for example, moving from the 3D session to a 2D session, because the new equipment no longer supports 3D. The change of location involves changes of access network. This change could then be reflected in new QoS requirements of application flows. The NaaS must recompose the network service following these new requests.

The NaaS must adapt the new network service request and support the new application flow. The NaaS architect must orchestrate the new composition and deploy the necessary network services. In our example, the purple network represents the suitable mobile network service.

Thus, the NaaS provider plays the role of orchestrator by provisioning adequate and adaptable network service to deal with the media delivery flows.

The 'Usage Level' of our example provides the logic of application service to the user—the composition of the application services through VPSN.

The 'Service Network Level' provides the network service composition related to the application of cloud gaming—the Network Service Delivery. The 'control plane' view at this level provides the assembly of network resources (VPCN) required to support and process the traffic of the media corresponding to requests from selected application services—the Media Service Delivery—which will be placed in the data plane.

7.9 Conclusion

For nearly all actors, digital transformation is a necessary precursor to the hyper-connected Internet of Everything, envisioned in the future. This requires virtualization, programming and 'as a Service' of IT as well as network functions and their automation. It is this convergence that we wanted to show in this paper.

For this purpose, we studied the evolution of cloud networking, which led us to synthesize an architecture based on the *cloudification network* and describe a virtualized (NFV) and programmable (SDN) NaaS.

By integrating all of these approaches, the QoS-based NaaS—virtualized and programmable—must:

(A) Be able to provide network services on demand and to provide appropriate and adaptable network services to meet the requirements of data flows.

(B) Apply the QoS approach to reflect an agreement between the required QoS (SaaS/PaaS) and the offered QoS (IaaS), and

(C) set the media delivery service.

Then it seems that it is important to separate the functional plane and the organizational plane for the purpose of composing them according to the requirements and deploying them in accordance with the most appropriate strategy.

For this reason, we propose a solution where we treat the functional and organizational aspects of personalized on-demand services. The use Case shows an example where the architecture builds its application from NPaaS, PaaS and IaaS and deploys on NIaaS depending on the desired QoS.

Glossary

APL—Application Level Gateway
CCDN—Cloud Content Delivery Networks
CDN—Content Delivery Networks
CNS—Cloud Networking Services
DHCP—Dynamic Host Configuration Protocol
GCM—Grid Component Model
IaaS—Infrastructure as a Service
IP—Internet Protocol
IPSec—Internet Protocol Security
MPLS—Multi Protocol Label Switching
MVNO—Mobile Virtual Network Operator
NaaS—Network as a Service
NAT—Network Address Translation
NFV—Network Functions Virtualization
NFVIaaS—Network Functions Virtualization Infrastructure as a Service
NIaaS—Network Infrastructure as a Service
NOS—Network Operating System
NPaaS—Network Platform as a Service
NSaaS—Network Software as a Service
OWL-S—Ontology Web Language for Services
PaaS—Platform as a Service
QoS—Quality of Service
ROI—Return on Investment

SaaS—Software as a Service
SAL—Service Abstraction Layer
SCC—Self Controlled Component
SDN—Software Defined Networking
SLA—Service Level Agreement
SOA—Service Oriented Architecture
USDL—Unified Service Description Language
VLAN—Virtual Local Area Network
VN—Virtual Network
VN-CoS—Virtual Network Classes of Services
VNF—Virtualized Network Function
VNFaaS—Virtual Network Function as a Service
VNOS—Virtual Network Operating System
VNPaaS—Virtual Network Platform as a Service
VC—Virtual Circuit
VPC—Virtual Private Cloud
VPCN—Virtual Private Connectivity Network
VPN—Virtual Private Network
VM—Virtual Machine
WLAN—Wireless Local Area Network
WSDL—Web Services Description Language
XML—eXtensible Markup Language

References

[1]   Amazon EC2. *Amazon Elastic Compute Cloud (Amazon EC2)*. Available from http://aws.amazon.com/ec2. [Accessed March 2015].

[2]   Microsoft Azure. *Windows Azure: Microsoft's Cloud Platform*. Available from http://www.windowsazure.com. [Accessed March 2015].

[3]   Bari M. F., Boutaba R., Esteves R., *et al.* 'Data center network virtualization: A survey'. *IEEE Communications Surveys & Tutorials*. 2013; **15**(2):909–928.

[4]   AKAMI. *Akami Solutions*. Available from http://www.akamai.fr. [Accessed March 2015].

[5]   Amazon. *Amazon CloudFront*. Available from http://aws.amazon.com/fr/cloudfront/. [Accessed March 2015].

[6]   Galán-Jiménez J., Gazo-Cervero, A. 'Overview and challenges of overlay networks: A survey'. *International Journal of Computer Science and Engineering Survey (IJCSES)*. 2011; **2**(1):19–37.

[7]   Hass R., Droz P., Stiller B. 'Autonomic service deployment in networks'. *IBM Systems Journal*. 2003; **42**(1):150–164.

[8]   GENI. *Global Environment for Network Innovations*. Available from http://www.geni.net/. [Accessed March 2015].

[9]   4AWARD. *The FP7 4WARD Project*. Available from http://www.4ward-project.eu/index.php?s=overview. [Accessed March 2015].

[10]   PlanetLab. *An Open Platform for Developing, Deploying, and Access-ing Planetary-Scale Services.* Available from http://www.planet-lab.org/. [Accessed March 2015].

[11]   Mosharaf Kabir Chowdhury N. M., Boutaba R. 'A survey in network vir-tualization'. *Computer Networks: The International Journal of Computer and Telecommunications Networking. Elseiver Computer Networks.* 2010; **54**(5):862–876.

[12]   Ayadi I., Simoni N., Diaz G. 'NaaS: QoS-aware cloud networking services'. *In 2013 12th IEEE International Symposium on Proceedings of the 12th IEEE International Symposium on Network Computing and Applications (NCA),* August 2013. Cambridge, MA: IEEE; pp. 97–100.

[13]   ETSI TR 102 805-3 V1.1.1 (2010-04). *Part 3: QoS Informational Struc-ture.* On-line ETSI report. Available from http://www.etsi.org/deliver/etsi_tr/102800_102899/10280503/01.01.01_60/tr_10280503v010101p.pdf. [Accessed February 2013].

[14]   ETSI TR 102 805-1 V1.1.1 (2009-11), *Part 1: User's E2E QoS. Analysis of the NGN.* On-line ETSI report. Available from http://www.etsi.org/deliver/etsi_tr/102800_102899/10280501/01.01.01_60/tr_10280501v010101p.pdf. [Accessed February 2013].

[15]   Simoni N., Yin C., Du Chene G. 'An intelligent user centric middleware for NGN: Infosphere and AmbientGrid'. *Communication Systems Software and Middleware and Workshops. IEEE COMSWARE 2008. Third International Conference.* Bangalore, India: IEEE; Jan. 2008. pp. 599–606.

[16]   Ayadi I., Simoni N., Diaz G. 'QoS-aware component for cloud comput-ing'. *Proceedings of the Ninth International Conference on Autonomic and Autonomous Systems (ICAS), 2013.* March 2013. Lisbon, Portugal. IARIA; pp. 14–20.

[17]   Ayadi I., Diaz G., Simoni N. 'QoS-based network virtualization to future net-works: An approach based on QoS network constraints'. *Network of the Future (NOF), 2013 Fourth International Conference.* October 2013. Pohang, South Korea. IEEE; 2013, pp. 1–5.

[18]   ETSI. *Network Function Virtualization. White paper.* Available from http://portal.etsi.org/NFV/NFV_White_Paper.pdf. [Accessed March 2015].

[19]   ETSI. *Network Function Virtualization (NFV); Architectural Frame-work.* Available from http://www.etsi.org/deliver/etsi_gs/NFV/001_099/002/01.01.01_60/gs_NFV002v010101p.pdf [Accessed March 2015].

[20]   Batalle J., Ferrer Riera J., Escalona E., Garcia-Espin J. 'On the implementation of NFV over an OpenFlow infrastructure: Routing Function Virtualization'. *In Proceeding of Future Networks and Services (IEEE SDN4FNS), 2013 IEEE SDN.* November 2013. Trento, Italy: IEEE; pp. 1–6.

[21]   Monteleone G., Paglierani P. 'Session Border Controller virtualization towards service-defined net-works based on NFV and SDN'. *In Proceedings of Future Networks and Services. (IEEE SDN4FNS), 2013 IEEE SDN* November 2013. Trento, Italy: IEEE; pp. 11–13.

[22] ONF. *Software-Defined Networking: The New Norm for Networks. White paper.* Available from https://www.opennetworking.org. [Accessed March 2015].

[23] McKeown N., Anderson T., Balakrishnan H., *et al.* 'Openflow: Enabling innovation in campus networks'. *ACM SIGCOMM Computer Communication Review.* 2008; **38**(2):69–74.

[24] Nunes Astuto B., Mendonça M., Nam Nguyen X., Obraczka K., Turletti T. 'A survey of software-defined networking: Past, present, and future of programmable networks'. *Communications Surveys and Tutorials. IEEE Communications Society, Institute of Electrical and Electronics Engineers (IEEE).* 2014; **16**(3):1617–1634.

[25] Kreutz D., Ramos F. M. V., Veríssimo P. F., Rothenberg C. E., Azodolmolky S., Uhlig S. 'Software-defined networking: A comprehensive survey'. *Proceedings of the IEEE.* 2015; **103**(1):14–76

[26] Xia W., Wen Y., Foh C. H., Niyato D., Xie H. 'A survey on software-defined networking'. *IEEE Communications Surveys & Tutorials. IEEE.* 2015; **17**(1):27–51.

[27] OpenDaylight. *OpenDaylight: Open Source SDN Platform.* Available from https://www.opendaylight.org. [Accessed March 2015].

[28] Sherwood R., Gibb G., Yap K. K., *et al. Flowvisor: A network virtualization layer.* Technical report Openflow-tr-2009-1, Stanford University, 2009.

[29] Blenk A., Basta A., Reisslein M., Kellerer W. 'Survey on network virtualization hypervisors for software defined networking'. *IEEE Communications Surveys & Tutorials.* 2016; **18**(1):655–685.

[30] Duan Q., Yan Y., Vasilakos A. V. 'A survey on service-oriented network virtualization toward convergence of networking and cloud computing'. *IEEE Transactions on Network and Service Management.* 2012; **9**(4):373–392.

[31] Open Cloudware. *Open Cloudware Project. Web page.* Available from http://www.opencloudware.org/bin/view/Main/. [Accessed March 2015].

[32] Aubonnet T., Ludovic Henrio L., Kessal S., *et al.* 'Management of service composition based on self-controlled components'. *Journal of Internet Services and Applications.* 2015; **6**(15):1–17.

Chapter 8
Mobile Cloud Networking: future communication architecture for mobile cloud

Zhongliang Zhao[1] and Torsten Braun[1]

Abstract

Commoditization and virtualization of wireless networks are changing the economics of mobile networks to help network providers, e.g. Mobile Network Operators (MNOs), Mobile Virtual Network Operator (MVNO), move from proprietary and bespoke hardware and software platforms to an open, cost-efficient and flexible cellular ecosystem. In addition, rich and innovation local services can be efficiently materialized through cloudification by leveraging the existing infrastructure. The future communication architecture for mobile cloud services – Mobile Cloud Networking (MCN) is a EU FP7 large-scale integrating project (IP) funded by the European Commission that is focusing on how cloud computing and network function virtualization concepts are applied to achieve virtualization of cellular networks. It aims at the development of a fully cloud-based mobile communication and application platform, or more specifically, it aims to investigate, implement and evaluate the technological foundations for the mobile communication system of long-term evolution (LTE), based on mobile network with decentralized computing and smart storage offered as one atomic service: on-demand, elastic and pay-as-you-go.

MCN will investigate, implement and evaluate the technological foundations for that system to meet real-time performance and support efficient and elastic use and sharing of radio access and mobile core network resources between operators. Mobile network functionalities—such as baseband unit processing, mobility management and quality of service (QoS) control—will run on the enhanced mobile cloud platform leveraging commodity hardware, which requires extensions toward higher decentralization and enhancing them to elastically scale up and down based on load. The end-to-end control and management orchestrates infrastructure and services across several technological domains: wireless, mobile core and data centres, providing guaranteed multiple end-to-end service, mobility through the Follow-Me Cloud concept. In this chapter, we present three enabling components of such a future MCN

---

[1]University of Bern, Bern, Switzerland

architecture: Radio Access Network as a Service (RANaaS), Information-Centric Networking as a Service (ICNaaS) and Mobility Prediction as a Service (MOBaaS).

8.1 Introduction

Nowadays continuously increasing number of mobile users, devices and new mobile applications has resulted in the mobile traffic growing at an unprecedented rate, which alternatively has a significant impact on the complexity of processes that are required to provide reliable cellular networks. Over the last few years, mobile operators are struggling to cope with challenges, though still not taking full advantages of new technologies, e.g. long-term evolution (LTE), that are helping to increase the capacity of Radio Access Network (RAN). This shortfall may be because of relying on highly centralized and custom hardware components. In addition, the current network components are not designed with elasticity in mind: e.g. during peak hours, traffic patterns can be highly dynamic and unpredictable, and any component can easily be overloaded with traffic. In this situation, it is difficult to distribute active sessions to other less loaded components in real-time and even more difficult to increase the components' capacity based on real-time demand. However, it is difficult to optimize resource (e.g. energy, bandwidth). These wastages are costly and directly influence the mobile operators' OPEX (operational expenses). For those reason, virtualization of cellular networks by means of cloud computing can be seen as a way of tackling these challenges.

However, today's cloud computing is confided to data centres and one of the weakest points of the established cloud computing value proposition is that it does not support the seamless integration of cloud computing services in the mobile ecosystem. Another observation is that infrastructure sharing, enabled by virtualization, and seen as one of the most fundamental enables of cloud computing, does indeed exist in the mobile Telco industry ever since the emergence of the Mobile Virtual Network Operator (MVNO). MVNOs are operators that can adopt different models. In one extreme (full MVNO), they might operate almost all parts of the cellular network, except the radio spectrum license. In the other extreme (light MVNO), the MVNO outsources all the operation, including the marketing and selling force. Despite these two observations, the principles of cloud computing are neither used nor supported by mobile operator and this regardless of fact that many of today's network components are intrinsically cloud-ready and could be hosted on top of a cloud computing platform. By means of virtualization, in near future all these facts together are expected to create a unique opportunity for both the cloud computing and the mobile Telco industry, as the virtualization is likely to exert a decisive influence on technical and business innovation of such networks. This unique opportunity enables Telco industry to provide a novel, distinct and atomic (i.e. all services as one bundle) Mobile Cloud Networking (MCN) service, that is Mobile Network with Decentralized Computing plus Smart Storage offered as One Service in on-demand fashion. The MCN project will investigate, implement and evaluate a fully cloud-based mobile communication and application platform for MCN services.

8.2   The MCN project

The MCN project [1] is an EU-funded research project, which is focusing on the integration of cloud computing and network function virtualization concept into cellular network. To be more specific, it is one of the pioneering projects in this domain, aiming at the development of a fully cloud-based mobile communication and application platform. The basis of the MCN system resolves around two main principles as mentioned before. The first principle is of cloud computing as defined by the U.S. National Institute of Standards and Technology (NIST), which states that a could computing service must display the characteristics of resource polling, broad network access, rapid elasticity and measured services (pay-as-you-go). The second and complimenting principle the MCN adopts is that of service-oriented architecture. These principles set out the characteristics any MCN service should have. The topmost motivations of the MCN project are to:

- Extend concept of cloud computing beyond data centres toward mobile end-user (as shown in Figure 8.1)
- Design a 3rd Generation Partnership Project (3GPP) compliant mobile
- Cloud networking architecture that exploits and supports cloud computing
- Design a novel virtualization layer, monitoring systems and general provisioning across the various domains with MCN
- Design a mobile platform for future mobile cloud-enabled services and applications
- Deliver and exploit concept of an end-to-end (E2E) MCN for novel applications and services

8.2.1   Objectives and scope

The technological approach of the MCN project is structured around several segments: cloud computing infrastructure foundation, wireless cloud, mobile core network cloud and mobile platform services. These serve as fundamental inputs for the design and evaluation of the MCN architecture. The necessary foundational, cloud computing infrastructure resources and services required to create, enable and deliver fully virtualized E2E MCN services are based on the popular OpenStack [2] framework and respective extensions to be developed. On top of this framework, mobile network enhancements are foreseen, namely to enable the concept of wireless cloud and mobile core cloud. The project will also design and develop a Mobile Application Platform for E2E mobility-aware service deployment including SLA (Service Level Agreement) management, AAA, Content Distribution Services (CDS) and Rating, Charging and Billing (RCB). The ultimate objective is thus to specify, implement, evaluate and standardize a fully cloud-based mobile communication and application platform.

The basic assumption of MCN is the existence of an infrastructure composed of RAN, Mobile Core Networks and Services (MCNS), Micro- and Macro Data Centres (DCs). Macro data centres are standard large-scale computing farms deployed and operated at strategically selected locations. Micro data centres are medium-to

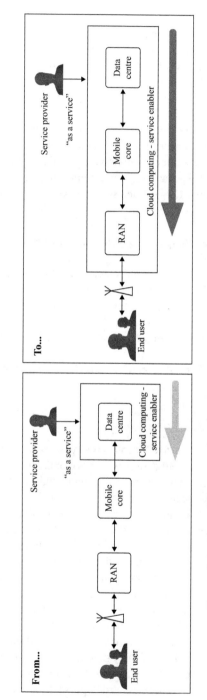

Figure 8.1   Extend the concept of cloud computing beyond data centre toward mobile end users

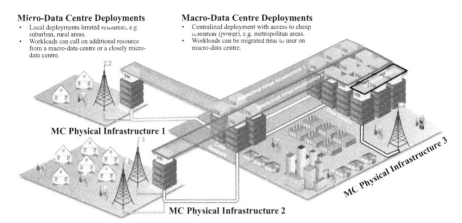

Micro-Data Centre Deployments
- Local deployments limited resources, e.g. suburban, rural areas.
- Workloads can call on additional resource from a macro-data centre or a closely micro-data centre.

Macro-Data Centre Deployments
- Centralized deployment with access to cheap resources (power), e.g. metropolitan areas.
- Workloads can be migrated near to user on macro-data centre.

MC Physical Infrastructure 1

MC Physical Infrastructure 2

MC Physical Infrastructure 3

Figure 8.2   From macro to micro data centres and its usages for Mobile Cloud Networking

small-scale deployments of server clusters across a certain geographic area, e.g. covering a city or a certain rural area and as part of a mobile network infrastructure (as shown in Figure 8.2). The data centres are the physical infrastructure and MCN enables deployments of RAN or any other mobile core networks or services on top of virtual infrastructures offered by the data centres. Virtual machines (VMs) build a platform layer on top of DCs that enable the instantiation of RAN and MCNS virtual infrastructures.

8.2.2   Project architecture

The MCN architecture follows a service-oriented architecture. In the MCN architecture, all functional elements are modelled as services. From the abstract view, the entities used to represent aspects of service orientation are a service, a service instance (SI) and service instance components (SICs). In particular, MCN services are implemented by resources, and these resources operate service-related instances. These resources can be both physical and virtual. Services in MCN are categorized into two main types: atomic services and composed services. Composed services are then further categorized into main services and supporting services. Service-related entities are managed by key MCN architectural entities. They are managed in a common consistent fashion regardless of their category. This consistency is reinforced by a common management life cycle. Each architecture entity and service with MCN shares a common life cycle model, which includes the following stages:

- Design: carry out service's technical design.
- Implement: implement the service manager (SM) and service orchestrator (SO).
- Deployment: deployment of the implemented elements, e.g. cloud controllers (CCs). Provide anything such that the service can be used, but do not provide

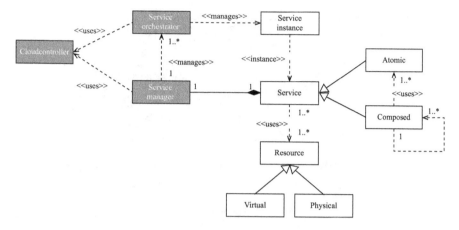

Figure 8.3   MCN architectural entities and relationships

access to the service, such as placing a VM image on the IaaS provider and create an instance from it, and installing machines from the VM images.

- Provisioning: provisioning of the service environment (e.g. interfaces and networks). Activation of the service such that the user can actually use it, such as installing Apache HTTP server, configure/activate it, and bringing in policies and whatever is needed to make the service run.
- Operation and runtime management: in this stage, SI is ready and running. Activities such as scaling, reconfiguration of SICs are carried out here.
- Disposal: release of SICs.

With this management life cycle, services can then be managed in a common uniform fashion. What enable the services are key architectural entities of the MCN architecture, and their relationships are shown in Figure 8.3. By using these core MCN architectural entities and the introduced life cycle, many services of different functionality can then be implemented, integrated and composed.

- SM: It provides an external interface to the user both programmatic and/or visual. It offers multitenant capable services to that user. The SM has two dimensions: the business that encodes business agreements, and the technology that manages different SOs of a particular tenant.
- SO: It embodies how the service is actually implemented. Generally, one SO per SM domain is instantiated per tenant. It oversees the complete (end-to-end) orchestration of a SI. It is implemented as a domain-specific component and manages the SI, which it creates, including scaling of the instance. The SO is managed by the SM and the SO monitors SI-specific metrics related to the SI. Although SIs are domain specific, they are composed of SICs.
- CC: It supports the deployment, provisioning and disposal of SOs. To the SOs, it also provides both atomic and support services through a service development kit (SDK).

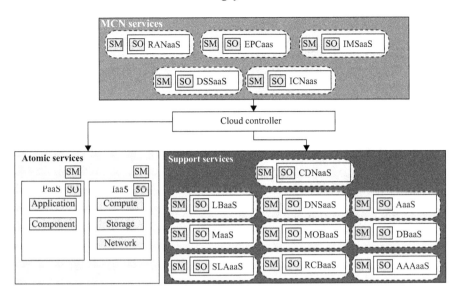

Figure 8.4   MCN services

8.2.3   Project services

MCN can enable wide and varied range of possible services. Such services can be (1) typical cloud computing atomic service, such as computing, storage and networking; (2) supporting services, which can be a full set of integrated services for E2E management solutions, such as monitoring as a service; (3) virtualized network infrastructure services, such as RANaaS, Evolved Packet Core as a Service (EPCaaS); (4) novel virtualized application and services, e.g. Content Delivery Network as a Service (CDNaaS); and (5) an E2E service that is comprised of (any of the above) subservices that are combined together with additional logic and configuration in order to provide a new service offering.

Using its conceptual architecture, an overview of the services that enable MCN is discussed below. The initial MCN service portfolio is shown in Figure 8.4.

As shown in Figure 8.4, MCN delivers a big list of services, including main services and supporting services:

- MCN services:
  o RANaaS
  o EPCaaS
  o IP-Multimedia-Subsystem as a Service (IMSaaS)
  o Digital Signage System-as-a-Service (DSSaaS)
  o ICNaaS
- Supporting services:
  o MOBaaS
  o Service Level Agreement as a Service (SLAaaS)
  o Load Balancing as a Service (LBaaS)

- o Monitoring as a Service (MaaS)
- o Domain Name System as a Service (DNSaaS)
- o Rating Charging and Billing as a Service (RCBaaS)
- o Authentication, Authorization and Accounting as a Service (AAAaaS)

In the rest of this chapter, we focus on explaining details of three services, which are RANaaS, ICNaaS and MOBaaS.

8.3 Radio Access Network as a Service

This section describes the details on the architecture and implementation of the RANaaS. This service focuses on the design, implementation and test of a set of elements for a system that can be used to manage RAN for an organization that specializes in providing on-demand RAN to customers.

8.3.1 Architecture reference model

While the concept of Cloud-RAN has been clearly defined, more research is needed to find an optimal architecture that maximizes the benefit behind C-RAN, and based on which a true proof-of-concept could be built. From the perspective of the operator such architecture has to meet the scalability, reliability/resiliency and cost-effective requirements. However, from the perspective of the software radio application, two main requirements have to be met: (1) strict hard deadline to maintain the frame and subframe timing and (2) efficient/elastic computational resources (e.g. CPU, memory) to perform intensive digital signal processing for different transmission modes (beam forming, CoMP, etc.).

Broadly, three main choices are possible to design a C-RAN, each of which provides a different cost–power–performance–flexibility trade-off.

- **Full GPP:** where all the processing (L1/L2/L3) is performed on the host/guest systems. According to China Mobile, the power consumption of the OpenAirInterface full GPP LTE soft modem is around 70 W per carrier.
- **Accelerated:** where only certain functions, such as Fast Fourier Transform/Inverse Fast Fourier Transform (FFT/IFFT), are offloaded to a dedicated hardware (e.g. Field Programmable Gate Array (FPGA) and Digital Signal Processor (DSP)), and the remaining functions operate on the host/guest. The power consumption is reduced to around 13–18 W per carrier.
- **System-on-Chip:** where the entire L1 is performed on a System-on-a-Chip (SoC) and the reminder of the protocol stack runs on the host/guest. The power consumption is reduced to around $\sim$8 W per carrier.

As shown in Figure 8.5, the hardware platform can either be full GPP or a hybrid. In the latter case, all or part of the L1 functions might be offloaded to dedicated accelerators, which can be placed locally at the cloud infrastructure to meet the real-time deadline and provide a better power-performance trade-off or remotely at Remote Radio Head (RRH) to reduce the data rate of fronthaul. Different service compositions

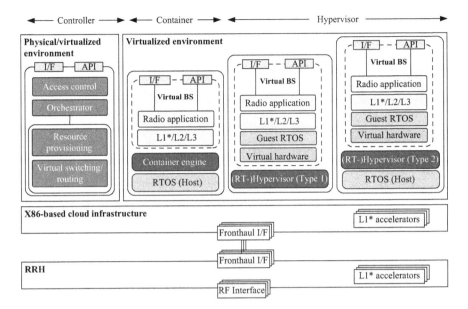

Figure 8.5   Candidate C-RAN architectures

can be considered, ranging from all-in-one software radio application virtualization to per carrier, per layer or per function virtualization as mentioned earlier. The virtualization is performed either by a container engine or a hypervisor, under the control of a cloud OS, which is in charge of life cycle management of a composite software radio application (orchestrator) and dynamic resource provisioning.

8.3.2   RANaaS life cycle

RANaaS describes the service life cycle of an on-demand, elastics and pay-as-you-go 3GPP RAN on the top of cloud infrastructure. Thus, life cycle management is a key for successful adoption and deployment of C-RAN and related services (e.g. MVNO as a service) (see Figure 8.6). It is a process of network design, deployment, resource provisioning, operation and runtime management and disposal.

8.3.3   RANaaS performance analysis

8.3.3.1   Base-Band Unit processing

Figure 8.7 illustrates the main RAN functions in both TX and RX spanning all the layers, which has to be evaluated to characterize the base-band unit (BBU) processing time and assess the feasibility of a full GPP RAN. Since the main processing bottleneck resides in the physical layer, the scope of the analysis in this chapter is limited to the BBU functions. From Figure 8.7, it can be observed that the overall processing is the sum of cell- and user-specific processing. The former only depends on the channel bandwidth and thus imposes a constant base processing load on the system,

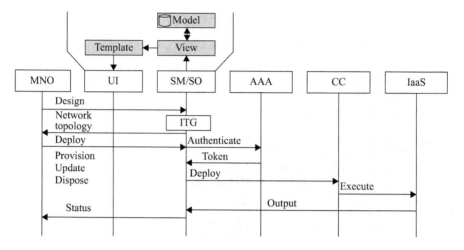

Figure 8.6   RANaaS life cycle

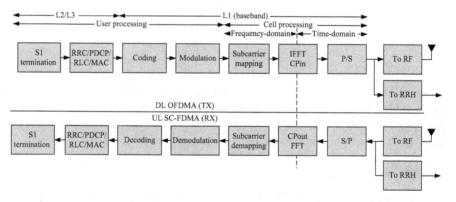

Figure 8.7   Functional block diagram of LTE eNB for DL and UL

whereas the latter depends on the modulation and coding scheme (MCS) and resource blocks allocated to users as well as signal noise ratio (SNR) and channel conditions. Figure 8.7 also shows the interfaces where the functional split could happen to offload the processing either to an accelerator or to a RRH.

To meet the timing and protocol requirements, the BBU processing must finish before the deadlines. One of the most critical processing that requires deadline is imposed by the Hybrid Automatic Repeat Request protocol (HARQ) in that every received MAC PDU has to be acknowledged (ACK'ed) or nonacknowledged (NACK'ed) back to the transmitter within the deadline. In frequency division duplex (FDD) LTE, the HARQ round trip time (RTT) is 8 ms. Each MAC PDU sent at subframe N is acquired in subframe $N + 1$, and must be processed in both RX and TX chains before subframe $N + 3$ allowing ACK/NACK to be transmitted in subframe

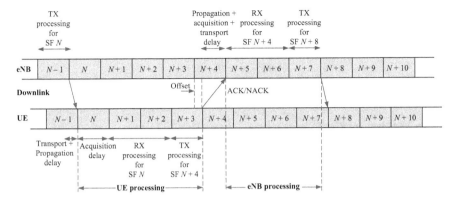

Figure 8.8   FDD LTE DL HARQ timing

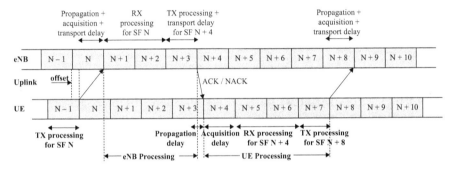

Figure 8.9   FDD LTE UL HARQ timing

$N + 4$. On the receiver side, the transmitted ACK or NACK will be acquired in subframe $N + 5$, and must be processed before subframe $N + 7$, allowing the transmitter to retransmit or clear the MAC PDU sent in subframe N. Figures 8.8 and 8.9 show an example of timing deadlines required to process each subframe in downlink and uplink respectively.

It can be observed that the total processing time is 3 ms, out of which 2 ms is available for RX processing and 1 ms for TX. Thus, the available processing time for an eNB to perform the reception and transmission is upper-bounded as follows:

$$TRx + TTx \leq \frac{\text{THARQ}}{2} - (\text{TPropagation} + \text{TAcquisition} + \text{TTransport} + \text{Toffset})$$

where THARQ $= 8$, TPropagation $+$ TAcquisition $+$ TTransport ≤ 1 ms and Toffset $= 0$ in DL.

Depending on the implementation, the maximum tolerated transport latency depends on the eNodeB processing time and HARQ period. As mentioned earlier, NGMN adopted a 250 μs for the maximum one-way fronthaul transport latency.

Hence, the length of a BBU-RRH link is limited to around 15 km to avoid too high round-trip-delays (given that the speed of light in fibre is approximately 200 m/μs). At maximum distance of 15 km, the remaining overall processing time will be between 2.3 and 2.6 ms.

8.3.3.2   Evaluation setup

Four set of different experiments are performed. The first experiment analyses the impact of different x86 CPU architecture on BBU processing time, namely, Intel Xeon E5-2690 v2 3 GHz (same architecture as IvyBridge), Intel SandyBridge i7-3930K at 3.20 GHz and Intel Haswell i7-4770 3.40 GHz. The second experiment shows how the BBU processing timescale with the CPU frequency. The third experiment benchmarks the BBU processing time in different virtualization environments including LXC, Docker and KVM against a physical machine (GPP). The last experiment measures the I/O performance of virtual Ethernet interface through the guest-to-host RTT.

All the experiments are performed using the OpenAirInterface DLSCH and ULSCH simulators designed to perform all the baseband functionalities of an eNB for downlink and uplink as in a real system. All the machines (hosts or guests) operate on Ubuntu 14.04 with the low latency (LL) Linux kernel version 3.17, x86-64 architecture and GCC 4.7.3. To have a fair comparison, only one core is used across all the experiments with the CPU frequency scaling deactivated except for the second experiment. The benchmarking results are obtained as a function of allocated physical resource blocks (PRBs), MCS and the minimum SNR for the allocated MCS for 75% reliability across four rounds of HARQ. Note that the processing time of the turbo decoder depends on the number of iterations, which is channel dependent. The choice of minimum SNR for a MCS represents the realistic behaviour and may increase number of turbo iterations. Additionally, the experiments are performed using a single user with no mobility, 8-bit log-likelihood ratios turbo decoder, SISO mode with AWGN channel and a full buffer traffic ranging from 0.6 Mbps for MCS 0 to 64 Mbps for MCS 28 in both directions. Note that if multiple users are scheduled within the same subframe in downlink or uplink, the total processing depends on the allocated PRB and MCS, which is lower than a single user case with all PRBs and highest MCS. Thus, the single user case represents the worst case scenario.

The processing time of signal processing module is calculated using timestamps at the beginning and at the end of each BBU function. OAI uses the rdtsc instruction implemented on all x86 and x64 processors to get a very precise timestamps, which counts the number of CPU clocks since reset. To allow a rigorous analysis, total and per function BBU processing time are measured as shown in Table 8.1. For statistical analysis, a large number of processing_time samples (10,000) are collected for each BBU function to calculate the average, median, first quantile, third quantile, minimum and maximum processing time for all the subframes in uplink and downlink.

CPU architecture analysis

Figure 8.10 depicts the BBU processing budget in both directions for the considered Intel x86 CPU architecture. It can be observed that the processing load increases

Table 8.1 OAI BBU processing time decomposition in downlink and uplink

RX function	Timing(μs)	TX function	Timing(μs)
OFDM demodulation	109.695927	OFDM modulation	108.308182
ULSCH demodulation	198.603526	DLSCH modulation	176.487999
ULSCH decoding	624.602407	DLSCH scrambling	123.744984
⌊ *interleaving*	12.677955	DLSCH encoding	323.395231
⌊ *demultiplexing*	117.322641	⌊ trubo encoder	102.768645
⌊ *rate matching*	15.734278	⌊ rate matching	86.454730
⌊ *turbo decoder*	66.508104	⌊ interleaving	86.857803
⌊ *init*	11.947918		
⌊ *alpha*	3.305507		
⌊ *beta*	3.377222		
⌊ *gamma*	1.018105		
⌊ *ext*	2.479716		
⌊ *intl*	5.441128		
Total RX	931	Total TX	730

with the increase of PRB and MCS for all CPU architectures, and that it is mainly dominated by the uplink. Furthermore, the ratio and variation of downlink processing load to that of uplink also increases with the increase of PRB and MCS. Higher performance (lower processing time) is achieved by the Haswell architecture followed by SandyBridge and Xeon. This is primarily due to the respective clock frequency, but also due to a better vector processing and faster single threaded performance of Haswell architecture. For the Haswell architecture, the performance can be further increased by approximately a factor of two if AVX2 (256-bit SIMD compared to 128-bit SIMD) instructions are used to optimize the turbo decoding and FFT processing.

CPU frequency analysis

Figure 8.11 illustrates the total BBU processing time as a function of different CPU frequencies (1.5, 1.9, 2.3, 2.7, 3.0 and 3.4 GHz) on the Haswell architecture. The most time consuming scenario is considered with 100 PRBs and downlink and uplink MCS of 27. In order to perform experiments with different CPU frequencies, Linux ACPI interface and cpufreq tool are used to limit the CPU clock. It can be observed that the BBU processing time scales down with the increasing CPU frequency. The figure also reflects that the minimum required frequency for 1 CPU core to meet the HARQ deadline is 2.7 GHz.

Based on the above figure, the total processing time per subframe, Tsubframe, can be modelled as a function of CPU frequency:

$$\text{Tsubframe}(x)[\mu s] = \alpha/x$$

where $\alpha = 7{,}810 \pm 15$ for the MCS of 27 in both directions, and x is CPU frequency measured in GHz.

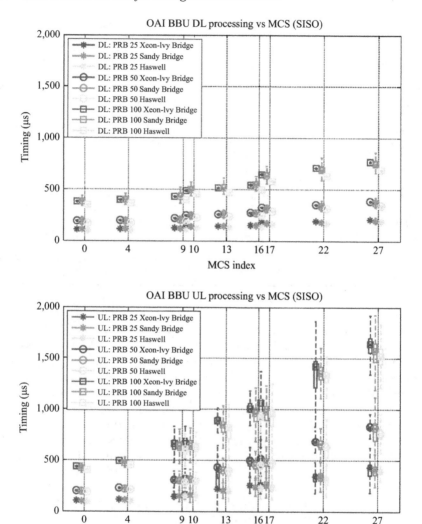

Figure 8.10   BBU processing budget in downlink (top) and uplink (bottom) for different CPU architecture

Virtualization technique analysis

Figure 8.12 compares the BBU processing budget of a GPP platform with different virtualized environments, namely, Linux Containers (LXC), Docker and KVM, on the SandyBridge architecture (3.2 GHz). While on average the processing time is very close for all the considered virtualization environments, it can be observed that GPP and LXC have slightly lower processing time variations than that of DOCKER and KVM, especially when PRB and MCS increase.

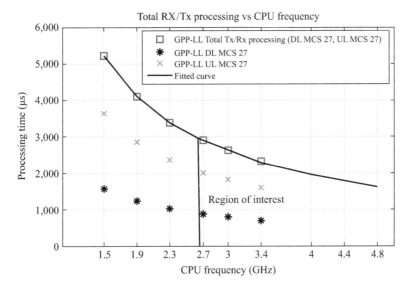

Figure 8.11   Total processing time as a function of CPU frequency

Figure 8.13 depicts the Complementary Cumulative Distribution Function (CCDF) of the overall processing time for downlink MCS 27 and uplink MCS 16 with 100 PRB. The CCDF plot for a given processing time value displays the fraction of subframes with execution times greater than that value. It can be seen that the execution time is stable for all the platforms in uplink and downlink. The processing time for the KVM (hypervisor based) has a longer tail and mostly skewed to longer runs due to higher variations in the non-native execution environments (caused by the host and guest OS scheduler). Higher processing variability is observed on a public cloud with unpredictable behaviours, suggesting that cares have to be taken when targeting a shared cloud infrastructure.

Discussion
By analysing the processing for a 1 ms LTE sub-frame, the main conclusion that can be drawn for the considered reference setup (FDD, 20 MHz, SISO, AWGN) is that with the CPU frequency of 3 GHz (on average), 1 processor core for the receiver processing assuming 16 QAM in uplink and approximately 1 core for the transmitter processing assuming 64-QAM in downlink are required to meet the HARQ dead-lines. Thus a total of 2 cores are needed to handle the total processing of an eNB in 1 ms (one subframe). With the AVX2 optimizations for this latest architecture, the computational efficiency is expected to double and thus a full software solution would fit with an average of 1 × 86 core per eNB. When comparing the results for different virtualization environments, the main conclusion that can be drawn is that containers (LXC and Docker) offer near bar metal runtime (native) performance while preserving the benefits of VMs in terms of flexibility, fast runtime and migration.

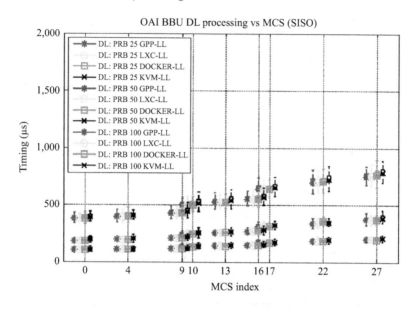

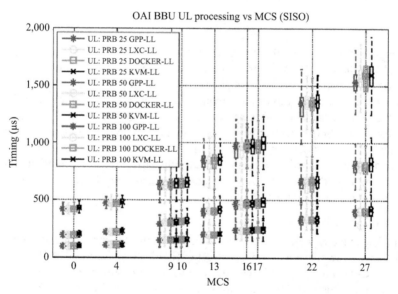

Figure 8.12   BBU processing budget in downlink (top) and uplink (bottom)

Furthermore, they are built on modern kernel features such as groups, namespace, chroot and sharing the host kernel and benefit from the host scheduler, which is a key to meet the real-time deadlines. This makes containers a cost-effective yet light-weight solution for RANaaS without compromising the performance. In summary, the LXC/Docker proved to provide a bar metal performance as they exploit native

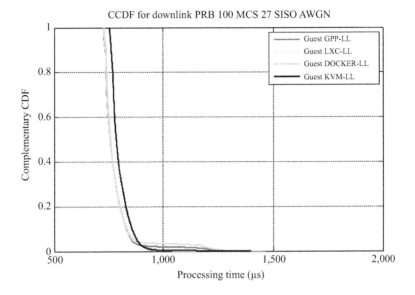

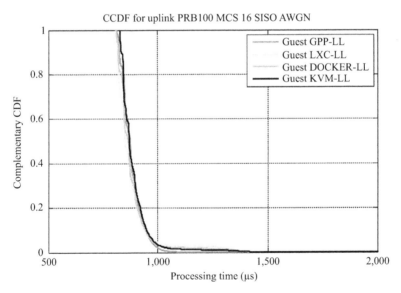

Figure 8.13 *BBU processing time distribution for downlink MCS 27 and uplink MCS 16*

Linux Features, and do not require a hypervisor. Furthermore, they are becoming more and more popular and well-integrated with OpenStack. However, they do not support multi-OS and require supports from the host OS. More details can be found in our publication [2].

8.3.4   Conclusions

The lesson learnt from the above and other experiments can be summarized as follows:

- Processing time deadline
  - o   FDD LTE HARQ requires a RTT of 8 ms that imposes an upper-bound for the sum of BBU processing time and the fronthaul transport latency. Failing to meet such a deadline has a serious impact on the user performance.
  - o   Virtualized execution environment of BBU pool must provide the required runtime.
- Containers (LXC and Docker) versus hypervisor (KVM)
  - o   Containers are more adequate for GPP RAN as they offer near-bar metal performance and provide direct access to the RF hardware. KVM performance is also good, but requires pass through mechanisms to access the RF hardware to reduce the delay of the hardware virtualization layer.
  - o   In case of containers, RAN requires low latency kernel in the host.
  - o   In case of full virtualization (KVM), hypervisor has to support real-time/low latency task (different techniques requires for types 1 and 2), and also the guest OS requires low latency kernel.
  - o   There is need for a dynamic resource provisioning/sharing, load balancing to deal with the cell load variations (scale out and in).
- Hardware
  - o   Probe the existing RF for their capabilities (FDD or time division duplex (TDD), frequencies, transmit power, etc.) and select the one that is required for the target RAN configuration.
  - o   Support of RRH:
    - –   Fronthaul (BBU to RRH link): a full-duplex 10 Gbps link is required. There is certain ETH configuration to be done here.
    - –   Either EXMIMO 2 (PCI-x if), and/or NI/ETTUS (USB3 if)
- Flexible configuration, build, run and monitoring
  - o   (Semi-)automatic generation of the RAN configuration file through the UI or selection and editing of predefined configuration files. The same holds for the compilation (e.g. Rel 8 or Rel 10) and execution (e.g. enable disable hooks).
  - o   Dynamic monitoring of the status of RAN.

8.4   Information-Centric Networking as a Service

ICNaaS can be used to improve the performance, efficiency and ease of retrieving content in mobile networks. With the approach of providing ICN components in the cloud, benefits such as the close integration with other cloud services could be obtained. This makes it easy for Enterprise End Users (EEUs) to request and deploy them according to their requirements.

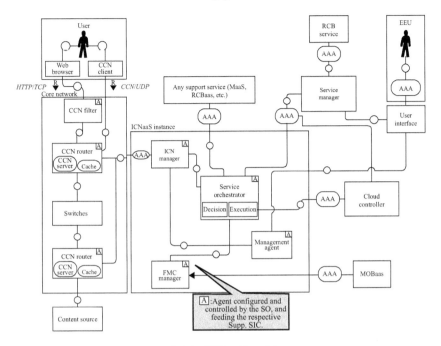

Figure 8.14   ICNaaS architecture

8.4.1   ICNaaS architecture design

Figure 8.14 presents the internal components of the ICNaaS architecture, the roles of different elements are listed below:

- **Service manager**: The service manager is the component responsible for handling the initial requests from EEUs for the service. Whenever an EEU wants the service, it requires the SM that will deliver the available Service Catalog, and asks the SM for the SI.
- **SO**: It is responsible for the orchestration of the service.
- **ICN manager**: The ICN manager is the component responsible for the coordination of the ICN layer, controlling and configuring all the CCNx routers with a global vision of the network topology. Whenever a new CCNx router joins the network, it contacts the ICN manager to register itself and obtains its initial configuration.
- **ICN filter**: The ICN filter is a proxy and packet classification filter. The latest and final version of this ICN filter supports the following features/functionalities:
  - Proxy used to translate HTTP GET request to CCNx Interest and CCN data packet (content) to HTTP traffic.
  - Packet classification filter that is used to filter CCNx-related traffic from HTTP-based traffic.

o   Multiuser support, meaning the ICN filter is able to handle request from different clients simultaneously.
o   HTTPS support, enabling users to use encrypted web traffic.
o   Large file transfer support.

- **CCNx router**: This component includes several daemons to manage requests and, the repository, running on a machine. The main goal of this component is to handle content requests (interest messages), which corresponds to routing these requests, or resolving requests if the content is available at the local cache. The content is provided from the cache of daemons or from a repository. When a content request is resolved, either by direct reply of the router or by routing the requested content objects, the router logs locally that request was processed at that moment. The other main daemon has two responsibilities: to send the local log information to the centralized database and receive commands from the MCN CCN Manager performing then the required actions. The two daemons and the changes made are described in the following subsections.

- **CCN server**: Developed in Java, this daemon is part of the CCNx router. It consists of multiple threads that perform different actions and as stated above, it has two responsibilities, send the log information to the database and receive and perform instructions coming from the MCN CCN Manager. For the first part, it has two threads, one that is continuously reading from the pipe, and another that periodically checks if it has data to upload, and if it has, sends it going back to sleep again after. For the second, it has a thread listening for incoming connections on a *ServerSocket*, and for every new incoming connection, instantiates a new thread that will perform all the necessary steps and end after everything is done.

- **Follow Me Cloud (FMC) manager**: This is a centralized processor of information and will be the brain of the content relocation mechanism. It will perform two main actions, receive the log data sent by every router storing it into the database, and receive messages from MOBaaS about users that are going to move, specifying some details about source, destination and number of users. The decision mechanism of FMC Manager will then take into account this information, the data stored in the database and other data that can come from other components and decide if it should and what it should migrate.

8.4.2   ICNaaS performance evaluation

The Normal, the YouTube and the Webserver scenarios were used to evaluate the ICNaaS/FMC decision algorithms. To save space, we skip the detailed description of evaluation scenarios, which can be found in [3]. In the experiment, the request simulator component was configured to perform the requests according to the configurations. A machine was dedicated to the FMC Manager, where the optimization algorithms are placed. The CCN Server receives the commands to perform the migration of content according to the decision of the FMC Manager.

In Figure 8.15 (top), CPU load is compared for the different number of CCNx Interests/s and for each of the three platforms. Clearly, the physical machine delivers

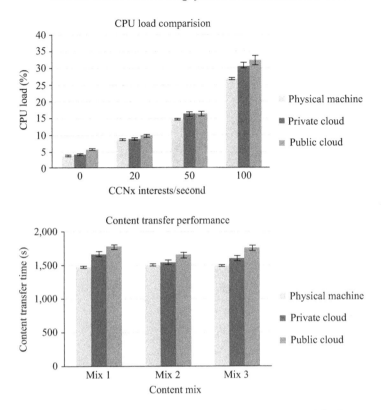

Figure 8.15   ICNaaS cloudification CPU load (top) and cloudified ICN content transfer time (bottom)

the best performance and the CPU is less loaded, followed by the private and then the public clouds. However, the differences are small and sometimes within the error margins, proving that the service can be cloudified without much impact on performance. In the same figure, one can also observe that even considering a single instance without scaling, the number of interests handled by the daemons for a given time unit is quite big without even reaching 50% of CPU load. From Figure 8.15 (bottom), we see the impact of the different platforms in terms of content transfer performance experienced by the end user. The difference is higher when the content mix consists of a majority of smaller objects, which by the quantity create more overhead and the usage of more processing resources. However, differences are not much higher for the cloud platforms when comparing to the physical machine, and even for Mix 1 they do not go much above 2 min for 1 GB of small content objects.

In Figure 8.16 (top), the total scaling time is compared for both scaling out and scaling in operations, considering a number of new instances of 1, 2, 5 and 10. Scaling in is a generally smooth operation, with the deletion of resources happening fairly quickly (between 10 and 15 s), regardless of the number of new instances. As for scaling out, there is a clear increase on the time taken because new resources have

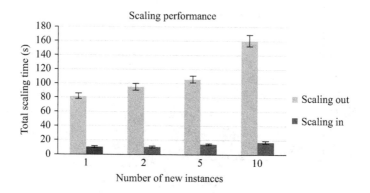

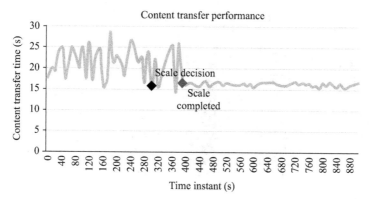

Figure 8.16   ICN scaling performance (top) and ICN content transfer time with scaling (bottom)

to be allocated, instances booted, service reconfigured, etc. However, the increase is not linear and it is smooth for the different number of new instances being deployed simultaneously, not increasing to values that would affect much the behaviour of the service before the scaling operation is finished. In Figure 8.16 (bottom), the advantage of scaling the service is clearly seen from the point of view of the end user. While during the first 6 min the service was having a highly unstable performance, ranging from good to fairly bad very quickly, after the scaling operation is triggered (upon 5 min of high load) and completed (about 85 s later) the performance becomes stable and at good levels in terms of content transfer times.

8.5   Mobility Prediction as a Service

End user's mobility prediction over historical movement traces has big potentials in optimizing urbanization activities, such as applications in navigation, traffic optimization, traffic monitoring and smart city. A prediction service hosted on the cloud enables consumers to request the prediction on-demand and response accordingly.

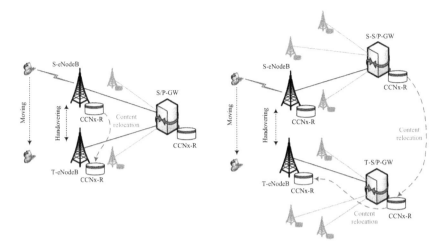

Figure 8.17   ICNaaS content relocation based on MOBaaS prediction

In this section, we describe MOBaaS, which is a network function of mobility avail-ability prediction cloudified over the OpenStack cloud infrastructure. MOBaaS can provide future location prediction in order to generate required triggers for on-demand deploying, provisioning, disposing of virtualized network components. This infor-mation can be used for self-adaptation procedures and optimal network function configuration during runtime operation, as well.

Predicting mobile user's locations at any time in the future is essential for a wide range of mobile applications, including location-based services, mobile access control, mobile multimedia QoS provision, crowd monitoring, as well as resource management for mobile communication and storage. In a cloudified LTE network, different virtualized service might also need the location awareness to optimize net-work performance. For instance, ICNaaS could benefit from knowing the location prediction results to place the users' interested content closer to locations that users will visit in the future, as shown in Figure 8.17. This information could also be used by RANaaS for the efficient deployment, upgrading and scaling of the radio resources dynamically. For EPCaaS, information about users' future locations assists in the mobility management procedure to support optimal traffic forwarding and seamless mobility handover operations.

8.5.1 Architecture reference model

Figure 8.18 presents the internal components of the MOBaaS implementation archi-tecture, its interfaces to the consumer and support service, as well as to the cloud platform and EEU. The roles of different elements are listed below:

- *Frontend*: It ties all internal components of MOBaaS together, and it handles prediction requests and starts algorithm threads. If there is a prediction request whose required input information is currently not available, the frontend stores the

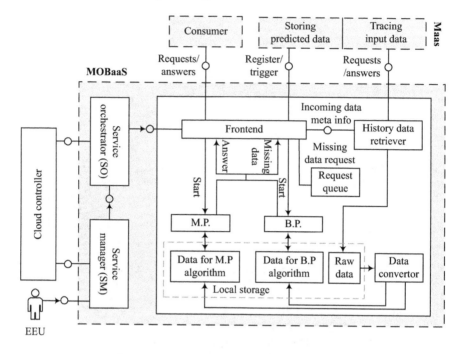

Figure 8.18   MOBaaS software architecture

request to the *Request Queue*, and that request will be postponed until the missing data become available. It also handles the trigger-based prediction. It periodically calculates future states of users' movements and stores the prediction information in MaaS. This data is used later to help decide if the registered consumers are informed correctly or not.

- *Mobility prediction*: This module includes the mobility prediction algorithm and predicts the location of an individual/a group of end users in a future moment of time. The improved version of Dynamical Bayesian Network (DBN) was used as the mobility prediction mechanism in this architecture. The rationale behind using DBN is that, the next location (cell) visited by a user only depends on its current location, the current time, and the day of the week that the user is in the movement. This approach utilizes the trace of mobile user trajectories to predict the next location that maybe visited.
- *History data retriever*: This module continually retrieves end users' historical movement traces from MaaS. This data later is processed and used as the input data for the prediction algorithms.
- *Data converter*: It processes the raw monitored data and converts it to the format usable by the algorithms.

- *Request queue*: It queues the requests when either a specific data to perform a prediction is missing in the input trace data, or more than one prediction is requested by consumers.
- *Data for M.P. algorithm*: These are the local databases storing the data processed by the *Data Converter* module. In order to perform estimations on the user(s) mobility or the available bandwidth in a certain network link, a significant amount of user(s) movement or link usage history information is required. This information can be acquired from MaaS.

MOBaaS consists of two types of notification mechanisms to provide the prediction results: request-based and trigger-based approaches. In the request-based approach, a web server running at the *Frontend* constantly waits for prediction requests from the consumers. Whenever a consumer requires a mobility prediction, it sends the request information to MOBaaS using a JSON (Java Script Object Notation) message. The request message includes the User ID (for single user), the current time and date, the current Cell ID and the time period of the required prediction. For multiusers prediction, the User ID is set to '0000' in the request message. Given this information, *Mobility Prediction* module makes the mobility prediction of which the cell specified user(s) will be located at a certain future time. Next, *Frontend* returns the prediction results in the JSON message to the consumers, including multiple pairs of <Cell ID, Probability>.

In the trigger-based approach, the web server running at the *Frontend* gets the prediction requests from consumers including two parameters: the maximum number of users (N) in each cell (for mobility prediction) as a threshold, and the prediction interval. Knowing this information, the *Mobility Prediction* module periodically performs mobility predictions, respectively. If the number of users per cell exceeds the threshold, a trigger message is generated by the *Frontend* and results are pushed into MaaS. This information could later be used by consumers for a specific action (e.g. scaling the resources, instantiating the VMs).

8.5.2   *MOBaaS cloudification architecture*

As shown in Figure 8.18, there are some components that are responsible for the cloudification operations of MOBaaS. In this subsection, we detail how mobility prediction service has been cloudified, which allows MOBaaS to be running in the cloud platform with all the benefits brought by cloud principles. Namely, EEU only needs to provide the network topology for the deployment and a few settings. Afterwards, it will get a running SI that manages itself for a typical operation. The key components for MOBaaS cloudification are the SMs, SO, CC and SDK. In the following, there components along with other cloudification-related components are described.

The technical implementation architecture of the cloudification procedure of MOBaaS follows the general MCN cloudification architecture. In principle the architectural elements are influenced by Start of Authority (SOA) principles, and constructs

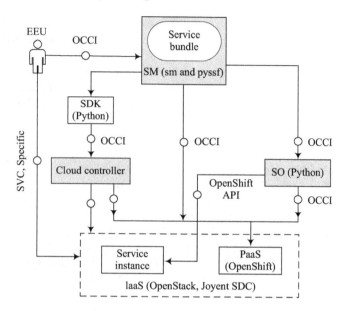

Figure 8.19   Cloudification architecture of MOBaaS

are in place to facilitate the service life cycle management. The detailed representation of the architecture element is shown in Figure 8.19, and the roles of different elements are listed below:

- *SO:* It is responsible for the orchestration of the service, consisting of two internal modules: The Service Orchestrator Decision (SOD) and Service Orchestrator Execution (SOE). The decision module is the component responsible for handling inputs from other services, mainly monitoring information that makes decisions on whether a scaling decision is needed. The execution module is responsible for the interaction with the CC using a SDK, ensuring that SIs are mapped to cloud resources and deployed, scaled and disposed whenever requested. It is also responsible for handling requests from SM to instant or dispose the MOBaaS SI through the SDK.
- *SM:* It provides an external interface to the EEU and a list of available services (Service Catalog). It is responsible for deploying the SO, and forwarding requests to SO for MOBaaS SI deployment and disposals. Upon receipt of a service creation request, the SM creates a Docker container in OpenShift and when the container is ready, pushes the service bundle into the container and deploys it. Once the SO is ready, the various life cycles of the services are activated.
- *Open cloud computing interface (OCCI)*: It enables to maintain interoperability and a common interface and model between different services. It forms the core of the interfaces exposed by the SM, CC and the SO. It is also the specification used by the northbound interface of the CC.

- *CC*: It is implemented through a set of modules to support service deployment, provisioning and disposal. Each of the modules is modelled as a service itself.
- *SDK*: It is designed to support developers to manage services. It serves two main purposes: (i) it supports the basic functionalities for the life cycle management of any service by allowing the SO to interact with the various internal services of the CC. Therefore, any changes to the internal implementation of the CC will have no impact on the code written, as the SDK abstracts the technologies used; (ii) it provides access to the support services, which allows for the developer of a certain service to interact with another service in a certain way, regardless how the other service is implemented.
- *Service bundle*: In every service while implementing, the SM has to provide the Service Bundle which has all necessary details for the deployment. It contains the Service Manifest details, which comprises of Service Template Graph (STG) where dependencies on other external services are encoded, and Infrastructure Template Graph (ITG), which is realized as a Heat Template that encodes all the details for the deployment of the service itself. The service bundle also contains the SO logic (the application code).
- *OpenStack infrastructure*: OpenStack is the cloud management framework that allows the life cycle management of VMs. It provides infrastructure cloud services, namely, compute, storage and networking. OpenStack is chosen as the reference implementation framework. The OpenStack Heat orchestration module allows service deployment as one logical unit instead of a collection of a number of VMs.
- *OpenShift*: A Platform as a service (PaaS) solution called OpenShift is adopted to enable the CC deployment module to host the SO instance. OpenShift provides quick and effective means to run SO workloads, and OpenShift v3 (Docker container based) builds the application container for every SO instantiation efficiently.

8.5.3   Sequence diagram of service management

The sequence diagrams are interaction diagrams that show the different steps that have to be done by the system to realize a certain operation requested from an external actor, either a person or an external system. It shows the different classes that are involved in the operation and the sequence of messages exchanged between the objects needed to carry out the functionality of the scenario.

In the following, we describe the general MOBaaS cloudification operations, including the sequence diagrams of service deployment, provisioning and disposal. Through the SM and the SO, the MOBaaS system can request the CC to deploy, dispose, provision and management operations of the MOBaaS instance. It allows the EEU to get a MOBaaS instance, deploy and dispose it in an on-demand fashion.

Figure 8.20 (top) shows the deployment of the service. The main users of the system (Service Provider) can log into the administrator panel and execute a command for deployment of an instance. As a start, the EEU requests through MOBaaS SM the deployment and provisioning of a MOBaaS instance. The MOBaaS SM module will

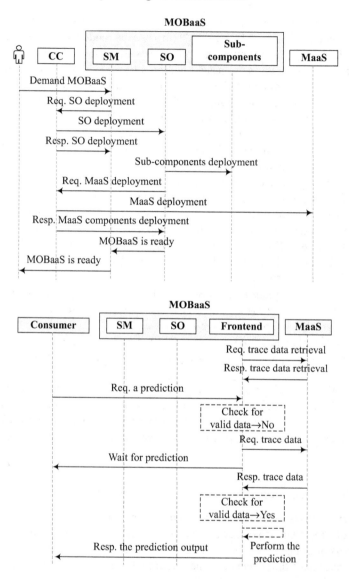

Figure 8.20   MOBaaS service instance deployment (top) and previsioning (bottom)

then send the request including the OpenStack Heat template graph of the service to the CC to start an instance of the MOBaaS SO. Heat is the main project in the OpenStack Orchestration program. It implements an orchestration engine to launch multiple composite cloud applications based on templates in the form of text files that can be treated as a code. In general, a Heat template describes the infrastructure for a cloud application in a text file that is readable and writable by humans. Once the MOBaaS SO is instantiated, the MOBaaS SM requests the SO to deploy the MOBaaS. Afterward, MOBaaS SO requests the CC to instantiate the ITG of the MOBaaS. Next, the CC instantiates the internal components of the MOBaaS instance (Frontend, M.P., B.P., Historical Data Retriever, Data Converter and Request Queue). After this, if everything is correct, as a response, an endpoint to the created instance will be returned, which makes us able to configure and manage the created instance. Figure 8.20 (bottom) shows provisioning phase of the service. After deployment, provisioning should have the actual VMs configured and running. To achieve this, after getting the MOBaaS SO instance from the CC, MOBaaS SM commands SO to provision required VMs according to the provided configuration. In this way, we have an MOBaaS instance deployed and ready to use. Similarly, when the EEU does not need the SI anymore, the disposal phased will be called, in which all the VMs that have been created before will be disposed, and all the resources that have been allocated to MOBaaS will be released.

8.5.4 *Mobility prediction algorithm*

One of the intuitive methods to determine a mobile user's movement pattern, leading to predict its future behaviour, is the attempt to trace and capture some sort of regularity in the user mobility. Many studies in this field show that most often a mobile user has the tendency to regularly behave the same way. Such behaviour regularity could be considered as user's profile and utilized to estimate places a user may visit in the future. The prediction schemes, relying on history of mobile node trajectories, are classified in *Temporal Dependency* category. In context of the LTE system, the Mobility Management Entity (MME) is one of the key components of the EPC in the LTE system, which is responsible for variety of important key functions (e.g. subscriber's authentication, keeping location information of users, gateway selection during registration process, mobility management, roaming and handovers). By monitoring and tracing information from MME, it would be possible to derive movement history of users. In this section, we describe the proposed mobility prediction algorithm, which utilizes the trace of mobile user trajectories to predict the future locations of users.

The proposed prediction algorithm benefits from the DBN model. The rationale behind using DBN is that, the next location (cell) visited by a user depends on: (i) its current location, (ii) the movement time and (iii) the day that user is in the movement. The proposed DBN model is illustrated in Figure 8.21.

In Figure 8.21, C_i represents a cell with ID i, and T_i defines the time of a day, and D shows the day of a week (e.g. Monday) and δ determines the future time. As it is shown in Figure 8.21 (left), the conditional distribution of the next location (cell) visited by a user comprises the current location, time and week day.

$$P(C(t + \delta)) = C_{i'}|C(t) = C_i, T(t) = T_t, D$$

Equation shown above can be considered as a location-dependent distribution, providing a given time and day, and can be modelled as a simple first order Markov

Chain (M.C.) that encodes the frequency (probability) of transitions between the cells. The DBN model can be simplified by integrating the transition time step (e.g. each minute) and the cell ID to derive a customized M.C. model, as shown in Figure 8.21 (right). In order to derive the transition probability of moving from one cell to others for each individual user, by counting the number of transitions from one cell to another, in regular intervals (e.g. each minute) on the given days of a week (e.g. all Monday and all Tuesday) in the trace files. Figure 8.22, as an example, shows two various transitions from one cell to other cells derived in two different times in the customized M.C. model. The spatial granularity of this algorithm is at the cell level, which means the algorithm outputs the possible future cells of a user with particular probabilities. The temporal granularity of the algorithm is dependent on the application requirements and could be tuned in scale of minutes. In order to evaluate performance of the proposed algorithm, we used the mobility data trace provided by Nokia for academic research. From this dataset, we picked data from 100 users ranging over 2–6 months of time. For each user, we separated available data into two parts: the learning data set (L) and the testing data set (T). Data set L is the first 70% of the user's trace and is used to derive the M.C. states and to calculate its transition probability matrix. Data set L contains the rest 30% of the data trace, which is utilized to validate the performance of the prediction accuracy. For instance, if the length of a data trace is 2.5 months (i.e. it includes trace data for ten Mondays), we use the data trace of the first seven Mondays during the learning phase and the rest for the testing phase.

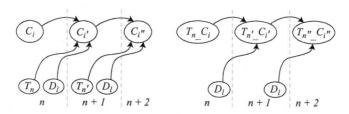

Figure 8.21 The DBN model used to drive the Markov chain states

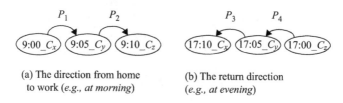

(a) The direction from home
to work (*e.g., at morning*)

(b) The return direction
(*e.g., at evening*)

Figure 8.22 Examples of customized M.C. states

8.5.5 Evaluation

The evaluation of MOBaaS includes two parts: one part is about the accuracy evaluation of the prediction algorithm and another part is about the performance of the cloudification operation of SI management.

8.5.5.1 Evaluation of the prediction algorithm accuracy

In order to evaluate accuracy of the proposed algorithm, we selected, for each user, 50 random states (representing the random times and IDs of the cells that user has been there) out of the Markov Chain states derived for each particular day of a week from the data set of L, and performed prediction calculation to find the future possible cells in next 20 minutes. We repeated the prediction for the same random states in the data set of T. Afterward, the Mean Absolute Error (MAE) for the corresponding test points, chosen from the learning and testing data sets, is computed in order to obtain accuracy of the prediction for each user in a particular day of a week.

$$\text{MAE} = \frac{1}{M} \sum_{i=1}^{M} |P_{iL} - P_{iT}|, \text{Accuracy} = (1 - \text{MAE}) * 100 \tag{8.1}$$

in which M represents the numbers of all possible transitions in the selected states, P_{iL} and P_{iT} define the calculated probability for each possible transition in the test points chosen from data set L, and the checked probability of possible transitions for the same selected states in the data set T, respectively.

Figure 8.23 shows the accuracy of the prediction: the left one is about the accuracy of some users, calculated per weekday and the right one is the overall accuracy of all the selected 100 users. As we can see from the results, prediction accuracy varies significantly from user to user. This is because the accuracy effectively pertains to the quality of data trace used to derive the transition probability matrix. Therefore, we

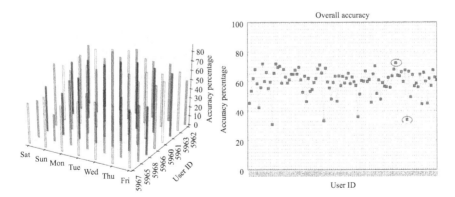

Figure 8.23   Accuracy of prediction algorithm for certain users per day (left) and all users/days (right)

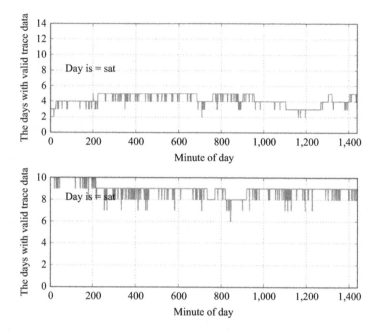

Figure 8.24 Quality of data trace for two different users (top: user ID 6025, bottom: user ID 5960)

pick up two users (user ID 6025 and 5960) from Figure 8.23 (right) with good quality and bad quality. Figure 8.24 illustrates data traces of the selected two users, leading to low (for user ID = 6025) and high (for user ID = 5960) prediction accuracy. As we can see, for user 5960, it has a high number of days where valid trace data is recorded, while user 6025 has very few numbers of days with valid data trace recorded.

8.5.5.2 Evaluation of the cloudification operations

This subsection describes how we evaluate the performance of the designed cloudification platform. The evaluation test beds, scenario and methodology and performance results are discussed next.

To support this evaluation, a performance evaluation framework is needed. The devised tool Graylog is used as a log aggregator to collect information and it is included in the SM library. Data collected by the monitoring system from SM and SO will show the timing of deployment, provisioning and disposal of a SI. We evaluate the MOBaaS cloudification implementation through two different cloud infrastructures with different resources allocated.

- Test bed 1: It offers all essential services, including OpenStack Nova with KVM, Glance, Cinder, Neutron and Heat. The OpenStack version is Juno running on Ubuntu 12.04. The test bed consists of a CC, computing node (Dell PowerEdge

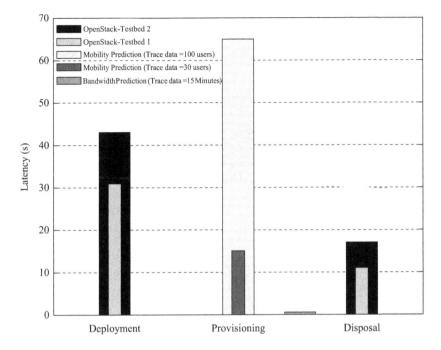

Figure 8.25   MOBaaS cloudification performance

R520) and a storage device (Dell Power Vault MD3800i). It operates 32 cores, Intel(R) Xeon(R) CPU E5-2450 v2@2.50 GHz, 192 GB RAM, 2.3 TB HDD for nova-computing file system, 9.0 TB HDD for the cinder block storage, and 120 floating IPs. We have committed the following resources for MOBaaS testing: 100 instances, 100 VCPU, 50 GB RAM, 50 public IPs and 1 TB of storage.

- Test bed 2: It offers also all the basic OpenStack services as Test bed 1. The OpenStack version is Kilo. It consists of multiple servers and has a total of 64 cores, 377 GB RAM, 5.5 TB disk and 44 floating IPs available. Additionally, Test bed 2 is an Identity Provider, a service that allows users to authenticate and get a token to interact with other services. The Identity Provider is implemented with OpenStack Keystone. The MCN PaaS, OpenShift v3, is hosted in Bart also and is used by the orchestration. Resources available to test MOBaaS are: 10 instances, 20 VCPU, 20 GB RAM, 8 public IPs and 1 TB of storage.

Figure 8.25 shows the results of MOBaaS cloudification performance, including the service deployment, provisioning for both the mobility and bandwidth predictions and disposal latency. Based on the obtained results, it can be observed that the deployment and disposal times are shorter in Test bed 1 (31 s for deployment and 11 s for disposal) than in Test bed 2 (42 s for deployment and 18 s for disposal). This is due to the fact that more resources are allocated at Test bed 1 during the evaluation procedure. However, in both cases, the delays are short and could fit on-demand instantiation requirements. From these results, we conclude that available amount of cloud resources has direct impact on the performance of cloudified services. In the provisioning phase, the mobility prediction latency depends on two factors: the day of the week and size of the trace file. The results show that predictions of weekdays take less time than prediction in weekends. This is due to the fact that user mobility is more regular in a weekday than in the weekend. The size of the trace file also significantly influences the prediction calculation time, and a prediction using a trace with more users takes a longer time than the predictions using a trace with a few number of users. From these results, we could learn that for time-critical smart city applications, the cloudified service could split the requests with large computation overhead, which might lead to a longer response delay, into multiple parallel lightweight computation procedures. Further details can be found in [4].

8.6 Related works

This section analyses related work, focusing on some projects that are with high relevance for the MCN project. T-NOVA [4] is probably the project with more similarities with MCN. It aims at designing an E2E architecture for NFV services, targeting all the layers including applications and infrastructure components. Different test beds are employed to evaluate T-NOVA, but the RAN service is not included. Considering that MCN claims to be compliant with ETSI NFV MANO [5] specification, it is also important to list existing open source projects. At the moment, several new projects provide an implementation of the ETSI NFV MANO specification. ETSI launched Open Source MANO initiative providing a reference architecture framework that implements MANO functionalities. Open Baton [6] provides a comprehensive reference implementation of ETSI NFV MANO, which has been further extended for supporting OCCI as northbound interface and can be integrated with MCN as an additional service orchestrator. The underlying MCN architecture distinguishes from the related work in the area of ETSI NFV MANO by assessing the mobile network infrastructure in diverse IaaS using a hierarchical orchestration approach.

8.7 Conclusion

The vast commercial advantages of digital business and cloud computing in particular have become apparent during the last three and a half years. This development has also set the benchmark for today's telecommunications industry. The MCN project has addressed the lack to exploit the unique opportunity that cloudified mobile telecommunication services yield for Europe. The basic assumption of MCN is the existence of RAN, MCN, Micro- and Macro Data Centres (DC). Macro data centres are standard large-scale computing farms deployed and operated at strategically selected locations. Micro Data Centres are medium- to small-scale deployments of server clusters across a certain geographic area, e.g. covering the area of a city or a certain rural district; they are part of a mobile network infrastructure. Ownership and operation of these infrastructure elements can vary. For instance, we can assume a mobile network operator that operates several RANs, mobile core networks, as well as DCs and thus enjoys full control of all technology domains. In addition, the technology also supports other scenarios such as that of a company that acts as E2E MCN provider without owning and operating any physical infrastructure. Such company would sign wholesale agreements with physical infrastructure owners, e.g. mobile network carriers in those geographic areas (e.g. big cities of entire countries), for which it wishes to provide access to MCN services. The same would be the case for contracting data centre operators in strategic locations and in order to complete a full MCN offering (RAN, Mobile Core, Data Centre). The distinctive aspect of this approach is that a MCN provider exploits the MCN architecture to compose and operate a virtual E2E infrastructure and platform layer on top of a set of fragmented physical infrastructure pieces provided by different mobile network and data centre owners/operators, thus providing a differentiated E2E MCN service (mobile network + compute + storage) that is not limited to a certain geographic area.

Aiming at these scenarios, the MCN project has investigated, implemented and evaluated the fundamental technological services for a mobile networks augmented by decentralized computing and smart Storage; all offered as one atomic service that is on-demand, elastic and pay-as-you-go. To realize this objective, the project extends the concept of cloud computing beyond data centres toward end users. It has designed 3GPP-compliant MCN architecture that exploits and supports cloud computing as well as a novel virtualization layer and general provisioning across the various domains within MCN.

In this chapter, we briefly describe three MCN enabling services: RANaaS, ICNaaS and MOBaaS, which were mostly responsible by University of Bern during the project period. Their design architecture, implementation details, cloudification procedure and performance evaluation are included and analysed.

References

[1]   MCN project website, 2006: http://www.mobile-cloud-networking.eu/
[2]   N. Nikaein, E. Schiller, R. Favraud *et al.* Network Store: Exploring Slicing in Future 5G Networks, ACM Mobi Arch Workshop. 2015. http://dx.doi.org/10.1145/2795381.2795390
[3]   A. Gomes, B. Sousa, D. Palma *et al. Edge caching with mobility prediction in virtualized LTE mobile networks.* Future Generation Computer Systems, Elsevier. http://dx.doi.org/10.1016/j.future.2016.06.022
[4]   Z. Zhao, Z. Zhao, L. Hendriks *et al.* Mobility and Bandwidth prediction as a Service in Virtualized LTE System. IEEE Cloud Net 2015. http://dx.doi.org/10.1109/ CloudNet.2015.7335295
[5]   T-NOVA project website, 2016: http://www.t-nova.eu/
[6]   ETSI NFV MANO specification, 2014: http://www.etsi.org/

Chapter 9

Composition of network and cloud service in next generation Internet

Jun Huang[1], Qiang Duan[2], and Ching-Hsien Hsu[3]

Abstract

Recent rapid development of both networking and cloud computing technologies is transforming the Internet from an infrastructure for data transportation to a general platform for service provisioning upon which a wide spectrum of cloud computing applications can be deployed. Such transform requires networking and cloud computing to be integrated in the next generation Internet, which calls for federated management of both networking and computing resources for service provisioning. The Service-Oriented Architecture (SOA), which has been widely adopted in cloud computing through the IaaS, PaaS, and SaaS paradigms, has been applied in networking to enable the Network-as-a-Service (NaaS) paradigm. The NaaS in networking together with IaaS in cloud computing offers a promising approach toward converged network–Cloud service provisioning in the next generation Internet. A key challenge to achieve high performance converged network–cloud services lies in composition of network and Cloud services with end-to-end performance guarantee. In this chapter, we present our recent research progress in Quality of Service (QoS)-aware network–cloud service composition, including system model and problem formulation for network–cloud service composition and our proposal of an efficient algorithm to solve this problem. We also report the experimental results for evaluating the performance of the proposed algorithm in this chapter.

9.1 Introduction

Cloud computing has been envisioned as a novel computing paradigm of the next generation Internet. Cloud infrastructure comprises a pool of virtualized computing functions and resources that are delivered on demand over networks [1]. In the Cloud environment, service delivery to an end user relies on communication functions

[1]Chongqing University of Post and Telecommunications, China
[2]Pennsylvania State University, USA
[3]Chung Hua University, Taiwan

offered by networks as well as computing functions provided by Cloud infrastructure [2]. Recent research has shown that networking has a significant impact on Cloud service performance [3–5]. The end-to-end performance evaluation between network and Cloud systems conducted in [6] showed that networks may form a bottleneck that limits Cloud service performance. The delay and cost introduced by inter-Cloud communications also have a strong impact on application performance [7]. Therefore, networking plays a crucial role in Cloud computing and becomes an indispensable ingredient for single, hybrid, and inter-Cloud service provisioning [8–11].

An illustrative example given in [12] is presented here to show the impact of networking on Cloud service performance. Considering a cloud environment in which a user has 100 GB of raw data that need to be processed in Amazon EC2 Cloud [13]. If the user obtains 10 EC2 VM instances and each instance can process 20 GB data per hour. Then the total processing time of the Cloud service is only 30 min. However, if the network between the user and EC2 supports 200 Mb/s throughput for data transmission, then the data transmission delay from the user to EC2 servers will be more than an hour. In this case, network delay for round-trip data transmission contributes to more than 80% of the total service delay.

Recent advances in networking technologies, especially network virtualization, Software-Defined Network (SDN), and application of the Service-Oriented Architecture (SOA) in networking, have significantly enhanced network control and management for supporting Cloud service provisioning [14]. Network virtualization decouples service functions from network infrastructure and is expected to be a key attribute of future networking [15]. The emerging Network Function Virtualization (NFV) technology allows network functions to be virtualized and implemented on general purpose data center hardware [16]. As a potential enabler of profound changes in both computing and networking, virtualization may bridge the gap between these two fields.

On the other hand, recent progress in network service provisioning has added more and more computing functions into the Internet infrastructure, which is transforming the Internet from an infrastructure for data transportation to a general platform for service provisioning upon which a wide spectrum of cloud computing applications can be deployed. The SOA, which has been widely adopted in cloud computing through the IaaS, PaaS, and SaaS paradigms, has been applied in networking to enable the Network-as-a-Service (NaaS) paradigm. The NaaS paradigm enables networking resources to be utilized on-demand by end users through standard service interfaces; much like computing capacities in Clouds being accessed as services by users. Therefore, NaaS is transforming tradition network service provisioning toward a Cloud IaaS service model. Such transform enables network services to be composed with computing services in a Cloud environment; thus allowing a convergence of network–Cloud service provisioning [17].

A key challenge to realizing converged network–Cloud service provisioning lies in Quality of Service(QoS)-aware service composition across the networking and Cloud computing domains. The objective of network–Cloud service composition is to select an appropriate sequence of network and Cloud services to meet end-to-end service performance requirements while optimizing networking and computing resource utilization. Although service composition has been extensively studied, most of the currently available approaches focus on composition of web and/or Cloud services instead of composition across network and Cloud services. Due to the real-time requirements of various network protocols, network–Cloud services typically need much shorter response time for composition requests; therefore requiring more efficient and effective algorithms that can not only meet diverse QoS constraints but also achieve much shorter response time.

In this chapter, we report our recent research progress toward addressing the challenging problem of QoS-aware network–Cloud service composition. We first present a model for QoS-aware network–Cloud service composition. This model allows multiple QoS metrics to be considered in network–Cloud services composition with guaranteed end-to-end service performance. We then formulate the problem of QoS-aware network–Cloud service composition as a variant of Multi-Constrained Optimal Path (MCOP) problem and develop an exact algorithm to solve the problem. We analyze the theoretical properties of the proposed algorithm to show that the proposed algorithm is effective and resilient in selecting network and Cloud services and able to guarantee end-to-end performance for composite service provisioning. We also provide experimental results to evaluate performance of the proposed algorithm. The obtained experimental results indicate that the proposed algorithm is efficient and accurate for service composition.

9.2 Modeling composite network–cloud service provisioning

The end-to-end services provisioned by future Cloud infrastructure will comprise data communication functions offered by network infrastructure as well as computing functions provided by data centers. Therefore, selection and composition of network and Cloud services play a key role in Cloud service provisioning for achieving the level of service performance required by end users. In this section, we will present a model and formulation for QoS-aware network–Cloud service composition.

9.2.1 A framework for network–cloud service composition

The SOA and virtualization becomes technical foundation for both Cloud computing and next generation networking, thus providing a promising mechanism for integrating networking and Cloud computing for composite service provisioning. Figure 9.1 shows the layered structure of a service-oriented framework for composing networking and Cloud computing systems presented in [14]. This framework consists of physical infrastructures for both networking and Cloud computing, the virtualization layer that encapsulates both networking and computing resources into SOA-compliant services, and the service provisioning layer that selects and composes the appropriate network/Cloud services for meeting user requirements.

NaaS is a key enabler in this framework for composite network-Cloud service provisioning. NaaS allows networking resources to be virtualized and encapsulated

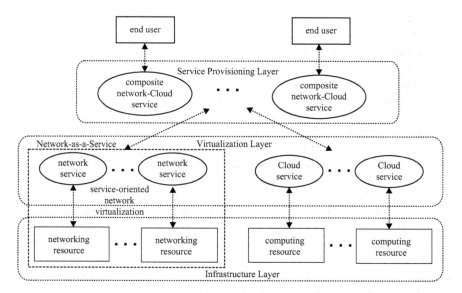

Figure 9.1   The layered structure of a service-oriented framework for networking and Cloud computing composition

in SOA-compliant services that can be combined in almost limitless ways with other service components that abstract both computing and networking resources. Therefore, NaaS may greatly expand the service spectrum that can be offered by composite network-Cloud service providers with the ability to match Cloud requirements to communication services through network service selection. NaaS enables provisioning of network services and Cloud services, which used to the responsibilities of separate providers, now to merge into one layer of composite services provisioning. This convergence leads to a new service delivery model in which the roles of traditional network service providers, like AT&T and Verizon, and computing service providers, such as Amazon and Google, merge together into one role of composite network–Cloud service provider. This new service delivery model may stimulate innovations in service development and create a wide variety of new business opportunities.

Figure 9.2 shows an illustrative example of composite network–Cloud service provisioning. In this example, the user uses two cloud services in sequence for processing a big data set and receives the results obtained by the second cloud service. In order to access the Cloud services for data processing and receive the process results back, the user also needs network services for data transmission as well. In this example, network services 1, 2, and 3 respectively provide data communications from the user to Cloud service 1, between Cloud services 1 and 2, and from Cloud service 2 back to the user. Therefore, the end-to-end service provided to this user is a composite network–Cloud service comprising three network services and two Cloud services. If the user wants a certain level of performance guarantee, for example, the

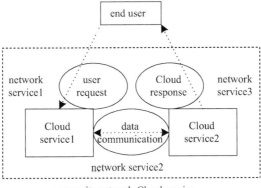

Figure 9.2   An example of composite network-Cloud service provisioning

end-to-end service delay, each component of the composite network–Cloud service must provide a maximum delay upper bound. In general, such QoS expectation can be defined in Service Level Agreements (SLAs) between the user and service providers.

9.2.2   Model and formulation for composite network–cloud service provisioning

The service composition process is invoked by the service provider upon receiving a service request from a user. A series of atomic service components are combined to form a composite service (path) that meets the user's QoS expectation. Figure 9.3 shows a model for composite network–Cloud service provisioning [18]. Suppose H service components S_i, $1 \leq i \leq H$, are needed by a given service request and there are l_i candidate services S_{ij}, $1 \leq j \leq l_i$, that meet the functional requirements for S_i. Note that a service component could be either a Cloud service or network service. To meet the requirement of the service request, a path traversing one candidate service of each service component needs to be determined, namely, $S_0 \to S_{11} \to S_{22} \to \cdots \to S_{Hl_H} \to S_l$. Therefore, selecting optimal service paths from a pool of available network and Cloud services while satisfying various QoS requirements is the main optimization objective for composite network–Cloud service provisioning.

A service network formed by interconnecting service components can be modeled as a directed graph $G(V, E)$ with n vertices and m edges. Each vertex (for example v) is associated with a weight c (c_v for vertex v) denoting its serving capability in service S_h, $1 \leq h \leq H$, and each edge $e \in E$ is associated with K weights $w = (w_1, w_2, \ldots, w_K)$ representing QoS parameters. Let $Req = \{C, W\}$ be a user's request, where $C = (C_1, C_2, \ldots, C_H)$ denotes the function capacity requirements on H services, $W = (W_1, W_2, \ldots, W_K)$ denotes the K non-functional end-to-end QoS constraints, then a series of selected service components is represented as a path p in the service network.

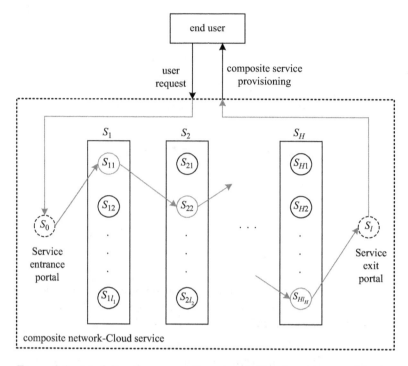

Figure 9.3   Modeling for composite network-Cloud service provisioning

In general, QoS parameters can be categorized to be either *positive* (quality increases as parameter value increases, e.g. reliability) or *negative* (quality decreases as parameter value increases, e.g. delay). In this model we assume all QoS parameters are negative and concentrate on the *additive* QoS parameters.

Definition 9.1. *Feasible Service Composition. A composed service, i.e., a path p in the service network, is said to be feasible if for $\forall v \in p$ there is $c_v \geq C_h$ where $v \in S_h, 1 \leq h \leq H$, and $w_k(p) \leq W_k$ for all $1 \leq k \leq K$.*

Denote $\{p^f\}$ as all feasible service compositions in $G(V, E)$, to each $p_i^f \in \{p^f\}$, there exists a smallest value $\eta_i \in (0, 1]$ such that $w_k(p_i^f) \leq \eta_i \cdot W_k$, $1 \leq k \leq K$, and for $\forall v \in p^f$ such that $c_v \geq C_h, 1 \leq h \leq H$ respectively.

Problem 9.1. *QoS-aware Service Composition (QSC). QSC is to select an optimal composition of services p^{opt} among feasible service compositions in $G(V, E)$ and the corresponding smallest value η^{opt} among all η_i such that $w_k(p^{opt}) \leq \eta^{opt} \cdot W_k$, $k \in [1, K]$ where $K \geq 2$.*

Theorem 9.1. *QSC problem is NP-hard.*

Proof. Since the constraint on vertex can be satisfied beforehand by pruning the topology in advance, QSC maps directly to the special case of MCOP, thus QSC is NP-hard.   □

Definition 9.2. *Path dominant. Given two paths p_1, p_2 in $G(V, E)$, path p_1 is said to dominate path p_2 (or path p_2 is dominated by path p_1) if $w_k(p_1) \leq w_k(p_2), k \in [1, K]$, and $\exists j \in [1, K], w_k(p_1) < w_k(p_2)$.*

Definition 9.3. *Pareto Minimum Paths Set. A Pareto minimum paths set consists of a set of paths that are not dominated by other paths.*

9.3 An algorithm for QoS-aware composition of network and cloud services

9.3.1 Algorithm description

In this section we present an algorithm named EXACT, shown in Algorithm 1, for solving the problem of QSC. This algorithm comprises five steps. First step (from lines 1 to 6) blocks any invalid request according to the end-to-end constraints on the path. Dijkstra algorithm is applied in this step to calculate the shortest path p_k^l according to each weight $w_k, 1 \leq k \leq K$. If the weight of the shortest path is greater than the requested weight, i.e., $W_k < w_k(p_k^l)$, it implies that the constraints imposed by a user request are too rigid to be satisfied; then this request would be treated as invalid and should be blocked in advance.

The second step (line 7 in Algorithm 1) prunes the topology according to the constraints on the vertex. This is a required step because c_v denotes the serving

---

Algorithm 1 EXACT

Input:
Graph: $G(V, E, w, W, c, C)$;
Output:
Path set: Pareto minimum path set MP;
1: **for** $k = 1$ to K **do**
2:     Apply Dijkstra to calculate the shortest path p_k^l according to weight $w_k(e)$ on each edge in $G(V, E)$;
3:     **if** $W_k < w_k(p_k^l)$ or $\max_{\substack{1 \leq i \leq K \\ i \neq k}} \left\{ \frac{w_i(p_k^l)}{W_i} \right\} > 1$ **then**
4:         **return** Invalid request, Exit;
5:     **end if**
6: **end for**
7: To each vertex $v \in S_h$, prune v and its connected edges if $c_v < C_h, 1 \leq h \leq H$;
8: Compute a new weight $w_M(e) = \sum_{k=1}^{K} \frac{w_k(e)}{W_k}$ for each edge $e \in E$;
9: Apply κ-shortest paths algorithm in terms of $w_M(e)$ on each edge to find the first κ paths $p_j^M, 1 \leq j \leq \kappa$ from source to destination, $MP \leftarrow \{p_j^M | 1 \leq j \leq \kappa\}$;
10: To all paths in MP, remove the path which is dominated by any other;
11: **return** MP;

---

capability of a vertex $c_v \in S_h$ in the service network and $c_v < C_h$ means such a node cannot meet the service requirement; therefore it should be eliminated in the initial stage. In addition to this, the notation $G(V, E)$ is still used here to denote the pruned topology.

The third step (line 8) of Algorithm 1 aggregates the weights on each edge through a linear combination manner, i.e., $w_M(e) = \sum_{k=1}^{K} \frac{w_k(e)}{W_k}$ for each edge $e \in E$.

In the fourth step (line 9), an existing κ-shortest path algorithm is applied in terms of new weight $w_M(e)$ on each edge to find the first κ shortest paths from source (service entrance portal) to destination (service exit portal). The obtained κ shortest paths form the *MP* initially.

The fifth step examines *MP* generated from the fourth step, and eliminates the path which is dominated by any other. That is, if there is a path p_j^M in *MP* such that each weight of it (for example delay) is inferior (equal or greater) than that of any other path in *MP*, then p_j^M should be removed from *MP*.

9.3.2 Algorithm analysis

Theorem 9.2. *Algorithm EXACT produces the Pareto minimum paths set MP within* $O((K + \kappa) \cdot (m + n \log n))$ *time.*

Proof. The first step of EXACT from lines 1 to 6 takes $O(K \cdot (m + n \log n))$ time to block the invalid requests. The second step prunes the topology that spends $O(n + m)$ time. Computing new weight for each in the third step takes $O(K \cdot m)$ time, and in the fourth step, a best-known κ-shortest paths algorithm [19] spends $O(\kappa \cdot (m + n \log n))$ time. The fifth step of the algorithm takes $O(K \cdot \kappa)$ time. Therefore, the time complexity of EXACT is $O((K + \kappa) \cdot (m + n \log n))$. □

Theorem 9.2 indicates that our proposed algorithm is light-weight and cost-effective for composing services to optimize the composite network–Cloud service provisioning. However, to the case of $\kappa = 1$, algorithm EXACT is essentially a K-approximation algorithm for QSC. The following theorem gives the detailed analysis for EXACT when $\kappa = 1$.

Theorem 9.3. *EXACT finds a K-approximation to QSC when $\kappa = 1$, i.e., $w_k(p_1^M) \leq K \cdot \eta^{opt} \cdot W_k, 1 \leq k \leq K$.*

Proof. For the optimal path p^{opt}, we have $w_k(p^{opt}) \leq \eta^{opt} \cdot W_k, 1 \leq k \leq K$, that is

$$\sum_{e \in p^{opt}} w_k(e) \leq \eta^{opt} \cdot W_k \tag{9.1}$$

Equation (9.1) can be rewritten as

$$\sum_{e \in p^{opt}} \frac{w_k(e)}{W_k} \leq \eta^{opt} \tag{9.2}$$

Summing (9.2) over K weights, then

$$\sum_{e \in p^{opt}} \sum_{k=1}^{K} \frac{w_k(e)}{W_k} \leq K \cdot \eta^{opt} \tag{9.3}$$

This implies

$$\sum_{e \in p^{opt}} w_M(e) \leq K \cdot \eta^{opt} \tag{9.4}$$

i.e.,

$$w_M(p^{opt}) \leq K \cdot \eta^{opt} \tag{9.5}$$

Since the path p_1^M in EXACT is calculated based on the new weight w_M, we have

$$w_M(p_1^M) \leq w_M(p^{opt}) \leq K \cdot \eta^{opt} \tag{9.6}$$

i.e.,

$$\sum_{e \in p_1^M} w_M(e) \leq K \cdot \eta^{opt} \tag{9.7}$$

On the other hand, $w_M(e) = \sum_{k=1}^{K} \frac{w_k(e)}{W_k} \geq \frac{w_k(e)}{W_k}$, hence

$$\sum_{e \in p_1^M} \frac{w_k(e)}{W_k} \leq K \cdot \eta^{opt} \tag{9.8}$$

It suggests

$$\sum_{e \in p_1^M} w_k(e) \leq K \cdot \eta^{opt} \cdot W_k \tag{9.9}$$

i.e.,

$$w_k(p_1^M) \leq K \cdot \eta^{opt} \cdot W_k \tag{9.10}$$

This proves that the algorithm EXACT is a K-approximation algorithm when $\kappa = 1$. □

Though EXACT finds a set of "coarse-resolution" paths, it is able to generate a Pareto set. Next we prove that *MP* generated by the proposed algorithm is a Pareto minimum paths set.

Lemma 9.1. *Path p_1^M determined by EXACT is a Pareto minimum path.*

Proof. Let us suppose that there exists a path p^* that dominates p_1^M, then

$$w_k(p^*) \leq w_k(p_1^M), \quad 1 \leq k \leq K \tag{9.11}$$

This can be rewritten as

$$\sum_{k=1}^{K} \frac{w_k(p^*)}{W_k} \leq \sum_{k=1}^{K} \frac{w_k(p_1^M)}{W_k} \tag{9.12}$$

i.e.,

$$w^M(p^*) \leq w^M(p_1^M). \tag{9.13}$$

This is contradictory with the fact that p_1^M is a shortest path calculated by new weight. Therefore, p_1^M is a Pareto minimum path. $\quad\square$

Theorem 9.4. *The set MP generated by the proposed algorithm is the Pareto minimum paths set.*

Proof. Suppose $p^* \notin MP$ dominates a path p_j^M in MP, we can prove that $w^M(p^*) \leq w^M(p_j^M)$ by following the same reasoning with Lemma 9.1. This is contradictory with the fact that $p^* \notin MP$, therefore, Theorem 9.4 holds. $\quad\square$

Theorem 9.4 indicates that the proposed algorithm is fully capable of generating a Pareto minimum paths set, which resolves QSC exactly. Furthermore, we notice that if κ is set to be a larger enough value, the complete Pareto set can be thus obtained by algorithm EXACT.

9.4   Experimental results for performance evaluation

In this section we present experimental results for evaluating the performance of the proposed algorithm. FPTAS [20] is the currently best-known algorithm for MCOP problem that can be applied for solving QSC problem with minor modification. Therefore, we compared the performance of EXACT against that of the modified FPTAS through numerical experiments to validate the efficiency of EXACT.

The performance metrics used in the rest of this section for evaluating EXACT and FPTAS are defined as follows.

Definition 9.4. *Average Execution Time (AET). AET indicates the average running time of an algorithm by its 20 independent runs. This metric is used for evaluating time cost performance of an algorithm, which reflects the time that a user has to wait before receiving the response from Cloud.*

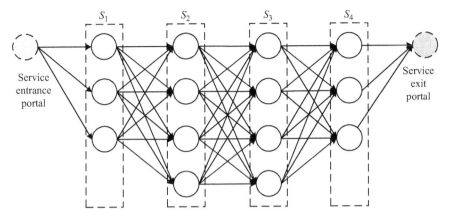

Figure 9.4   An example of generated network

Definition 9.5. *Path Weights Distance (PWD). PWD denotes the distance between the weights of a path returned by the algorithm and path constraints; that is, for a path p returned by an algorithm*

$$PWD(p) = \sqrt{\sum_{k=1}^{K} \left(1 - \frac{w_k(p)}{W_k}\right)^2}$$

This metric reflects the QoS that can be guaranteed by the service composition to end users.

We conducted experiments that generate a set of random service networks with node numbers ranged from 20 to 200. Each link in a generated network has three uniformly distributed QoS parameters. Figure 9.4 shows an example of the generated network, in which there are 14 nodes and the number of virtualized service H is equal to 4. In order to evaluate the performance under different parameter configurations, we set $H \in \{5, 10\}$ for all randomly generated networks, $\kappa \in \{1, 2, 3\}$ for EXACT, and $\varepsilon \in \{0.1, 0.5, 1.0\}$ for FPTAS. Our experiments were conducted in two ways. First, we tested the impact of parameter H variation on algorithms' performance by setting constant κ and ε for EXACT and FPTAS, respectively. Second, we examined the performance of both algorithms with different parameters settings.

In the first set of experiments, we set $\kappa = 2$ for EXACT, $\varepsilon = 0.5$ for FPTAS, then compared *AET* and *PWD* of two algorithms when parameter $H = 5$ and 10, respectively. Figure 9.5 shows the comparative results about *AET* (in millisecond) of both algorithms with different H, where the left subfigure is the *AET* of EXACT, the right one is the *AET* of FPTAS. From this figure we can see that the *AET* of two algorithms are almost unaffected by the parameter H, which follows our previous theoretical analysis that the time complexity of both algorithms are independent of H. In addition to this, we observed that EXACT runs much faster than FPTAS. As shown

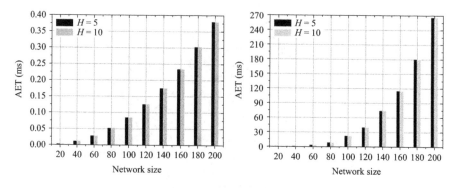

Figure 9.5   AET comparisons with different H

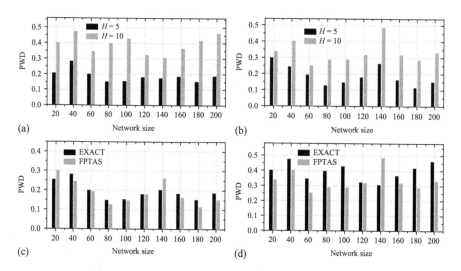

*Figure 9.6   PWD comparisons: (a) PWD comparison for EXACT with different H,
(b) PWD comparison for FPTAS with different H, (c) PWD comparison
between EXACT and FPTAS when H = 5, and (d) PWD comparison
between EXACT and FPTAS when H = 10*

in the figure, the *AET* of EXACT is close to a linear function of network size, while *AET* of FPTAS is almost an exponential function of network size. Particularly, to the case of network with 200 nodes, the EXACT runs more than 600 times faster than FPTAS. Therefore, our proposed algorithm is better than FPTAS in terms of running time.

Figure 9.6 shows the PWD comparison of both algorithms with different *H* settings, in which Figure 9.6(a) and (b) plots the *PWD* comparison for EXACT and FPTAS, respectively, when *H* varies, Figure 9.6(c) and (d) compares the *PWD* between

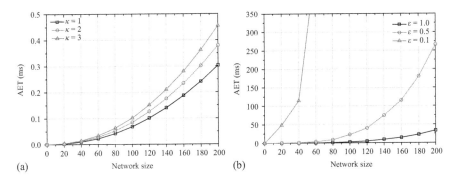

Figure 9.7   AET comparisons with different κ and ε: (a) AET comparison for EXACT with different κ and (b) AET comparison for FPTAS with different ε

EXACT and FPTAS when $H = 5$ and 10. As the definition of *PWD* indicated, the larger *PWD* is, the better path finds. Since our proposed algorithm EXACT would generate two candidate solutions, we took a greater value for comparison, i.e., $PWD = \max\{PWD(p_1^M), PWD(p_2^M)\}$ for EXACT. The results presented in Figure 9.6(a) and (b) show that both algorithms find smaller *PWD* with smaller H. This is because weights on each edge are uniformly distributed in the given intervals. The hop counts of obtained paths decrease when H becomes smaller. Moreover, we can observe from Figure 9.6(c) and (d) that *PWD* of EXACT is better than that of FPTAS except for the cases of 20 nodes when $H = 5$ and 140 nodes when $H = 5$ and $H = 10$. In other words, EXACT can find a better path than FPTAS in most cases. Based on the results shown in Figures 9.5 and 9.6, we claim that the proposed EXACT outperforms FPTAS in most cases under the parameter configurations of $\kappa = 2$ and $\varepsilon = 0.5$.

In the second set of experiments, we set $H = 10$ and tested the impact of variation in κ and ε on the performance of both algorithms, which were also measured by *AET* and *PWD*. Figure 9.7 gives the comparative results of *AET* with different κ and ε values for EXACT and FPTAS respectively, in which Figure 9.7(a) shows the *AET* of EXACT, Figure 9.7(b) draws the *AET* of FPTAS. It can be seen from this figure that *AET* of EXACT increases approximately in a linear manner when κ increases, which conforms to the theoretical analysis in Theorem 9.2. On the other hand, *AET* of FPTAS tends to increase with ε exponentially. This figure also shows that EXACT runs much faster than FPTAS even for the case of EXACT when $\kappa = 3$, FPTAS when $\varepsilon = 1.0$. It indicates that EXACT has better time performance than FPTAS.

The comparative results about *PWD* of both algorithms are shown in Figure 9.8, in which we took the greatest *PWD* again for EXACT's results presentation. Figure 9.8(a) and (b) gives the *PWD* comparison for EXACT with different κ and FPTAS with various ε, respectively, Figure 9.8(c) and (d) compares the *PWD* between EXACT and FPTAS in the condition of $\kappa = 1, 2, \varepsilon = 1.0$ and $\kappa = 3, \varepsilon = 0.1, 0.5$. As can be seen clearly in Figure 9.8(a), the greater κ sets, the greater *PWD* finds, i.e., EXACT finds better path with increasing of κ. Similarly, Figure 9.8(b) indicates that FPTAS is able

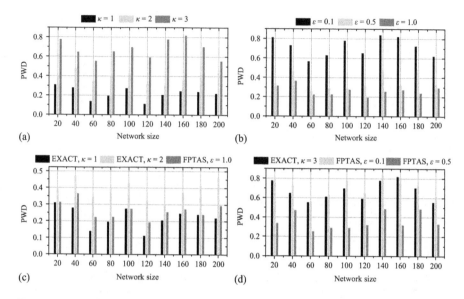

Figure 9.8   PWD comparisons: (a) PWD comparison for EXACT with different κ, (b) PWD comparison for FPTAS with different ε, (c) PWD comparison between EXACT and FPTAS when κ = 1, 2, ε = 1.0, and (d) PWD comparison between EXACT and FPTAS when κ = 3, ε = 0.5, 0.1

to obtain better path when ε tends to be small. Observations on $\kappa = 1$ and $\varepsilon = 1.0$ data in Figure 9.8(c) show that EXACT can find a better path than FPTAS; that is to say, in this case EXACT is superior to FPTAS in both running time and finding path. Also, we can see from Figure 9.8(d) that *PWD*s of EXACT are greater than those of FPTAS when $\kappa = 3$ and $\varepsilon = 0.5$, but they are smaller than those of FPTAS when $\kappa = 3$ and $\varepsilon = 0.1$. This implies that the *PWD* of EXACT tends to increase with κ increasing. Although *PWD*s of EXACT with $\kappa = 3$ setting are little worse than that of FPTAS with $\varepsilon = 0.1$, EXACT takes much less time than FPTAS. One can expect that EXACT could find a better path than FPTAS with increasing of κ while taking less amount of running time. Therefore, we claim that our proposed algorithm has better performance than FPTAS both in *AET* and *PWD*.

9.5   Conclusion

In this chapter, we discussed the significant role of the emerging Network-as-a-Service (NaaS) paradigm in next generation Internet for enabling the transform of Internet from a data communication infrastructure to a general platform for diverse service provisioning. Through virtualization and abstraction of network resources and capabilities, NaaS provides a promising mechanism to bridge the gap between networking and Cloud computing, which leads to composite network–Cloud service provisioning.

QoS-aware service composition plays a crucial role in such a service environment for meeting various user requirements. In this chapter, we presented a model and formulation for the QoS-aware network–Cloud service composition problem and described an algorithm for solving this problem. Theoretical analysis and experimental evaluation show that the proposed algorithm is effective and efficient in selecting services for composite network–Cloud service provisioning.

References

[1]   I. Foster, Y. Zhao, I. Raicu, and S. Lu, "Cloud computing and grid computing 360-degree compared," in *Proceedings of GCE'08*, Nov. 2008, pp. 1–10.

[2]   M. Armbrust, A. Fox, R. Griffith, *et al.* "A view of cloud computing," *Communications of the ACM*, vol. 53, no. 4, pp. 50–58, 2010.

[3]   K.-T. Chen, Y.-C. Chang, H.-J. Hsu, D.-Y. Chen, C.-Y. Huang, and C.-H. Hsu, "On the quality of service of cloud gaming systems," *IEEE Transactions on Multimedia*, vol. 16, no. 2, pp. 480–495, Feb. 2014.

[4]   W.-H. Hsu and C.-H. Lo, "QoS/QoE mapping and adjustment model in the cloud-based multimedia infrastructure," *IEEE Systems Journal*, vol. 8, no. 1, pp. 247–255, Mar. 2014.

[5]   K. R. Jackson, K. Muriki, S. Canon, S. Cholia, and J. Shalf, "Performance analysis of high performance computing applications on the Amazon Web services Cloud," in *Proceedings of the 2010 IEEE International Conference on Cloud Computing Technology and Science*, Nov. 2010, pp. 159–168.

[6]   Q. Duan, "Modeling and delay analysis for converged network-cloud service provisioning systems," in *The 2013 International Conference on Computing, Networking and Communications (ICNC2013)*, 2013, pp. 66–70.

[7]   S. S. Woo and J. Mirkovic, "Optimal application allocation on multiple public Clouds," *Computer Networks Journal*, vol. 68, pp. 138–148, 2014.

[8]   M. Fiorani, S. Aleksic, P. Monti, J. Chen, M. Casoni, and L. Wosinska, "Energy efficiency of an integrated intra-data-center and core network with edge caching," *IEEE/OSA Journal of Optical Communications and Networking*, vol. 6, no. 4, pp. 421–432, Apr. 2014.

[9]   C. Assi, S. Ayoubi, S. Sebbah, and K. Shaban, "Towards scalable traffic management in cloud data centers," *IEEE Transactions on Communications*, vol. 62, no. 3, pp. 1033–1045, Mar. 2014.

[10] C. Shan, C. Heng, and Z. Xianjun, "Inter-Cloud operations via NGSON," *IEEE Communications Magazine*, vol. 50, no. 1, pp. 82–89, 2012.

[11] Y. Xia, M. Zhou, X. Luo, Q. Zhu, J. Li, and Y. Huang, "Stochastic modeling and quality evaluation of infrastructure-as-a-service clouds," *IEEE Transactions on Automation Science and Engineering*, vol. 12, no. 1, pp. 162–170, 2015. [Online]. Available: http://dx.doi.org/10.1109/TASE.2013.2276477

[12] J. Huang, Q. Duan, S. Guo, Y. Yan, and S. Yu, "Converged network–cloud service composition with end-to-end performance guarantee," *IEEE Transactions on Cloud Computing*, 2015. DOI: 10.1109/TCC.2015.2491939.

[13]   Amazon, Elastic Compute Cloud, http://aws.amazon.com/ec2/.

[14]   Q. Duan, Y. Yan, and A. V. Vasilakos, "A survey on service-oriented network virtualization toward convergence of networking and Cloud computing," *IEEE Transactions on Network and Service Management*, vol. 9, no. 4, pp. 373–392, 2012.

[15]   N. M. M. K. Chowdhury and R. Boutaba, "A survey of network virtualization," *Computer Networks*, vol. 54, no. 5, pp. 862–876, 2010.

[16]   ETSI NFV Industry Specification Group (NFV-ISG), "Network Function Virtualization (NFV) – network operator perspectives on industry progress," white paper published in Oct. 2014.

[17]   Q. Duan, "Modeling and performance analysis on network virtualization for composite network-cloud service provisioning," in *Proceedings of 2011 IEEE World Congress on Services*, Aug. 2011, pp. 548–555.

[18]   J. Huang, G. Liu, and Q. Duan, "On modeling and optimization for composite network–cloud service provisioning," *Journal of Network and Computer Applications*, vol. 54, pp. 35–43, 2014.

[19]   M. M. J. Hershberger and S. Suri, "Finding the *k* shortest simple paths: a new algorithm and its implementation," *ACM Transactions on Algorithms*, vol. 3, no. 4, pp. 851–864, 2007.

[20]   J. Huang and Y. Tanaka, "QoS routing algorithms using fully polynomial time approximation scheme," in *Proceedings of the ACM/IEEE International Workshop on Quality of Service 2011*, Jun. 2011, pp. 1–3.

Chapter 10

NaaS-enabled service composition
in SDN environment

Yongyi Ran[1], Jian Yang[1], Enzhong Yang[1], and
Shuangwu Chen[1]

10.1 Introduction

Service-Oriented Architecture (SOA), service composition, and Next Generation Service Overlay Network (NGSON) have been intensively investigated in the past decade, producing a plenty of contributions concerning novel models and solutions [1,2]. As the emerging fundamental advances in Everything-as-a-Service (XaaS), cloud computing and Software Defined Networking (SDN), service composition also evolves and encounters several challenges. One important concern is to provide Network as a Service (NaaS) in terms of network connectivity, virtual private networks, bandwidth on-demand as well as routing in the context of service composition.

Most of previous research [1–3] related to service composition have been conducted by assuming that the service composition process does not involve the network resource of routers and switches, only considering the application-level services residing in computing devices. Consequently, previous contributions are predominantly devoted to the solution for composing multiple available computation-intensive services concerning node mapping, composability, Quality-of-Service (QoS) awareness, and load balancing. Although the simple connectivity may be considered in the context of traditional service composition, customized network provisioning between compute-related components has not been fully investigated, and still remains challenging due to imposing a fundamental network upgrade upon the existing network infrastructure.

In the conventional networks having decentralized structure, the control plane is coupled with the data planes, both of them coexisting in the same device [4]. Furthermore, the network devices are closed systems, which are bound to specific products and their versions. That is to say, the operation of the products from different vendors relies on their vendor-dependent configuration and management through private low-level commands. This eventually induces an intolerable cycle for updating/upgrading the network devices, in turn ossifying the network as well as negating

[1]Department of Automation, University of Science and Technology of China, China

network evolution and its innovation. All of these obviously pose severe restrictions to the implementation of NaaS-enabled service composition. SDN [5] as an emerging networking paradigm creates an opportunity for solving these problems. The core mechanism of SDN is separating the physical binding of the control plane and the forwarding plane. Specifically, a centralized network controller is built by extracting control of each network device and fusing them together. In order to enable the network controller to efficiently manage and manipulate the network devices, a network operating system (NOS) is necessary and deployed to provide the basic functions of dynamically configuring the network devices and acquiring the network information for constructing an abstract network model.

Decoupling the control plane with the data plane is well achieved by defining programming interface for the controller to interact with the switches. OpenFlow is a well-known available protocol stack [6], which defines a standard invoking interface for controlling forwarding plane in the context of SDN. It provides the capability of identifying, manipulating, and isolating the flows passing switches (routers). This capability further facilitates network virtualization that enables the physical network resources shared in the form of independent virtualized networks. These fundamental natures of OpenFlow are beneficial for achieving a holistic service composition where network connectivity, virtual private network, bandwidth on-demand as well as routing are encapsulated as basic service components.

Due to the powerful computing paradigm of cloud computing, many computation-intensive service components are implemented in cloud computing center. Against this background, it is highly desirable to integrate networking and computing resources as a set of virtualized resources, for the sake of coordinately managing, controlling, aggregating, and optimizing utilization of joint resource of network and computing domains. Hence, merging networking services and cloud computing services triggers a new important topic of NaaS-enabled service composition [7]. Naturally, considering the transmission schema and routing rules in the service composition, rather than the connectivity between the compute-related components, may induce additional complexity.

In addition, to reduce the complexity of the service management, a logical and abstract method for describing services, network topology, workflows, etc. is widely applied by service providers. However, the physical network or servers cannot operate directly with this logical-level approach. Hence, the logical description should be translated into the script executable for the substrate layer. For instance, the multicasting with a multicast tree can be mapped into specific forwarding configuration of the OpenFlow switches.

Given the challenges described above, this chapter focuses on conceiving a service composition framework supporting NaaS with the aid of software-defined networking. Specifically, we present the framework and software necessary for implementing the NaaS-enabled service composition. It provides a comprehensive solution for jointly utilizing the resource of the networking and computing domain, which relies on the techniques of coordinate management, aggregation, controlling, and optimization. Then, several specific mechanisms are induced in the framework for the sake of providing *multipath*, *multicast*, and *multi-domain* routing service as a service component. A prototype system is constructed by integrating OpenFlow network of moderate scale and OpenStack. Experiments based on this prototype are conducted to demonstrate the applicability of the conceived framework.

The remaining of this chapter is structured as follows. Service composition with the aid of SDN is discussed in Section 10.2. In Section 10.3, the NaaS-enabled framework is conceived for implementing service composition, which elaborates the framework, software tools as well as design for encapsulating a network service as a basic service component. Section 10.4 presents two use cases to show the success of the conceived framework. Finally, Section 10.5 offers the conclusion of this chapter.

10.2 Overview of service composition in SDN

The basic idea of service composition is to compose atomic service components automatically into diverse complex services based on dynamic service requirements. Many key aspects have been focused in the previous works, e.g.: composability [8], QoS-awareness [9], function mapping [10], and load balancing [3,11], etc. In the context of cloud computing, the elasticity of cloud resources benefits a lot to both SaaS service providers and common users who instantiates their services in cloud computing based on service-oriented architecture (SOA). Ye *et al.* [12] proposed a cloud service composition framework that selects the optimal composition based on an end user's long-term QoS requirements. The authors in [13] proposed a network-aware approach to service composition in a fat-tree cloud data center, consisting of a network resource consumption function, a network QoS computation, and a network-aware service composition model. Recently, network function virtualization (NFV) [14] emerges as a solution to provide flexible allocation of network resources and accelerate the integration of heterogeneous network architecture and services. For a specific network service, the operators only need to specify a chain of Virtual Network Functions (VNFs) and the data flows within that service chain. There exists a plethora of NFV-related research such as unified control and optimization framework [15], VNF placement [16], VNF scheduling and resource optimization [17] as well as managing service chain [18].

Overall, most of the previous works only simply consider the network connectivity and network capacity in a service composition processing, the network transmission schema and routing policies are never considered as service components being composed into a compositional service. Actually, it is challenging in traditional networks to flexibly provide network services due to the drawbacks of fundamental infrastructure.

10.2.1 SDN and cloud in substrate layer

Recent years have witnessed a widespread increase of interest in Software Defined Networking (SDN), which is a promising network technology for building next-generation networks and network services. As described in Section 10.1, OpenFlow is proposed as a standard communication interface defined between the central control and the forwarding plane in an SDN architecture [5]. The natures of OpenFlow enable it to provide network connectivity, virtual private network, bandwidth on-demand as well as routing as a service for upper-layer users. In this chapter, therefore, OpenFlow

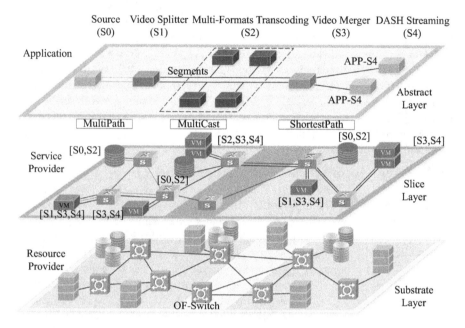

Figure 10.1   The layers of service composition in an SDN environment

is utilized to construct a software-defined network as the substrate layer shown in Figure 10.1. In order to make it more convenient to achieve network virtualization and provide network resource in the form of network slice [19], we employ FlowVisor [19] as an OpenFlow hypervisor enabling multiple SDN controllers to share the same physical network and network resources. The isolation features of FlowVisor comprise topology, bandwidth, traffic, and forwarding tables. More details about FlowVisor is described in [19]. Furthermore, OpenStack [20] is utilized to manage the compute and storage resources in our proposed framework and prototype.

With the aid of SDN technology and OpenStack, the resource provider in the substrate layer illustrated in Figure 10.1 can provide the compute-related components to the service provider in the form of virtual machines and can supply a network slice with an SDN controller to achieve the connectivity and routing between the compute-related components. In order to offer an integrated resource slice with both computing and network resources, we jointly invoke the application programming interfaces (APIs) of FlowVisor, OpenFlow and OpenStack. Finally, a slice layer is partitioned and isolated for service provider (as shown in Figure 10.1). How to share the resources of substrate layer and provide a virtual network for upper-layer users is also related to Virtual Network Embedding (VNE), which is widely studied in recent years [21–25].

10.2.2   NaaS in slice layer

As described above, resource provider can provide an integrated resource slice for each service composition request to service provider, while the service provider manages all the slices in the slice layer. The elements in slice layer mainly include abstract compute-related components and abstract network service components among the abstract compute-related components.

When applying SOA in the field of networking, the NaaS paradigm emerges, through which functionalities of various networking systems can be exposed and utilized as network services. NaaS supports encapsulation and virtualization of networking resources in the form of SOA-compliant networking services, which can be composed with computing services in a Cloud-SDN environment. Generally, a network service may denote any type of networking component at different levels, such as network connectivity, virtual networks, and network functions. In this chapter, we mainly discuss the following three categories.

10.2.2.1 Network connectivity

Resource provider in the substrate layer shown in Figure 10.1 offers a wide range of NaaS technologies, but network connectivity with on-demand bandwidth are especially common [26], which can provide the basic network path among computing components in a service composition processing. Additionally, we can determine the characteristics of the connection like security, reliability, and so on. Bandwidth on-demand is a service that allocates network resources based on the needs of particular devices on that network. More bandwidth is allocated to nodes experiencing peak demands, and, as demand subsides, bandwidth is scaled down.

10.2.2.2 Routing rules and optimization

Data traffics among compute-related components are common for a compositional service, thus it is essential to design appropriate transmission schema and routing rules for such data traffics with the purpose of improving network utility and efficiency of the compositional service. As illustrated in Figure 10.1, a DASH (Dynamic Adaptive Streaming over HTTP [27,28]) application is used as an example. In this application, the elementary compute-related components comprise video splitting, multi-formats transcoding, video merging as well as DASH streaming. Obviously, special network services can be employed among compute nodes to guarantee QoS and improve network utility. In particular, *multipath* can be applied between the video source and the video splitter to decrease the transmission latency, *multicast* can be used between the video splitter and the transcoding nodes to save the network bandwidth, also, *shortest-path* can be abstracted as a service between two compute nodes to reduce the transmission time.

10.2.2.3 Network functions

Generally, network functions refer to the edge network functions, including deep packet inspection [29], load balancing [30], firewall [31], etc. Recently, many studies have been conducted in the field of Network Functions Virtualization (NFV) [32,33] in both academia and industry. However, this is out of the scope of this chapter, we focus more on how to provide network services such as transmission schema, routing rules between compute-related components to support a more flexible and comprehensive service composition.

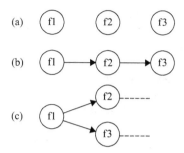

Figure 10.2   Examples of service request graph

10.2.3   Service request graph in abstract layer

By running different compute-related components in a composable way, more complex services could be composed and provided. In particular, the compute-related components would be connected by different network services. As shown in Figure 10.1, a service composition request can be submitted in the form of service request graph (as shown in Figure 10.2) to our proposed NaaS-enabled service composition system.

As illustrated in Figure 10.2, a service request graph contains one or more compute-related components as well as their network support as described in Section 10.2.2. Here we propose three basic types of service request graphs as follows. Moreover, we can obtain more complicated graphs by combining these three basic ones.

- *Separate graph*: In Figure 10.2(a), three compute-related components could be instantiated and executed independently and parallelly. There is no data traffic or control signal between each other. In this context, no special network service is needed.
- *Sequential graph*: In Figure 10.2(b), three compute-related components will be instantiated and executed sequentially. The function $f2$ cannot be started before $f1$ finishes its work. The solid arrow line represents a signal or data traffic between two components. The network service "ShortestPath" can be specified when such a request is submitted.
- *Branch graph*: In Figure 10.2(c), there exists a branch along the workflow, which means that the component $f1$ will trigger two other computing component after finishing its own work. "Multicast" can be specified when we submit such a request.

10.2.4   Service composition

First of all, service components should be well-designed before composition. Loose coupling and reuse are most common characteristics associated with SOA. However, even the most reusable service component would not be available, if this component cannot be discovered by those who are responsible for creating compositional services. In addition, if the most loosely coupled service component cannot be assembled into an effective composition, it would also have limits on reuse potential. Therefore, the principles of service discoverability and service composability come into play in the context of service composition.

The natures of discoverability basically assist SOA in avoiding the casual creation of redundant services or services that implement redundant logic, and enable the service components to become as discoverable as possible, regardless of the implementation environment. Composability is simply a form of reuse where a service component should be designed in a standardized way to maximize the composition opportunities.

In this chapter, as illustrated in Figure 10.1, the slice layer, which maintains the information of service components and network slice, is owned and managed by the service provider. When a specific application is launched in the form of request graph as described in Section 10.2.3, the service provider is to find the optimal compute-related components first, and then will design a proper transmission schema and routing policies as network services according to the submitted request graph and stitch these network services with compute-related components. For the sake of optimization, the service provider also can jointly consider the availability of compute components and the network resources.

To illustrate the basic principle of service composition with NaaS more clearly, we take a DASH-based streaming service as an example, as shown in Figure 10.1. The elementary compute-related components comprise S0 – video storage, S1 – video splitting, S2 – multi-formats transcoding, S3 – video merging as well as S4 – video streaming. The abstract service components {S0,S1,S2,S3,S4} are managed by the slice layer, and will be instantiated in the substrate layer when an application is launched. For the sake of improving network utility while guaranteeing the quality of transmission, *multipath* can be applied between the video source (storage) and the video splitter to decrease the transmission latency, *multicasting* can be employed between the video splitter and the transcoding nodes to reduce the bandwidth waste, also, *shortest-path* can be abstracted as a network service which can be placed between the video merger and the streaming servers to reduce the transmission delay. When the service provider intends to provide such a service, a service request graph as illustrated in the abstract layer in Figure 10.1 should be submitted first. Afterwards, the slice layer will select appropriate abstract compute components and design networking services for this service request graph, and then request the corresponding computing and network resource from the substrate layer to instantiate this requested compositional service.

10.3 NaaS-enabled framework for service composition

In this section, we present how to achieve the functions of the three layers described in Section 10.2 by proposing a framework of NaaS-enabled service composition.

10.3.1 Framework of service and resource management

As shown in Figure 10.3, the framework of service and resource management for service composition in SDN and cloud environment is designed. Two big parts are marked

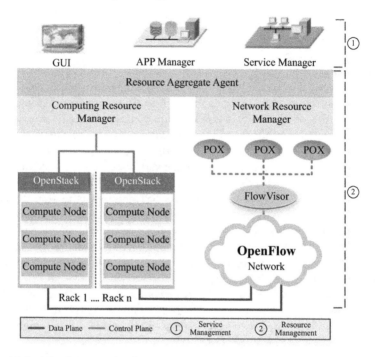

Figure 10.3   The framework of service and resource management for service composition in SDN

in the figure, the first part is for service management and the second part is for resource management. The resource provider is responsible for resource management, and manages the computing resource by OpenStack and network resource by OpenFlow, OpenVisor and POX. While the service provider is in charge of service management, and manages service components and compositional services/applications. Resource management owns a module named "Resource Aggregate Agent" which provides APIs for service management to instantiate service components in physical layer.

To present the framework in more detail, we illustrate the software modules of the proposed framework in Figure 10.4. First, in order to virtualize the computing and network resources in the substrate layer, we apply OpenStack to manage the physical servers and instantiate virtual machines with virtual internal network among them. At the same time, the switches and routers in the substrate layer are all OpenFlow-enabled, and FlowVisor is utilized to virtualize the physical network into network slices and acts as the hypervisor of OpenFlow-enabled switches and routers. In addition, for the sake of monitoring the status of system, Nagios is employed to measure the common performance metrics, such as the utilization of CPU, memory and disk. The APIs of OpenStack, FlowVisor, OpenFlow and Nagios are all encapsulated in the module "Resource Handler API" which will be used by upper-layer modules. Second,

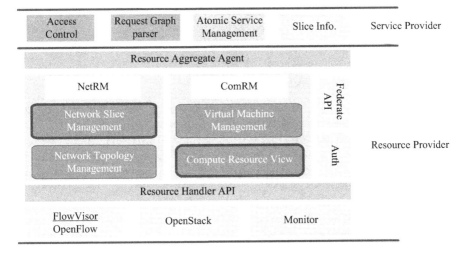

Figure 10.4   The software modules of the proposed framework

in order for resource provider to supply integrated resource slices with computing resource and network resource to service provider, we design two major modules – "NetRM" and "ComRM" to manage and instantiate compute-related components and network service components respectively, where "NetRM" stands for Network Resource Manager including two sub-modules – "Network Slice Management" and "Network Topology Management", and "ComRM" stands for Computing Resource Manager including "Virtual Machine Management" and "Compute Resource View". Furthermore, the module "Auth" is designed to testify the authority of using the substrate resources, and the module "Federate-API" is to share infrastructure resources crossing multi-domains. Third, to provide unified APIs for service provider to instantiate a request graph (i.e., a service composition request), the module "Resource Aggregate Agent" is designed.

The service provider is responsible for processing a service composition request and making this request instantiated in physical infrastructure. As shown in Figure 10.4, the module "Access Control" is designed to decide whether the service composition request can be accepted or not. "Request Graph Parser" module is to parse the submitted request graph (as described in Section 10.2.3, a service composition request is submitted in the form of request graph) to derive the requested compute-related components and the network dependencies among these compute-related components. Afterwards, the module "Atomic Service Management" will be executed to map the abstract service components parsed by "Request Graph Parser" to the corresponding concrete computing resource and transmission schema/routing policies via invoking the APIs provided by "Resource Aggregate Agent". Once the instantiation being completed by resource provider in the form of a resource slice, the information of such a slice will be managed in the module "Slice Info".

To facilitate the utilization of NaaS in service composition, we present several special designs for commonly used network services based on the proposed framework, including *multipath, multicast* as well as *multi-domain* routing services.

10.3.2 Design of multipath

Multipath routing (like ECMP) is capable of improving end-to-end transmission throughput and transmission reliability as well as reducing end-to-end transmission delay. As illustrated in Figure 10.1, for the DASH application, we can transmit the original video content to the video splitter via multiple paths to save transmission time. In this subsection, we present how to modify the network design of OpenStack to support the network service – "multipath routing" more flexibly.

When employing OpenStack to establish a cloud environment, one controller, many compute nodes as well as several services (nova, glance, neutron, etc.) should be deployed. Advanced Message Queuing Protocol (AMQP) is used as the communication mechanism among these nodes. Neutron supplies a way to alleviate the stress on the network to make it easier to provide NaaS in a cloud environment. Therefore, the most common NaaS comes from the Neutron of OpenStack, which let cloud users create network services at Level 2 or Level 3 and then associate their virtual machine (VM) instances with these network services. We know that Neutron itself is very flexible, but there are still some restrictions on utilizing multipath routing in internal networks of OpenStack. Here we list three main aspects: (1) two virtual Network Interface Cards (vNICs) which belong to the same virtual machine cannot be bind to the same internal network, (2) neutron assumes that only one physical interface on each compute node is used to access all networks, (3) and the number of interfaces of virtual machine are fixed and cannot be added or removed once the virtual machine is instantiated.

In this chapter, by taking some ideas from OVS-Plugin [34] and NEuca (a Quantum Plugin) [35] of ExoGENI, a multiple data plane is proposed to support multipath routing for service composition in an SDN and cloud based environment. As illustrated in the dashed box of Figure 10.5, the total number of physical links from the NICs of the host to the edge OpenFlow-enabled Switch depends on the hardware parameters (the number of NICs, the number of ports on switch, etc.). However, we can dynamically associate multiple physical interface with the OVS-Switch when OVS-Plugin is installed and used.

When a compute-related component should be instantiated in a certain OpenStack rack due to the upper layer request, OpenStack will firstly create an internal network with an IP address range and a VLAN ID (VID). Afterwards, a virtual machine (VM) will be instantiated to support this compute-related component, and this VM will be attached to this internal work. In our proposed design, we can bind multiple vNICs belonging to the same VM to the same internal network to support multipath routing between different VMs. At the same time, the vNICs can be newly added and removed dynamically. To do so, the "Agent" in Figure 10.5 should create another one or more virtual interfaces for one VM and associate these virtual interfaces with the OVS-Switch ports by using the same VLAN ID. Then, an injected daemon, which

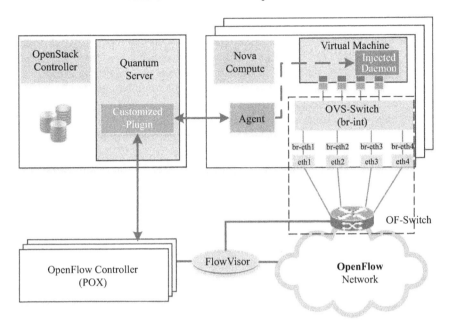

Figure 10.5   A design of multiple data plane to support multipath routing

is executed on each VM, will be run to configure the internal interfaces of this VM. Furthermore, OVS-Switch will be configured to forward data packets between the virtual interfaces and the external networks.

As described before, once instantiating and allocating VMs for each compute-related component in OpenStack-based cloud environment, the module "ComRM" will return the information of VMs and their network locations to the module "Resource Aggregate Agent", then "Resource Aggregate Agent" sends the information and the network service requirements from service provider to the module "NetRM" to isolate a network slice. Finally, "Resource Aggregate Agent" will aggregate the compute nodes (in the form of VMs) with network slice to get an integrated resource slice and send the final slice to the service provider. Once service provider gets this resource slice, the compositional service can be started.

10.3.3   Multicast in SDN

Multicast is a typical one-to-many transmission paradigm, which is commonly used in modern cloud computing data centers. The natures of multicast benefit a lot to reducing the load of sender and saving network bandwidth. For example, in data centers, multicast always can save the inter-rack bandwidth because chunk replication is usually bandwidth hungry. In addition, multicast can also accelerate the delivery of OS images and decrease the execution time of this type of tasks. Therefore, we present how to implement the multicast service as a service component for service composition in SDN in this section.

Taking the DASH application in Figure 10.1 as an example again, to achieve adaptive video streaming, we should prepare different formats with different resolutions for the same one video, thus transcoding is essential. In order to speed up the process of transcoding, we can send the same video segment to different computing nodes to transcode the video into different formats. As illustrated in Figure 10.1, the video splitter (S1) divides the video content into many segments, and then each video segment is delivered to several transcoding nodes (S2) via multicast routing. To instantiate such a multicast network service, the module "ComRM" will map the compute nodes into the OpenStack racks first, and then the multicast tree among these compute components is calculated by "multicast service component" implemented in the module "Atomic Service Management" according to the compute node information from "ComRM" and network topology information from "NetRM", and finally the module "NetRM" isolates a network slice with multicast routing rules by using the calculated information of "multicast service component".

Each divided video segment can be sent by a separate User Datagram Protocol (UDP) socket to the network. Thus, the number of UDP flows is identical to the number of video segments. Intuitively, we can allocate a logical ID (LID) to each UDP flow associated with a video segment to identify the flow and simplify the management. However, OpenFlow switch cannot identify a LID easily due to its fixed match-action tables. In our proposed multicast mechanism, LID is trickily mapped into a two-tuples (Src IP, Src Pt), where "Src IP" is source IP and "Src Pt" indicates source port. To achieve this, we can use the APIs of SDN controller to insert a new entry into the flow tables of OpenFlow switches, this entry can make the UDP flows of each segment be forwarded along the multicast tree via filtering (Src IP, Src Pt). It can be found that only source IP and source port are used to identify multicast flows on the OpenFlow switches, which means that the destination IP field of all packets of a segment would be set as the same IP value (this IP may be one of the destination IPs) during multicast transmission over OpenFlow network. Obviously, we cannot concurrently write all the destination IP addresses into the packet header. In this case, when packets reach the edge switches, they cannot be forwarded to the clients correctly due to a wrong destination IP. Thus, at the edge switches, we should add a Modify-Field action to filter the specific multicast packets and then modify the field of destination IP and port according to the IP address and port of the destination clients.

UDP has no retransmission mechanism, so we should take extra steps to achieve a reliable multicast service based on UDP. We can use the unique sequence number in UDP header to recognize packet loss. If there exists packet loss when the transmission of a video segment is completed, the client can issue retransmission requests to source node to obtain the lost packets, which will be retransmitted by TCP unicast.

Based on the design described above, a generic multicast function can be implemented in the modules "Atomic Service Management" (as shown in Figure 10.4). When a multicast paradigm is required among computing nodes (as the example in Figure 10.1) in a service composition process, the interfaces in the modules "Atomic Service Management" could be invoked to easily and quickly achieve a multicast as a service component.

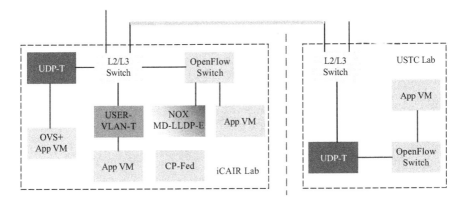

Figure 10.6   Layer-2 connection across two different domains

10.3.4   Multi-domain connectivity

In the context of multi-domains, a layer-2 connection across two different domains over the existing Internet is usually needed. If we can provide a layer-2 connection as a network service, this will benefit a lot to two compute components located in two different domains. In this section, we present how to provide a layer-2 network connection as a network service for service composition.

In this chapter, an open source code EthUDP [36] is applied to create transparent tunnel over IPv4/IPv6 network. First, we need two servers with two NICs each. The two NICs are connected to the Internet and the OpenFlow-enabled network respectively. Second, to run the EthUDP program, two parameters *<VID,Port>* should be set to create a VLAN-based layer-2 network connection. More detail can be found at the website [36]. EthUDP also has been used to interconnect the international networks between the Lab of University of Science and Technology of China (USTC) and the Lab of Northwestern University (United States) as illustrated in Figure 10.6.

Similarly, when a layer-2 network connectivity is needed in a service composition, the module "Atomic Service Management" will deploy the software-based UDP-T dynamically. For example, in Figure 10.1, the Video Merger (S3) could use the multi-domain connectivity to transfer the transcoded video content to a video streaming server located in a different domain.

10.4   Use cases

In order to verify the feasibility of our proposed NaaS-enabled service composition framework, two use cases with prototype system and some preliminary results are described as follows.

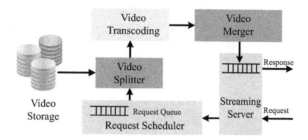

Figure 10.7   The architecture of DASH-based streaming service

10.4.1   Use case 1: DASH-based streaming service

The first example is deploying DASH-based streaming service in an SDN-Cloud environment.

10.4.1.1   Prototype system

Firstly, we implement a prototype system to support DASH-based streaming application over OpenFlow-enabled networks as shown in Figure 10.1. To establish an economical OpenFlow-enabled network, we apply Pantou [37] to rebuild the commercial NETGEAR 3800 switches to enable OpenFlow, and use 16 such OpenFlow-enabled switches to set up a OpenFlow-enabled network. Secondly, POX [38] is used as the SDN controller. We deploy POX, FlowVisor as well as Management Module on a physical server with Ubuntu 12.04 where the basic configuration of this server is: two Intel Xeon E5-2620 CPUs, 64 GB RAM and three NICs. We connect one NIC to a TP-Link switch (No.: TL-SG1024DT), and then connect this TP-Link switch to the console port of all NETGEAR 3800 switches (OpenFlow-enabled) to form an out-of-band control network (IP address 192.168.0.0/24). In order to enable this server to communicate with OpenStack nodes, another NIC is connected to the top of rack switches. Afterwards, OpenStack all-in-one package is applied to install OpenStack on six physical servers with Ubuntu 12.04 to enable instantiating compute components, and Python is then employed to develop the software modules shown in Figures 10.3 and 10.4.

After having the SDN and Cloud environment, we can deploy the DASH streaming application over such an environment in a compositional way as illustrated in Figure 10.1. The more detailed architecture of the DASH-based streaming application is illustrated in Figure 10.7. The elementary compute-related components for this DASH streaming application comprise S0 – video storage, S1 – video splitting, S2 – multi-formats transcoding, S3 – video merging as well as S4 – video streaming. The abstract service components {S0,S1,S2,S3,S4} are managed by the slice layer, and will be instantiated in the substrate layer when an application is launched. To implement video splitting, transcoding and merging, we use FFmpeg [39] here to process video contents. For the sake of improving network utility while guaranteeing the quality of transmission, *multipath* can be applied between the video source (storage) and the video splitter to decrease the transmission latency, *multicasting* can be

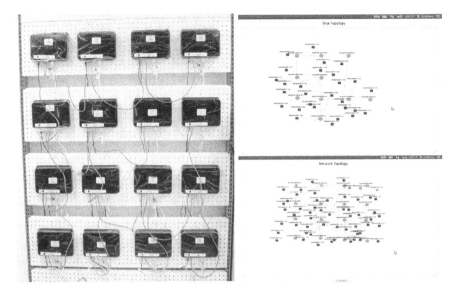

Figure 10.8 Pictures of our prototype system

employed between the video splitter and the transcoding nodes to reduce the bandwidth waste, also, *shortest-path* can be abstracted as a network service which can be placed between the video merger and the streaming servers to reduce the transmission delay. The left part of Figure 10.8 shows the physical devices we used, and the left part illustrate the substrate layer topology and the slice layer topology.

10.4.1.2 Preliminary result

Based on the above prototype system, we carried out several experiments to evaluate the effectiveness and performance of our proposed NaaS-enabled service composition framework in terms of the bandwidth utilization of the source node and the overall network with and without multicast service. The preliminary results are shown in Figures 10.9 and 10.10. It can be observed that enabling proper NaaS in a compositional service can achieve a better networking utility.

10.4.2 Use case 2: dynamic switching for multi-tree multicast

Besides providing a multicast path/tree for a one-to-many communication pattern, we also can integrate a routing optimization into this basic multicast network service, i.e., a multiple multicast tree switching mechanism [40].

10.4.2.1 Prototype system

As mentioned before, multicast is a typical one-to-many transmission paradigm, which is commonly used in modern cloud data centers. The natures of multicast benefit a lot to reducing the load of sender and saving network bandwidth. For example, distributed file system is widely used in data centers, such as GFS in Google,

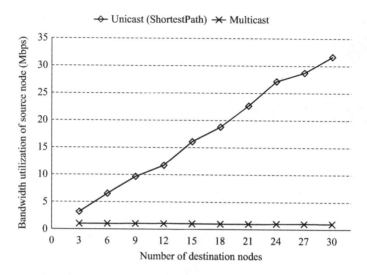

Figure 10.9   Bandwidth utilization of the source node

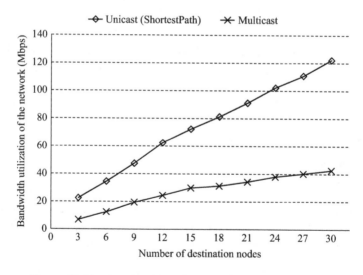

Figure 10.10   Bandwidth utilization of the overall network

HDFS in Hadoop, and COSMOS in Microsoft. Files are divided into many fixed-size chunks, like, 64 MB or 100 MB. Each chunk is replicated to several copies and stored in servers located in different racks to improve the reliability. In this case, multicast can save the inter-rack bandwidth because chunk replication is usually bandwidth hungry. In our first use case, multicast is also applied to accelerate the delivery of the video segments and save the bandwidth.

However, in data center networks, multicast paradigm with single multicast tree generally cannot cope with network congestion and failure well. Multi-tree multicast

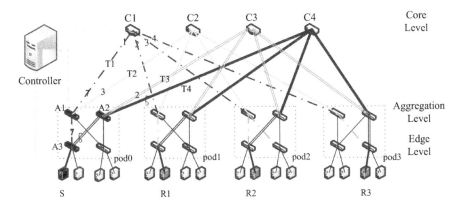

Figure 10.11   OpenFlow-based fat-tree data center network

is able to achieve load balance and failure recovery by applying dynamic multi-tree switching mechanisms, but it is hard to be deployed and implemented in the traditional networks. OpenFlow separates the control plane and data plane, and thus has better controllability and programmability. Therefore, here such a multi-tree switching mechanism is abstracted as a network service for multicast in OpenFlow based data center with a fat-tree architecture, shown as Figure10.11. The same way as in the first use case can be used to build this prototype system.

To achieve such a network service, the group table with group type "select" defined in OpenFlow protocol [41] is used to select different multicast tree dynamically. For each multicast session, we select several multicast trees as candidates and calculate a priority-level value for each multicast tree by using the monitored network status from OpenFlow controller. This priority-level value can be employed to configure the weight of action bucket in group table. Each multicast tree corresponds to a weight and a bucket. Then the select algorithm in group table can choose a proper multicast tree according to the weight values of action buckets. Of course, we can also design different select algorithm to achieve different optimization objective. Here, we adopt a round robin algorithm and an adaptive-filter [42] based prediction algorithm. For the latter, adaptive-filter is used to predict a priority-level value for each candidate tree.

10.4.2.2   Preliminary result

To evaluate this routing optimization service in a service composition process, we integrate this multi-tree multicast mechanism into the first use case to deliver the video chunks (the multicast stage shown in Figure 10.1) in a prototype of fat-tree data center network. In Figure 10.11, the node S is the source node, and the nodes $\{R1, R2, R3\}$ are destination nodes. The dynamic switching algorithm can be deployed and executed on switches $\{A1, A2, A3\}$. Here we compared three different mechanism for this multicast service. The first one is single multicast tree with Shortest Path Tree (SPT-Single Tree), the second one is multi-tree multicast with round robin algorithm (Round-Robin), and the last one is multi-tree multicast with adaptive-filter algorithm (AF).

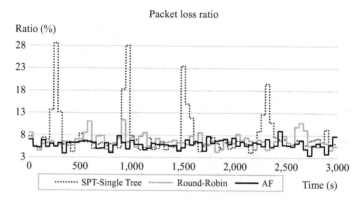

Figure 10.12   The average data loss ratio

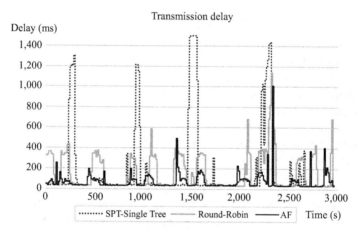

Figure 10.13   The average transmission delay

It is clear from Figures 10.12 and 10.13 that network service with a routing optimization can achieve a better performance in data loss ratio and transmission delay. However, hereby it is noted that a too frequent switching between different multicast trees may insult in an additional control overhead and introduce a switching delay.

10.5   Conclusion

This chapter discussed NaaS-enabled framework for composing services in the context of SDN, where networking services are encapsulated into basic service components, for the sake of obtaining flexible and efficient networking utility. We presented the system architecture and software necessary for enabling NaaS to drive the process of service composition. It is allowed to coordinately managing and aggregating the cross-layer resources concerning the network layer and application layer. Furthermore, the *multipath*, *multicast* and *multi-domain* routing services are provided as basic components for the service-composition process. Finally, two use cases are carried out to demonstrate the success of the conceived framework.

Acknowledgment

This work was supported by the National Natural Science Foundation of China (No. 61573329) and the State Key Program of National Natural Science of China (No. 61233003).

References

[1] H. Ma, F. Bastani, I.-L. Yen, and H. Mei, "QoS-driven service composition with reconfigurable services," *IEEE Transactions on Services Computing*, vol. 6, no. 1, pp. 20–34, 2013.

[2] D. Chiu and G. Agrawal, "Cost and accuracy aware scientific workflow composition for service-oriented environments," *IEEE Transactions on Services Computing*, vol. 6, no. 4, pp. 470–483, 2013.

[3] X. Gu and K. Nahrstedt, "Distributed multimedia service composition with statistical QoS assurances," *IEEE Transactions on Multimedia*, vol. 8, no. 1, pp. 141–151, 2006.

[4] D. Kreutz, F. M. V. Ramos, P. E. Veríssimo, C. E. Rothenberg, S. Azodolmolky, and S. Uhlig, "Software-defined networking: A comprehensive survey," *Proceedings of the IEEE*, vol. 103, no. 1, pp. 14–76, Jan. 2015.

[5] S. Sezer, S. Scott-Hayward, P. K. Chouhan, *et al.*, "Are we ready for SDN? Implementation challenges for software-defined networks," *IEEE Communications Magazine*, vol. 51, no. 7, pp. 36–43, Jul. 2013.

[6] N. McKeown, T. Anderson, H. Balakrishnan, *et al.*, "OpenFlow: Enabling innovation in campus networks," *SIGCOMM Computer Communication Review*, vol. 38, no. 2, pp. 69–74, Mar. 2008.

[7] Q. Duan, Y. Yan, and A. Vasilakos, "A survey on service-oriented network virtualization toward convergence of networking and cloud computing," *IEEE Transactions on Network and Service Management*, vol. 9, no. 4, pp. 373–392, 2012.

[8] B. Medjahed and A. Bouguettaya, "A multilevel composability model for semantic web services," *IEEE Transactions on Knowledge and Data Engineering*, vol. 17, no. 7, pp. 954–968, Jul. 2005.

[9] Q. Wu, F. Ishikawa, Q. Zhu, and D. H. Shin, "QoS-aware multigranularity service composition: Modeling and optimization," *IEEE Transactions on Systems, Man, and Cybernetics: Systems*, vol. 46, no. 11, pp. 1565–1577, Nov. 2016.

[10] B. Sahoo and P. Bhuyan, "A selection approach in service composition of SOA," in 2016 *International Conference on Recent Trends in Information Technology (ICRTIT)*, Apr. 2016, pp. 1–6.

[11] B. Raman and R. H. Katz, "Load balancing and stability issues in algorithms for service composition," in *INFOCOM 2003. Twenty-Second Annual Joint Conference of the IEEE Computer and Communications. IEEE Societies*, vol. 2, Mar. 2003, pp. 1477–1487.

[12] Z. Ye, S. Mistry, A. Bouguettaya, and H. Dong, "Long-term QoS-aware cloud service composition using multivariate time series analysis," *IEEE Transactions on Services Computing*, vol. 9, no. 3, pp. 382–393, May 2016.

[13] A. Klein, F. Ishikawa, and S. Honiden, "Towards network-aware service composition in the cloud," in *Proceedings of the 21st International Conference on World Wide Web*. New York, NY, USA: ACM, pp. 959–968, 2012.

[14] R. Mijumbi, J. Serrat, J. L. Gorricho, N. Bouten, F. D. Turck, and R. Boutaba, "Network function virtualization: State-of-the-art and research challenges," *IEEE Communications Surveys & Tutorials*, vol. 18, no. 1, pp. 236–262, First quarter 2016.

[15] Y. Li, F. Zheng, M. Chen, and D. Jin, "A unified control and optimization framework for dynamical service chaining in software-defined NFV system," *IEEE Wireless Communications*, vol. 22, no. 6, pp. 15–23, Dec. 2015.

[16] M. C. Luizelli, L. R. Bays, L. S. Buriol, M. P. Barcellos, and L. P. Gaspary, "Piecing together the NFV provisioning puzzle: Efficient placement and chaining of virtual network functions," in *2015 IFIP/IEEE International Symposium on Integrated Network Management (IM)*, May 2015, pp. 98–106.

[17] L. Qu, C. Assi, and K. Shaban, "Delay-aware scheduling and resource optimization with network function virtualization," *IEEE Transactions on Communications*, vol. 64, no. 9, pp. 3746–3758, Sep. 2016.

[18] H. Moens and F. D. Turck, "Customizable function chains: Managing service chain variability in hybrid NFV networks," *IEEE Transactions on Network and Services Management*, vol. 13, no. 4, pp. 711–724, Dec. 2016.

[19] R. Sherwood, G. Gibb, K.-K. Yap, *et al.*, "FlowVisor: A network virtualization layer," *OpenFlow Switch Consortium*, Tech. Rep., 2009.

[20] Openstack. [Online]. Available: http://www.openstack.org/

[21] A. Fischer, J. Botero, M. Till Beck, H. de Meer, and X. Hesselbach, "Virtual network embedding: A survey," *IEEE Communications Surveys & Tutorials*, vol. 15, no. 4, pp. 1888–1906, 2013.

[22] C. Papagianni, A. Leivadeas, S. Papavassiliou, V. Maglaris, C. Cervello-Pastor, and A. Monje, "On the optimal allocation of virtual resources in cloud computing networks," *IEEE Transactions on Computing*, vol. 62, no. 6, pp. 1060–1071, 2013.

[23] M. Chowdhury, M. R. Rahman, and R. Boutaba, "Vineyard: Virtual network embedding algorithms with coordinated node and link mapping," *IEEE/ACM Transactions on Network*, vol. 20, no. 1, pp. 206–219, Feb. 2012.

[24] F. Esposito, I. Matta, and Y. Wang, "Vinea: An architecture for virtual network embedding policy programmability," *IEEE Transactions on Parallel and Distributions Systems*, vol. 27, no. 11, pp. 3381–3396, Nov. 2016.

[25] S. Abdelwahab, B. Hamdaoui, M. Guizani, and T. Znati, "Efficient virtual network embedding with backtrack avoidance for dynamic wireless networks," *IEEE Transactions on Wireless Communications*, vol. 15, no. 4, pp. 2669–2683, Apr. 2016.

[26] R. Gouveia, J. Aparício, J. Soares, B. Parreira, S. Sargento, and J. Carapinha, "SDN framework for connectivity services," in *2014 IEEE International Conference on Communications (ICC)*, June 2014, pp. 3058–3063.

[27] MPEG-DASH. [Online]. Available: http://dashif.org/mpeg-dash/

[28] T. Stockhammer, "Dynamic adaptive streaming over HTTP: Standards and design principles," in *Proceedings of the Second Annual ACM Conference on Multimedia Systems*. ACM, 2011, pp. 133–144.

[29] C. Xu, S. Chen, J. Su, S. M. Yiu, and L. C. K. Hui, "A survey on regular expression matching for deep packet inspection: Applications, algorithms, and hardware platforms," *IEEE Communications Surveys Tutorials*, vol. 18, no. 4, pp. 2991–3029, Fourth quarter 2016.

[30] H. Uppal and D. Brandon, "OpenFlow based load balancing," *CSE561: Networking Project Report*, University of Washington, Seattle, WA, 2010.

[31] Firewall (dev), floodlight. [Online]. Available: http://www.openflowhub.org/display/floodlightcontroller/Firewall+(Dev)

[32] Network Functions Virtualization. [Online]. Available: http://www.etsi.org/technologies-clusters/technologies/nfv

[33] R. Mijumbi, J. Serrat, J. l. Gorricho, S. Latre, M. Charalambides, and D. Lopez, "Management and orchestration challenges in network functions virtualization," *IEEE Communications Magazine*, vol. 54, no. 1, pp. 98–105, Jan. 2016.

[34] Ovs quantum plugin. [Online]. Available: http://openvswitch.org/openstack/documentation/

[35] Neuca quantum plugin. [Online]. Available: http://groups.geni.net/geni/wiki/GENIRacksHome#ExoGENIRacks

[36] J. Zhang. EthUDP. [Online]. Available: https://github.com/bg6cq/ethudp

[37] Pantou : OpenFlow 1.0 for OpenWRT. [Online]. Available: http://archive.openflow.org/wk/index.php

[38] Open Networking Lab, POX Wiki. [Online]. Available: http://openflow.stanford.edu/display/ONL/POX+Wiki

[39] FFmpeg. [Online]. Available: http://www.ffmpeg.org/

[40] M. Sun, X. Zhang, L. Wang, H. Shi, and W. Zhang, "A multiple multicast tree optimization solution based on software defined network," in *2016 Seventh International Conference on Information and Communication Systems (ICICS)*, April 2016, pp. 168–173.

[41] S. Kerner, "OpenFlow protocol 1.3. 0 approved," *Enterprise Networking Planet*, 2012.

[42] S. S. Haykin, *Adaptive Filter Theory*. Pearson Education India, Delhi, India, 2008.

Chapter 11

Network-as-a-Service in software-defined networking for end-to-end quality of service provisioning

Qiang Duan[1]

Abstract

Software-Defined Network (SDN) is expected to have a significant impact on future networking. Although exciting progress has been made toward realizing SDN, application of this new networking paradigm in the future Internet to support end-to-end QoS provisioning faces some new challenges. The autonomous network domains coexisting in the Internet and the diverse user applications deployed upon the Internet call for a uniform Service Delivery Platform (SDP) that enables high-level network abstraction and inter-domain collaboration for end-to-end service provisioning. However, the currently available SDN technologies lack effective mechanisms for supporting such a platform. This chapter presents an SDP framework that applies the Network-as-a-Service (NaaS) principle to provide network abstraction and orchestration for end-to-end service provisioning in SDN-based future Internet. In order to address the new challenges brought in by resource abstraction enabled by NaaS to system modeling and analysis for QoS evaluation, a profile-based analysis method developed based on the network calculus theory is also presented in this chapter.

11.1 Introduction

Software-Defined Network (SDN) is emerging network architecture that may have a significant impact on the development of future networking technologies. SDN architecture decouples network control and data forwarding functions, thus enabling network control to become directly programmable and underlying network infrastructure to be abstracted for applications [1]. Key features of SDN include separation between control plane and data plane, logically centralized network control, and programmability of the control plane. These features combined together gives SDN

---

[1] Information Sciences and Technology Department, Pennsylvania State University Abington College, USA

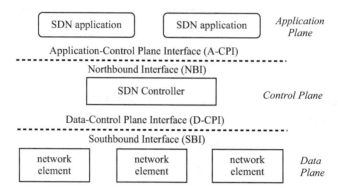

Figure 11.1   General architecture of Software-Defined Networking

some great advantages in networking, including simplified and enhanced network configuration and operation, flexible and efficient network control and management, and improved network performance for meeting various application requirements. Therefore, SDN is expected to play a crucial role in the future Internet.

A general architectural framework for SDN, as shown in Figure 11.1, consists of three planes: the data plane, control plane, and application plane; and the two interfaces: the interface between data and control planes (D-CPI/southbound interface) and the interface between application and control planes (A-CPI/northbound interface). The data plane comprises distributed network resources that perform functions of data transport and processing. Devices on the data plane expose their capabilities and resource states to the control plane via the southbound interface. The behaviors of data plane resources are directly controlled through this interface. The SDN controller manages distributed network resource states and provides a global abstract view of the data plane to the application plane via the northbound interface. The SDN applications specify their networking requirements to the controller and define operations of the abstracted network resources through this interface. The SDN controller translates applications requirements to low-level control instructions that may be performed by the network elements on the data plane.

SDN architecture and its enabling technologies recently formed an important research area that has attracted extensive attention from both academia and industry. Active research topics in this area include SDN-enabled switching devices, SDN controllers, network operating systems, various network control/management applications, protocols between the data and control planes (southbound interface), and Application Programming Interfaces (APIs) for programming the control plane (northbound interface). Exciting progress has been made on SDN development and numerous research results have been reported in literature [2–4].

Although the SDN architecture has been successfully applied in some networking systems such as enterprise networks, data center networks, and inter-data center communications, adoption of this new networking paradigm in a large scale inter-networking scenario such as the future Internet faces new challenges that must be further investigated. One of the key issues lies in end-to-end service provisioning across heterogeneous network domains with QoS guarantee for meeting diverse user requirements.

In an enterprise or data center network, the user applications, network controller, and data forwarding devices all belong to the same administration domain; therefore information of underlying network infrastructure can be made available to upper layer applications easily. However, in the Internet end users (computing applications) and network service providers often belong to different domains; therefore detailed information of network states may not be directly visible to applications. In addition, end-to-end communication paths in the Internet often traverse multiple autonomous systems operated by different organizations, each has the authority to choose its own networking technologies and management policies. The rapid development and wide adoption of cloud computing has made the end-to-end service provisioning more complex. The services requested by end users often demand not only data communication operations but also data process functionalities offered by computing facilities. Therefore, end-to-end service provisioning in such a heterogeneous networking scenario requires a higher-level network abstraction for flexible interaction between users and service providers and loose-coupling collaboration among the involved autonomous systems. This calls for an SDP that supports flexible and effective user-network interaction and inter-domain collaboration.

However, currently available SDN technologies lack an effective mechanism for building such an SDP. Although a variety of SDN controllers have been developed, there is no standard yet for achieving interoperability between these controllers. What resulted is that no single vendor could deliver a standard-based northbound API for application development, or a standardized interface between controllers. In a large scale inter-domain networking scenario, it is not feasible to require all autonomous network domains to adopt the same type of SDN controller. Therefore, lack of interoperability between SDN controllers prevents applications from functioning seamlessly across different controllers for inter-domain network service provisioning. Recent works on inter-domain networking in SDN mainly focused on distributed collaboration between SDN controllers for routing. End-to-end service delivery across heterogeneous SDN domains has not been sufficiently studied.

Recently, application of the service-orientation principle in SDN to address the challenging problem of end-to-end service delivery started attracting researchers' attention. The Service-Oriented Architecture (SOA) [5] offers an effective mechanism to enable flexible interactions among autonomous systems to meet diverse service requirements. SOA has been widely adopted in various areas, including cloud computing and Web services, as the main model for service delivery. Application of the SOA principle in networking leads to a Network-as-a-Service (NaaS) paradigm, which enables networking resources and functionalities to be utilized by users as services through a standard abstract interface, much like computational resources are utilized as services in cloud computing.

NaaS enables abstraction of networking systems into network services that can be discovered, selected, and accessed by users, thus offering a flexible mechanism for user–network interaction. Network abstraction enabled by NaaS allows flexible collaboration among autonomous network domains via loose-coupling service inter-actions. SOA-compliant network services may be composed with Cloud services through a unified service orchestration platform to provision composite network-cloud services to end users. Therefore, NaaS may greatly facilitate end-to-end service provisioning in the future Internet.

On the other hand, NaaS-based service provisioning also brings new challenges to system modeling and performance analysis for end-to-end QoS provisioning. Tra-ditional analytical approaches to QoS analysis are based on queueing theory and typically developed for specific network architecture with certain assumptions about service implementations. However, NaaS introduces network resource abstraction that allows end users to utilize network infrastructure resources without knowledge of spe-cific network implementations. Therefore, NaaS-based service provisioning calls for more flexible methods for QoS analysis that are agnostic to service implementations.

The research work presented in this chapter tackles the challenging problem of end-to-end service provisioning in SDN by exploiting the NaaS notion. A frame-work of an SDP that applies the NaaS paradigm in SDN is first proposed. The framework enables high-level network abstraction and inter-domain network service orchestration. Then a flexible approach for modeling and analyzing end-to-end QoS is presented. This method is based on network calculus theory and is agnostic to service implementations and thus is applicable to network services with heteroge-neous implementations for meeting diverse user requirements. SDN is becoming an important part of cloud infrastructure that allows composition of network and cloud services. The application of this new performance evaluation method to composite network-cloud services is presented at the end of this chapter.

11.2   End-to-end service provisioning in SDN—challenges and solutions

11.2.1   Challenges to end-to-end service provisioning in SDN

SDN is expected to be widely applied in future networking, including public carrier networks such as the Internet backbone and wireless mobile networks as well as data center and enterprise networks. These networks are typically under the administration of different organizations, thus forming different autonomous domains. Therefore end-to-end network service delivery in SDN-based future Internet often requires data communications traversing heterogeneous network domains.

Recent rapid advancement in SDN research has yielded diverse technologies for realizing this new network architecture. Various SDN-enable switches have been developed. Although OpenFlow [6] has been widely adopted for controlling switches in the data plane, it is not the only southbound interface for SDN. Possible protocols that may potentially play the same role include Forwarding and Control Element Sep-aration (ForCES) [7], Path Computation Element Communication Protocol (PCEP) [8], Protocol-Oblivious Forwarding (POF) [9], and OpFlex [10]. A wide variety of SDN controllers and network operating systems have also been developed. These include both centralized controllers such as NOX [11], Beacon [12], and Floodlight

[13], and distributed network operating systems such as ONIX [14], ONOS [15], and HyperFlow [16].

Diversity in available SDN technologies brings in challenges to end-to end service provisioning across multiple domains in SDN-based future Internet. Autonomous systems in the Internet should have the freedom to employ various SDN technologies, including switches, southbound protocols, and network controllers, that fit their particular networking needs. On the other hand, the objective of service provisioning is to deliver network services across the heterogeneous domains for meeting the diverse requirement of end users. Therefore, end-to-end service provisioning in the future Internet requires not only effective inter-domain collaboration but also flexible interaction between upper layer user applications and the underlying network domains.

However, the currently available SDN technologies lack sufficient capability of meeting this requirement for end-to-end service delivery. Development of SDN controllers often lacks consideration of interoperability with controllers from other vendors. Work on distributed deployment of SDN controllers mainly focuses on cooperation among multiple homogeneous controllers in the same domain and thus are insufficient to handle heterogeneity of the controllers in multi-domain cases. Moreover, despite rapid development on standard southbound interface, currently there is no common standard for the northbound API between SDN controllers and network control/management applications. These applications are often developed based on the API provided by a particular type of controller and thus are tightly coupled with the controller design. Such tight coupling between applications and controllers significantly limits the capability of service provisioning over heterogeneous controllers in a multi-domain SDN environment.

Recently some study on inter-domain issues in SDN has been reported in the literature. SDNi [17] is a protocol recently proposed by IETF for coordinating operations and exchanging information between SDN controllers in different domains. The implementation of SDNi suggested in [17] is to extend BGP for information exchange. However, the hop-by-hop nature of BGP makes routing among domains in a decentralized manner without knowledge of end-to-end routes, which may not be able to achieve a global optimal path for end-to-end QoS provisioning. Research reported in [18,19] employs the SDN principle to address the inter-domain routing problem. Both works are based on BGP and thus are limited by its decentralized feature to fully realize the SDN benefit of centralized control with a global network view. The inter-AS routing proposed in [18] assumes that homogeneous controllers, specifically the NOX-OpenFlow controller, are used in all domains and thus may not be applicable to large-scale multi-domain scenarios. The multi-AS routing control platform proposed in [19] assumes the existence of a mechanism to communicate with SDN domain controllers without detailed discussion on the realization of such a mechanism.

In [20] the authors argue that BGP is a poor candidate for inter-domain routing in SDN and propose decoupling between routing and policy control to facilitate interoperability among SDN domains. The distributed control plan proposed in [21] employs a message-oriented communication bus for information exchange among SDN domain controllers. The aforementioned research focuses on controller collaboration for inter-domain routing in SDN. End-to-end service provisioning needs more than just routing across multiple domains. Flexible interaction between user applications and the SDN controllers in different domains of the underlying network infrastructure is another important aspect that so far has received little attention. It requires a high-level network abstraction, loose-coupling interaction between applications and controllers, and flexible collaboration among heterogeneous controllers.

11.2.2   Network-as-a-Service in SDN—a promising solution

The SOA [5] offers a promising approach to addressing the challenges for end-to-end service provisioning in multi-domain SDN. The SOA can be described as architecture within which all functions are defined as independent services with invokable interfaces that can be called in defined sequences to form business processes. A *service* in SOA is a module that is self-contained (i.e., the service maintains its own states) and platform-independent (i.e., the interface to the service is independent with its implementation platform). Services can be described, published, located, orchestrated, and programmed through standard interfaces and messaging protocols. A key feature of SOA is "loose-coupling" interaction among heterogeneous systems, which allows entities to collaborate with each other while keep themselves independent. This feature makes SOA very effective architecture for coordinating heterogeneous systems to provide services that meet various application requirements.

Application of the service-orientation principle in networking provides a promising approach to addressing some challenges in the future Internet. Such a service-oriented networking paradigm is referred to as *Network-as-a-Service* (NaaS), in which networking resources are abstracted and utilized in form of SOA-compliant network services. In principle, a network service may represent any type of networking component at different levels, including an entire network domain, a single physical or virtual network, or an individual network node. Multiple network services can be combined into one composite inter-network service through a service orchestration mechanism.

Recently the NaaS paradigm has started attracting attention from the networking research community and interesting progress has been reported in the literature. Costa *et al.* proposed a NaaS model for data center networks in [22] for enabling cloud tenants to have direct access to network infrastructure for improving service performance. Cloud-based network architecture that combines the cloud service model with the network openness enabled by SDN was proposed in [23] in order to offer various network protocol services. An SDN control platform called Meridian was presented in [24], which provides a service-level network model with connectivity and policy abstractions for cloud networking. Bueno and his colleagues developed a NaaS-based Network Control Layer (NCL) that provides an abstraction layer to obtain homogeneous control over heterogeneous network infrastructure [25].

The above works made interesting progress of applying NaaS in SDN for network service provisioning; however, they mainly focus on single-domain cases or assume homogeneous SDN controllers. The framework proposed in [22] assumes that applications can directly acquire detailed knowledge of underlying network infrastructure, which is reasonable in a single data center environment but not realistic for the large scale Internet with multiple autonomous domains. The prototype given in [23] for realizing the proposed network architecture used NOX controller and OpenFlow protocol for controlling all switches. Both Meridian platform and NCL were implemented based only on Floodlight controller. Cooperation between SDN domains with heterogeneous controllers for end-to-end service delivery is still an opening issue that has not been sufficiently addressed yet.

Some researchers have employed the NaaS paradigm for inter-domain QoS provisioning in SDN. Zhu *et al.* proposed Software Service Defined Network (SSDN) architecture in [26], which employs SOA-based Enterprise Service Bus (EBS) to build a network software service layer that allows networking resources in multiple domains with different SDN controllers to be federated for end-to-end service delivery. However, some key technologies for achieving QoS guarantee with such a service layer, for example, abstraction of service capabilities and cross-domain resource allocation, were not addressed in [26]. The authors of [27] developed a distributed QoS architecture for SDN, which employs a hierarchical control plane where a super controller coordinates the local SDN controllers in multiple domains to support end-to-end multimedia streaming. However, [27] did not give any specific mechanism for the super controller to coordinate heterogeneous SDN controllers in different network domains for service delivery.

Preliminary study of NaaS-based inter-domain service delivery in SDN was reported in our previous work [28]. In this chapter, we further elaborate the idea of NaaS–SDN integration and present a framework of a NaaS-based SDP for a multi-domain SDN environment. This platform provides a high-level abstraction of each SDN domain as a network service and enables network service orchestration for end-to-end service delivery.

11.3 NaaS-based service delivery in SDN

The framework of a NaaS-based SDP in a multi-domain SDN environment is shown in Figure 11.2. In this framework each network domain may have its own choice of SDN technologies, including data plane switches, SDN controllers, and the southbound interface. A domain may also implement various control programs upon its own SDN controller to perform functions such as QoS routing and traffic engineering within the domain scope. Each network domain is abstracted as a network service through a NaaS interface, which provides a high-level abstraction of networking capabilities of the entire domain, including both forwarding and control functionalities, to the SDP. The NaaS interface also allows the SDP to specify its networking requests and policies to each domain. The NaaS-based network abstraction makes network infrastructure of each domain transparent to upper layer applications, thus enabling SDP to coordinate the resources provided by network domains for delivering network services to support diverse user applications.

The SDP serves as a middleware between upper layer user applications and the underlying network infrastructure consisting of heterogeneous domains.

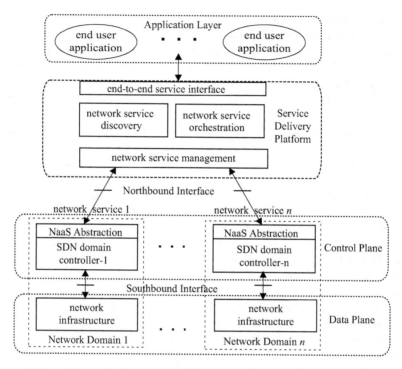

Figure 11.2 A NaaS-based service delivery platform in software-defined network

Key components of the SDP include a service interface and the modules for service management, service discovery, and service orchestration. The SDN controller of each network domain is responsible to publish and update an abstract model of the domain service capability at the service management module. The service interface allows upper layer user applications to specify their requests for end-to-end network services. Upon receiving a service request from an end user, the service discovery module searches the service registry maintained by the service management module to discover a network service for meeting the request. If no single network service provided by any individual domain can meet the requirement, the orchestration module will search for a service chain of multiple network services and orchestrate them for end-to-end service delivery. Then the service management module will send requests to the SDN controllers of all domains involved in service delivery for this user to allocate sufficient bandwidth for meeting user QoS requirement. In addition to these key components, the SDP may also perform some global network management functions, for example user authentication, service request authorization, end-to-end path computation, and traffic engineering.

The presented framework gives functional architecture for an SDP for inter-domain SDN, which may be realized with various implementations. Enabling technologies are required for implementing key functions in two categories: (i) internal modules of the SDP, mainly including the service management, discovery, and orchestration modules and (ii) interfaces for the SDP to interact with user applications and network domains, including the service interface and the network abstraction interface. Recent research on NaaS has yielded various technologies for network service description, discovery, and composition. A summary of these technologies can be found in the survey paper [29]. These technologies form the foundation for implementing the key modules in the SDP. Standard interfaces for network and service abstractions form the other key aspect for realizing the SDP. From an end user's perspective, the SDP plays the role of a service broker in the SOA architecture; therefore standard Web Service interfaces between service consumers and a service broker can be applied to realize the service interface between the SDP and the upper layer user applications. The network abstraction interface between the SDP and various network domains is essentially a SDN northbound interface. RESTful Web Service has been widely adopted for implementing a northbound interface in SDN. Application Layer Traffic Optimization (ALTO) [30] and Interface to Routing System (I2RS) [31] are two RESTful compatible protocols based on which a network abstraction interface may be realized.

The proposed SDP framework combines advantages of NaaS and SDN for improving end-to-end service provisioning in the future Internet. The separated data and control planes and logically centralized controlling enabled by SDN allows a global control mechanism over heterogeneous network infrastructure. NaaS provides a high-level abstraction of autonomous networking systems and enables loose-coupling collaboration among them. The proposed NaaS-based SDP offers a uniform platform upon which third party service providers can develop and deploy new end-to-end network services to meet various application requirements without knowing detailed implementations of underlying network infrastructure. Such an SDP enables a new business model in which a service provider can lease networking resources from various domains and orchestrate the resources for end-to-end network service provisioning. Such a business model is similar to the model for cloud service provisioning, which allows service providers to lease computing resources from infrastructure providers for offering cloud services to end users.

Please note that in many SDN literature the term *application* is often used to refer to software that programs the network through API to a SDN controller in order to perform network control and management functions, for example, applications for routing and traffic engineering. In this senses they are network control/management applications. The user applications in the framework shown in Figure 11.2 are different. They are software programs that utilize the network services provided by the SDP to perform their own computing functions; that is, end users of network services.

Another important advantage of the proposed SDP is to realize the benefit of logically centralized control promised by the SDN paradigm in large scale multi-domain networking environments. Such a centralized control with a global network view is particularly important for achieving end-to-end service delivery with QoS guarantee in the Internet consisting of various autonomous systems. Due to the heterogeneity of network protocols and technologies in these systems, exposure of networking capabilities to a central control unit without appropriate abstraction would

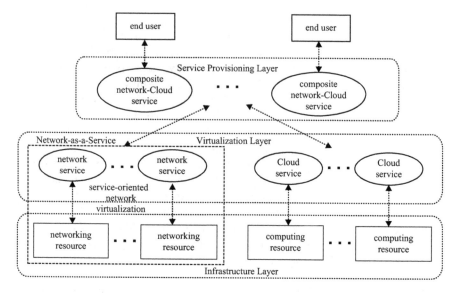

Figure 11.3 A NaaS-based framework for network-cloud convergence

lead to unmanageable complexity. The high-level abstraction enabled by the SDP addresses the diversity challenge, thus making centralized control for end-to-end QoS provisioning possible.

The NaaS-based SDP in SDN may also greatly facilitate the notion of converged networking and Cloud computing. Figure 11.3 gives a layered framework for enabling network-Cloud convergence, in which resources in both networking and computing domains are virtualized into SOA-compliant services. In this convergence framework, composition of network and computing services expands the spectrum of cloud services that can be offered to users. The loose-coupling feature of SOA provides a flexible and effective mechanism in this network-cloud convergence framework that supports interaction between networking/computing infrastructure and service provisioning functions as well as collaboration among heterogeneous networking and computing domains [29].

11.4 Modeling and performance analysis for NaaS-based QoS provisioning in SDN

11.4.1 Modeling and analysis

The network service performance that can be achieved by the SDP to an user applications is determined by the service capacities offered by the network domains involved in service delivery and the traffic load generated by the application as well. The NaaS paradigm adopted by the SDP make network domains transparent to user applications, which can utilize the services provided by network domains without detailed knowledge of internal implementations of the domains.

Therefore a key requirement for a performance analysis method for the SDP is being able to characterize service capabilities offered by network domains and being independent with domain implementations. The concept of *service curve* in network calculus [32] may be applied to develop a general profile that is independent with domain implementations thus applicable to various network services.

Let $R(t)$ and $D(t)$, respectively, be the accumulated amount of traffic of a flow that arrives at and departs from a network domain by time t. Given a non-negative, non-decreasing function, $\mathscr{P}(\cdot)$, where $\mathscr{P}(0) = 0$, we say that the domain guarantees a *Capability Profile* $\mathscr{P}(\cdot)$ for the flow, if for any $t \geq 0$ in the busy period of the system,

$$D(t) \geq R(t) \otimes \mathscr{P}(t) \tag{11.1}$$

where $\otimes$ denotes the min-plus convolution operation defined in network calculus as $h(t) \otimes x(t) = \inf_{s:0 \leq s \leq t} \{h(t-s) + x(s)\}$.

A service contract between the SDP and a network domain can be specified with a set of parameters, typically include the maximum latency and minimum bandwidth that the domain should guarantee. Therefore, the capability profile for such a network service is

$$\mathscr{P}(t) = P[r, \theta] = \max\{0, r(t - \theta)\}, \tag{11.2}$$

which is referred to as *Latency-Rate* (LR) profile and the parameters θ and r are respectively called the latency and service rate of the profile. The rate parameter r reflects the minimum available bandwidth that the domain promises for service delivery to a flow. The parameter θ is to characterize the latency introduced by transmission delay and switch processing delay. Note that the latency parameter is different from the service delay performance and the latter is impacted significantly by the queueing delay introduced by packet buffering at switches in the domain.

A model for NaaS-based SDP in SDN is shown in Figure 11.4. The system consists of a series of tandem network service components $S_1, S_2, \ldots, S_n$, each of which is an abstraction of the service capability provided by a network domain to the service delivery system [33].

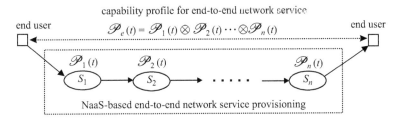

Figure 11.4   A model for NaaS-based end-to-end service delivery in SDN

It is known from network calculus theory that the service curve guaranteed by a series of tandem servers can be obtained from the convolution of the service curves guaranteed by these servers. Since the capability profile defined in (11.1) is essentially the service curve that a service component guarantees to a traffic flow, the capability profile of the end-to-end system for the same flow can be determined accordingly. Assume that service components $S_1, S_2, \ldots, S_n$ have the capability profiles $\mathscr{P}_1(t), \mathscr{P}_2(t), \ldots, \mathscr{P}_n(t)$, respectively, then the capability profile for the end-to-end service delivery system, denoted by $\mathscr{P}_e(t)$, can be obtained as,

$$\mathscr{P}_e(t) = \mathscr{P}_1(t) \otimes \mathscr{P}_2(t) \cdots \otimes \mathscr{P}_n(t). \tag{11.3}$$

Suppose each service component S_i has an LR profile, $\mathscr{P}_i(t) = P[r_i, \theta_i]$, then it can be proved that the capability profile of the end-to-end service delivery system is

$$\mathscr{P}_e(t) = P[r_1, \theta_1] \otimes \cdots \otimes P[r_n, \theta_n] = P[r_e, \theta_\Sigma] \tag{11.4}$$

where $r_e = \min\{r_1, r_2, \ldots, r_n\}$ and $\theta_\Sigma = \sum_{i=1}^{n} \theta_i$. Equation (11.11) implies that if each service component in an end-to-end service delivery system can be modeled by an LR profile, then the networking capability of the entire service system can also be modeled by a LR profile. The latency parameter of the end-to-end profile is the summation of latency parameters of all service components in the system, and the end-to-end service rate parameter is the minimum service rate of all the service components.

The end-to-end network QoS experienced by an end user is determined by the traffic load generated by the user as well as the service capacity provided to the traffic flow of this user. Therefore, traffic parameters are specified in the Service Level Agreement (SLA) between a service user and the service provider. A typical specification of user traffic load is to bound arrival traffic in a flow with peak rate p, sustain rate ρ, and the maximum burst size σ, which may be enforced by employing a token-bucket regulator at network boundary. According to network calculus, the accumulated amount of traffic in a token-bucket regulated flow that arrives at the entry of a service delivery system by any time instant t has an upper limit

$$L(t) = \min\{pt, \sigma + \rho t\}, \tag{11.5}$$

which is referred to as the *Load Profile* of the flow.

The maximum end-to-end delay performance is associated with both the networking capacity of the delivery system, which is modeled by a capability profile, and the traffic load characteristics, which is described by a load profile. It has been shown in network calculus that for a service delivery system with a capability profile $\mathscr{P}(t)$ under traffic described by a load profile $L(t)$, the maximum delay d_m for network service delivery can be determined as

$$d_m = \max_{t:t \geq 0} \{\min\{\delta : \delta \geq 0 \ L(t) \leq \mathscr{P}(t + \delta)\}\}. \tag{11.6}$$

For a service delivery system with an LR capability profile $P[r_e, \theta_e]$ and a load profile $L(t) = \min\{pt, \sigma + \rho t\}$ for a traffic flow, following (11.6) we can obtain that the maximum delay guaranteed by an end-to-end service delivery system to the flow can be determined as

$$
d_m = \begin{cases} \theta_e + \left(\frac{p}{r_e} - 1\right)\frac{\sigma}{p-\rho} & \text{for } p > \rho, \ r_e \geq \rho \\ \theta_e & \text{for } p = \rho, \ r_e \geq \rho \end{cases} \tag{11.7}
$$

Equation (11.7) implies that a delay upper bound d_m is achievable for a flow only when the $r_e \geq \rho$; that is, the minimum available bandwidth promised by all involved network domains to the SDP is no less than the sustained rate of traffic load. When sufficient bandwidth is available, the delay performance of variable rate traffic ($p > \rho$) is a function of both capability profile parameters (θ_e and r_e) and load profile parameters (p, ρ, σ). For constant rate traffic ($p = \rho$), the delay is just the total end-to-end latency θ_e. This is because service rate guarantee ($r_e \geq \rho$) removes extra queuing delay for constant rate traffic flows.

11.4.2 Numerical examples

A networking scenario of service provisioning across three network domains S_i, $i = 1, 2, 3$ is considered to illustrate the application of the modeling and analysis techniques. Suppose two traffic flows traverse the service delivery system. Flow f_1 transmits a stream of video packets while f_2 delivers audio traffic. Both flows require a small maximum end-to-end networking delay. The traffic parameters for f_1 are peak rate $p = 5.3$ Mb/s, sustained rate $\rho = 1.5$ Mb/s, and the maximum burst size $\sigma = 140$ kbits. The traffic parameters for f_2 are peak rate $p = 3.2$ Mb/s, sustained rate $\rho = 1.1$ Mb/s, and the maximum burst size $\sigma = 300$ kbits. These parameters are derived from the traffic analysis results reported in [34]. Each network domain is assumed to provide an LR capability profile to the end-to-end system with a maximum transmission unit $M = 1,000$ bytes.

The maximum end-to-end delays for f_1 and f_2 are evaluated with various amounts of bandwidth allocated in the three network domains. The obtained results are plotted in Figure 11.5, where d_m^1 and d_m^2 denotes the maximum delay for f_1 and f_2 respectively. This figure shows that the maximum delay for both flows decrease with increasing available bandwidth offered by network domains; that is, the more bandwidth is allocated in underlying network infrastructure the tighter is the end-to-end delay bound guaranteed by the SDP. Comparison between the curves of d_m^1 and d_m^2 shows that although both of them decrease with increasing available bandwidth, d_m^2 drops faster than d_m^1 does, which implies that the same amount of bandwidth increment can make more significant improvement in delay performance for f_2 than it does for f_1. Another observation we can make from Figure 11.5 is that different delays are achieved for these two flows with the same amount of available bandwidth. This indicates that the end-to-end delay performance guaranteed to a flow is impacted by the traffic characteristic of the flow as well as the networking capability provided by the service delivery system to the flow.

One of the key potential advantages offered by SDN is a logical centralized control vision enabled by de-coupling the data and control planes. The NaaS-based SDP

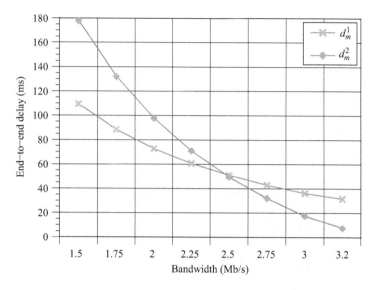

Figure 11.5   End-to-end delay performance for f_1 and f_2

facilitates realizing this advantage of SDN in the large scale Internet with heterogeneous domains. The proposed profile-based analysis technique enables end-to-end service performance evaluation with such a centralized control vision. Without the global control provided by the SDP, if one wants to analyze end-to-end service delay in a multi-domain SDN environment with traditional analysis techniques, she has to determine the delay bound guaranteed by each individual domain then sum them up.

In order to evaluate the end-to-end performance improvement enabled by the SDP with a centralized SDN controller, we also calculate the accumulated delay bounds for flows f_1 and f_2 crossing three network domains, denoted as d_T^1 and d_T^2 respectively, using the traditional domain-by-domain analysis technique. Figure 11.6 gives comparison between end-to-end delay and accumulated delay for these two traffic flows. The obtained results indicate that for each flow accumulated delay bound is apparently larger than the end-to-end delay obtained with the same amount of available bandwidth. Such comparison shows that the profile-based modeling and analysis method matches the NaaS-based SDN control for end-to-end service delivery; therefore, it may obtain a tighter delay bound that allows us to evaluate the maximum service delay performance more precisely.

11.5   Modeling and performance analysis for composite network-compute services

11.5.1   Modeling and analysis

A typical delivery system for composite network-compute services is shown in Figure 11.7, which consists of network services for data transmissions, compute

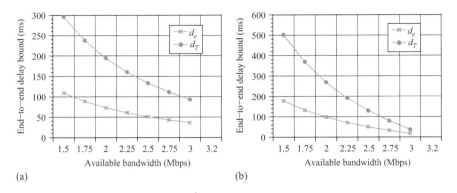

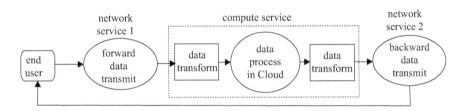

Figure 11.6 End-to-end vs. accumulated delay performance for f_1 and f_2: (a) flow f_1 and (b) flow f_2

Figure 11.7 A typical delivery system for composite network-compute services

services for data process, and data transform functions between the data transmissions and data process [35].

In order to analyze the composite service performance, one must examine the communication capability offered by the network service and data processing capability provided by the compute service. As we presented in last section, network service capability can be modeled by the network calculus-based capability profile. Similarly, a profile-based method may also be employed for modeling data processing capabilities of cloud services. Specifically, an *LR* profile can also characterize service capabilities of typical computing systems. Cloud service providers typically offer a certain service capacity units to users. For example, each type of virtual machine instance in Amazon EC2 provides a predictable amount of computing capacity and I/O bandwidth. The latency and rate parameters of the *LR* profile for a cloud service can be derived from the processing capacity and I/O bandwidth information specified by its provider.

In order to represent the data transformation effect of computing function provided by cloud infrastructure, the concepts of scaling function and scaling curve, which were originally developed in [36] as an extension to network calculus, are adopted in the model for composite network-compute service provisioning systems.

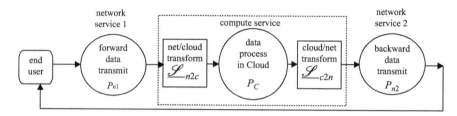

Figure 11.8   A profile-based capability model for composite network-compute service provisioning

A scaling function is defined as a function $S(\cdot)$ that assigns an amount of scaled data $S(a)$ to an amount of data a. Scaling function is a general concept for taking into account data transformation in a system model. It does not model any queuing effect—a scaling function is assumed to have zero delay. Queuing-related effect of Cloud computing is modeled by the service curve-based capability profile of the compute service component.

Given a scaling function S, the function $\mathscr{L}$ is called a (minimum) scaling curve of S iff $\forall b \geq 0$ it applies that $\mathscr{L}(b) \leq \inf_{a \geq 0}\{S(b+a) - S(a)\}$.

Applying the above defined capability profile and scaling curve, a composite network-compute service provisioning system can be modeled with capability profiles of network and compute service components and the scaling curves that represent data transform between networking and computing, as shown in Figure 11.8. In this system the network services for forward data transmission (from user to data center) and backward data transmission (from data center to user) are respectively modeled by the profiles $P_{n1}(t)$ and $P_{n2}(t)$. The compute service offered by the data center is modeled by the profile $P_C(t)$. The scaling curve $\mathscr{L}_{n2c}$ models data transform from forward transmission to computing server while the scaling curve $\mathscr{L}_{c2n}$ models data transform from computing server to backward data transmission.

Applying the alternative server method (Theorem 3.1 in [36]) in the model shown in Figure 11.8, network service 1 for forward data transmission and the scaling function for network/compute data transform can be switched without impacting the performance bound guaranteed by the system, if the capability profile of network service 1 is transformed to $P_{n1}^{\mathscr{S}} = \mathscr{L}_{n2c}(P_{n1})$. Similarly, the scaling function for compute/network data transform and network service 2 for backward data transmission can be switched, if the capability profile of network service 2 is transformed to $P_{n2}^{\mathscr{S}} = \mathscr{L}_{c2n}^{-1}(P_{n2})$. Then the composite service system has the alternative model shown in Figure 11.9.

Since the capability profile defined in (11.1) is essentially the service curve of a service component, the capability profiles for both network service components (for forward and backward data transmissions) and the compute service component can be integrated into one profile using the convolution operation defined in network calculus. Therefore, the end-to-end capability profile for the composite system, denoted by $P_{e2e}(t)$, can be determined as,

$$P_{e2e}(t) = \mathscr{L}_{n2c}(P_{n1}(t)) \otimes P_C(t) \otimes \mathscr{L}_{c2n}^{-1}(P_{n2}(t)). \tag{11.8}$$

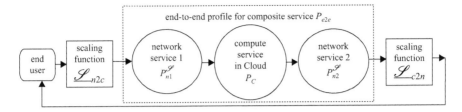

Figure 11.9   Alternative model for composite network-compute service provisioning

Suppose each service component in a composite network–compute system has a *LR* profile; that is, $P_{n1} = \mathscr{P}[r_1, \theta_1]$, $P_C = \mathscr{P}[r_C, \theta_C]$, and $P_{n2} = \mathscr{P}[r_2, \theta_2]$. Then the transformed profile for network service 1 is

$$P_{n1}^{\mathscr{S}} = \mathscr{L}_{n2c}(P_{n1}) = \mathscr{P}[\mathscr{L}_{n2c}(r_1), \theta_1] \tag{11.9}$$

and the transformed profile for network service 2 will be

$$P_{n2}^{\mathscr{S}} = \mathscr{L}_{c2n}^{-1}(P_{n2}) = \mathscr{P}[\mathscr{L}_{c2n}^{-1}(r_2), \theta_2]. \tag{11.10}$$

Then the end-to-end capability profile of the composite service provisioning system is

$$
\begin{aligned}
P_{e2e} &= \mathscr{L}_{n2c}(\mathscr{P}[r_1, \theta_1]) \otimes \mathscr{P}[r_C, \theta_C] \otimes \mathscr{L}^{-1}[\mathscr{P}(r_2, \theta_2)] \\
&= \mathscr{P}[\mathscr{L}_{n2c}(r_1), \theta_1] \otimes \mathscr{P}[r_C, \theta_C] \otimes \mathscr{P}[\mathscr{L}_{c2n}^{-1}(r_2), \theta_2] \\
&= \mathscr{P}[r_e, \theta_e]
\end{aligned}
\tag{11.11}
$$

where $r_e = \min \left\{ \mathscr{L}_{n2c}(r_1), r_C, \mathscr{L}_{c2n}^{-1}(r_2) \right\}$, $\theta_e = \theta_1 + \theta_C + \theta_2$.

The entry of the composite network–compute service system where user applications load the system is the boundary of the forward networking system; therefore the traffic load for forward data transmission is the load of the composite service system. Based on the alternative model given in Figure 11.9, service delay performance is determined by the end-to-end profile $P_{e2e}(t)$ because scaling functions do not introduce any delay. Due to the network/compute scaling curve $\mathscr{L}_{n2c}$ in front of the end-to end profile, the actual load that determines service delay performance should be characterized by a transformed load profile $\mathscr{L}^{\mathscr{S}}(t) - \mathscr{L}_{n2c}(\mathscr{L}(t))$. Therefore, given the end-to-end capability profile $P_{e2e}(t)$ of a composite network–compute service system, the maximum service delay d_{max} guaranteed by the system to the user can be determined as

$$d_{max} = \max_{t:t \geq 0} \left\{ \min \left\{ \delta : \delta \geq 0 \ \mathscr{L}_{n2c}(\mathscr{L}(t)) \leq P_{e2e}(t + \delta) \right\} \right\}. \tag{11.12}$$

Suppose a composite network–compute system has a *LR* profile for each service component and a leaky-bucket load profile $\mathscr{L}[p, \rho, \sigma]$, then the transformed load profile for the system is $\mathscr{L}_{n2c}(\mathscr{L}[p, \rho, \sigma])$. Following (11.11) and (11.12), the maximum service delay guaranteed by this system can be determined as

$$d_{max} = \theta_\Sigma + \left(\frac{\mathscr{L}_{n2c}(p)}{r_e} - 1\right)\frac{\mathscr{L}_{n2c}(\sigma)}{\mathscr{L}_{n2c}(p) - \mathscr{L}_{n2c}(\rho)} \tag{11.13}$$

where $\theta_\Sigma = \theta_{n1} + \theta_{n2} + \theta_C$ is the total service latency including round-trip network latency and latency of the compute service.

11.5.2 Numerical examples

Considering a service provisioning scenario in which an end user transmits data to a cloud data center for processing and then receives the processed data back from the cloud. Based on the measurement results reported in [37,38], traffic parameters of the load profile for this testing case are assumed to be 320 Mb/s, 120 Mb/s, and 200 Kbits for the peak rate, sustained rate, and burst size respectively. For simplicity, forward and backward data transmissions are assumed to be provided by the same network service with a 10 Gb/s link capacity and a packet length of 1500 bytes.

A communication intensive service scenario was first examined. In this scenario data transform from forward transmission to the computing server decreases the load with a scaling factor $\mathscr{S}_{n2c} = 1/4$ and data transform from computing server to backward transmission increases the load with a scaling factor $\mathscr{S}_{c2n} = 2$. In this scenario we considered two cases in which latency parameters of the computing server are 150 μs and 300 μs. The end-to-end delay upper bounds for both cases are denoted respectively as d_{c1}^e and d_{c2}^e in Figure 11.10. Both curves in this figure drop with

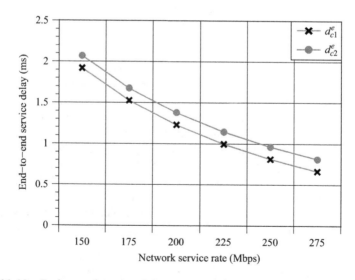

Figure 11.10   End-to-end service delay vs. network service rate for a communication intensive application

increasing network service rate, which indicates that leasing more bandwidth from the network service providers may significantly improve end-to-end delay performance for composite service provisioning. Comparison of the two curves of d_{c1}^e and d_{c2}^e shows that given the same network service rate, smaller server latency may give a tighter end-to-end delay bound but its impact is not as significant as that of increasing network bandwidth.

Then we examined a computing intensive service scenario in which data transforms between network and compute services increase the load for data processing with a scaling factor $\mathcal{S}_{n2c} = 2$ and decrease the load for data transmission with a scaling factor $\mathcal{S}_{c2n} = 1/4$. The end-to-end service delay bounds with various compute service capacities are given in Figure 11.11, in which d_{n1}^e, d_{n2}^e, and d_{n3}^e denotes the delay bounds obtained with 100, 150, and 200 Mbps network service rates. This figure shows that maximum delay of the composite service decreases significantly with increasing compute service capacity. Figure 11.11 also shows that the three delay curves are very close to each other although not completely overlap, which implies that given the same compute service capacity, different amounts of network bandwidth do not change the delay bound much. The results shown in Figure 11.11 indicate that in this scenario computing server capacity is the decisive factor for the maximum service delay and increasing network service rate only has minor contribution to delay performance improvement. This is because the compute service forms a bottleneck for this computing intensive scenario; therefore obtaining sufficient computing capacity in the data center is the key to achieving end-to-end delay guarantee for the composite service.

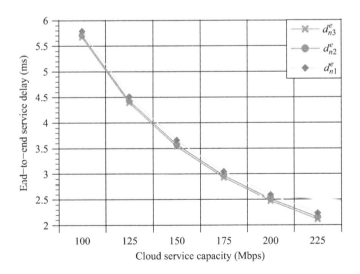

Figure 11.11   End-to-end service delay vs. compute server capacity for a computing intensive application

11.6 Conclusion

The autonomous network domains coexisting in the Internet and the diverse user applications deployed upon the Internet call for a uniform SDP that offers a high-level network abstraction and enables inter-domain collaboration for end-to-end service provisioning. Currently available SDN technologies still lack effective mechanisms for supporting such a platform. In order to address this important and challenging issue, in this chapter we presented an SDP framework that employs the Network-as-a-Service (NaaS) principle to provide a high-level network abstraction and enables inter-domain collaboration through service orchestration for end-to-end service provisioning. The NaaS-based SDP also facilitates the convergence of networking and cloud computing through composition of network and compute services via a unified service-oriented mechanism. On the other hand, the network abstraction and inter-domain service orchestration bring in new challenges to system modeling and performance analysis for evaluating the end-to-end QoS guarantee that can be offered to the NaaS-based SDP to end users. In order to address such challenges to QoS analysis, a profile-based analysis method developed based on the network calculus theory was presented in this chapter. This profile-based method is agnostic to the specific implementations of network services thus are applicable to evaluate end-to-end QoS of services comprising heterogeneous domains, including both networking and computing systems.

References

[1]   ONF, "Open Networking Foundation Software-Defined Networking (SDN) Definition," https://www.opennetworking.org/sdn-resources/sdn-definition, 2013.

[2]   W. Xia, Y. Wen, C. H. Foh, D. Niyato, and H. Xie, "A survey on Software-Defined Networking," *IEEE Communications Surveys and Tutorials*, vol. 17, no. 1, pp. 25–51, 1st quarter 2015.

[3]   Y. Jarraya, T. Madi, and M. Debbabi, "A survey and a layered taxonomy of Software-Defined Networking," *IEEE Communications Surveys and Tutorials*, vol. 16, no. 1, pp. 1955–1980, 4th quarter 2014.

[4]   D. Kreutz, F. Ramos, P. Verissimo, C. E. Rothenberg, S. Azodolmolky, and S. Uhlig, "Software-Defined Networking: A comprehensive survey," *Proceedings of the IEEE*, vol. 103, no. 1, pp. 14–76, 2015.

[5]   T. Erl, *Service-Oriented Architecture – Concepts, Technology, and Design*. Prentice-Hall, NJ, 2005.

[6]   N. McKeown, T. Anderson, H. Balakrishnan, *et al.*, "OpenFlow: Enabling innovation in campus networks," *ACM SIGCOMM Computer Communication Review*, vol. 38, no. 2, pp. 69–74, 2008.

[7]   A. Doria, J. H. Salim, R. Hass, *et al.*, "Forwarding and control element separation (ForCSE) protocol," *Internet Engineering Task Force Specification*, Mar. 2010.

[8]   J. Vasseur and J. L. Roux, "IETF RFC5440: Path Computation Element Communication Protocol (PCEP)," Mar. 2009.

[9]   H. Song, "Protocol-Oblivious Forwarding: Unleash the power of SDN through a future-proof forwarding plane," in *Proceedings of the Second ACM SIGCOMM Workshop on Hot Topics in Software Defined Networking (HotSDN'13)*, pp. 127–132, Jan. 2013.

[10] M. Smith, M. Dvorkin, Y. Laribi, V. Pandey, P. Garg, and N. Weidenbacher, "OpFlex control protocol," *Internet Research Task Force Internet-Draft*, Apr. 2014.

[11] N. Gude, T. Koponen, J. Pettit, *et al.*, "NOX: Toward an operating system for networks," *ACM SIGCOMM Computer Communication Review*, vol. 38, no. 3, pp. 105–110, 2008.

[12] D. Erickson, "The Beacon OpenFlow controller," in *Proceedings of the Second ACM SIGCOMM Workshop on Hot Topics in Software Defined Networking (HotSDN'13)*, Jan. 2013.

[13] "Floodlight OpenFlow Controller," http://www.projectfloodlight.org/floodlight/.

[14] T. Koponen, M. Casado, N. Gude, *et al.*, "ONIX: a distributed control platform for large-scale production networks," in *Proceedings of the Ninth USENIX Conference on Operating Systems Design and Implementation*, Oct. 2010.

[15] U. Krishnaswamy, P. Berde, J. Hart, *et al.*, "ONOS: an open source distributed SDN OS," http://www.slideshare.net/ON-LAB/onos-open-network-operating-system-an-opensource-distributed-sdn-os.

[16] A. Tootooonchian and Y. Ganjali, "HyperFlow: a distributed control plane for OpenFlow," in *Proceedings of the 2010 Internet Network Management Conference on Research on Enterprise Networking*, Apr. 2010.

[17] H. Yin, H. Xie, T. Tsou, D. Lopez, P. Aranda, and R. Sidi, "SDNi: a message exchange protocol for Software Defined Networks (SDNS) across multiple domains," *Internet Research Task Force Internet-Draft*, Jun. 2012.

[18] R. Bennesby, P. Fonseca, E. Mota, and A. Passito, "An Inter-AS routing component for Software-Defined Networks," in *Proceedings of the 2012 IEEE/IFIP Network Operations and Management Symposium (NOMS'12)*, Aug. 2012.

[19] V. Kotronis, X. Dimitropoulos, and B. Ager, "Outsourcing the routing control logic: better internet routing based on SDN principle," in *Proceedings of the 11th ACM Workshop on Hot Topics in Networks (Hotnets'12)*, Oct. 2012.

[20] P. Thai and J. C. de Oliveira, "Decoupling policy from routing with software defined interdomain management: interdomain routing for SDN-based networks," in *Proceedings of the 2012 IEEE International Conference on Computer Communications and Networks (ICCCN'12)*, Jul. 2012.

[21] K. Phemius, M. Bouet, and J. Leguay, "DISCO: distributed multi-domain SDN controller," in *arXiv preprint arXiv:1308.6138*, Aug. 2013.

[22] P. Costa, M. Migliavacca, P. Pietzuch, and A. L. Wolf, "NaaS: Network-as-a-Service in the Cloud," in *Proceedings of the Second USENIX Workshop on Hot Topics in Management of Internet, Cloud, and Enterprise Networks and Services*, Apr. 2012.

[23]   T. Feng, J. Bi, H. Hu, and H. Cao, "Networking-as-a-Service: a Cloud-based network architecture," *Journal of Networks*, vol. 6, pp. 1084–1090, Jul. 2011.

[24]   M. Banikazemi, D. Olshefski, A. Shaikh, J. Tracey, and G. Wang, "Meridian: an SDN platform for Cloud network services," *IEEE Communications Magazine*, vol. 51, pp. 120–127, Feb. 2013.

[25]   I. Bueno, J. Aznar, E. E. J. Ferrer, and J. A. Garcia-Espin, "An OpenNaaS based SDN framework for dynamic QoS control," in *Proceedings of the 2013 IEEE SDN for Future Networks and Services (SDN4FNS)*, Nov. 2013.

[26]   J. Zhu, W. Xie, L. Li, M. Luo, and W. Chou, "Software service defined network: centralized network information service," in *Proceedings of the 2013 IEEE SDN for Future Networks and Services (SDN4FNS)*, Nov. 2013.

[27]   H. E. Egilmez and a. M. Tekalp, "Distributed QoS architectures for multimedia streaming over software defined networks," *IEEE Transactions on Multimedia*, vol. 16, no. 6, pp. 1597–1609, Oct. 2014.

[28]   Q. Duan, "Network-as-a-Service in Software-Defined networks for end-to-end QoS provisioning," in *Proceedings of the 2014 IEEE Wireless and Optical Communications Conference*, May 2014.

[29]   Q. Duan, Y. Yan, and A. V. Valisakos, "A survey on service-oriented network virtualization toward convergence of networking and Cloud computing," *IEEE Transactions on Network and Service Management*, vol. 9, pp. 373–392, Dec. 2012.

[30]   R. Alimi, R. Penno, and Y. Yang, "Internet-Draft: Application Layer Traffic Optimiation (ALTO) protocol," Mar. 2014.

[31]   A. Atlas, J. Halpern, S. Hares, and D. Ward, "Internet-Draft: An Architecture of Interface to the Routing System," Jun. 2013.

[32]   J. L. Boudec and P. Thiran, *Network Calculus: A Theory of Deterministic Queueing Systems for the Internet*. London: Springer Verlag, Jun. 2001.

[33]   Q. Duan, M. Zeng, and J. Huang, "Performance analysis for a service delivery platform in software-defined network," in *Proceedings of the 2015 ACM Symposium of Applied Computing (SAC 2017)*, Apr. 2015.

[34]   J. W. Roberts, "Internet traffic, QoS, and pricing," *Proceedings of the IEEE*, vol. 92, no. 9, pp. 1389–1399, 2004.

[35]   Q. Duan, "Modeling and performance analysis for composite network-compute service provisioning in software-defined cloud environment," *Digital Communications and Networks*, vol. 1, pp. 181–190, 2015.

[36]   M. Fidler and J. B. Schmitt, "On the way to a distributed systems calculus: an end-to-end network calculus with data scaling," in *Proceedings of ACM SIGMetrics/IFIP Performance 2006*, pp. 287–298, Jun. 2006.

[37]   K. R. Jackson, K. Muriki, S. Canon, S. Cholia, and J. Shalf, "Performance analysis of high performance computing applications on the Amazon Web services Cloud," in *Proceedings of the 2010 IEEE International Conference on Cloud Computing Technology and Science*, pp. 159–168, Nov. 2010.

[38]   G. Wang and T. S. E. Ng, "The impact of virtualization on network performance of Amazon EC2 data center," in *Proceedings of the IEEE INFOCOM 2010*, pp. 1–9, Mar. 2010.

Chapter 12
Flow management and orchestration for virtualized network functions in software-defined networks
Po-Han Huang[1] and Charles H.-P. Wen[2]

Abstract

This chapter discusses network function virtualization (NFV) to enhance flexibility and reduce costs in the deployment of service networks. NFV utilizes virtualization (e.g., virtual machines (VMs)) to separate network functions (NFs) from hardware in the form of virtual network functions (VNFs), for placement within general-purpose host machines. This makes it possible for network operators to serve a larger number of users and meet service level agreements (SLAs). This necessitates the intelligent management of VNFs and the flow among them. Software-defined networking (SDN) is ideal for this, because the separation of control and data planes makes it possible to centralize network operations. To better understand the concept of Network as a Service (NaaS), this chapter describes how NFV works with SDN, and how the flow among service chains of VNFs in SDN networks can be managed. We outline issues crucial to the design of networks from various perspectives using a number of performance metrics. Experimental results illustrate how SLA affects network performance in NFV with SDNs. Thus, flow management for service chains of VNFs in SDNs is also covered. Two categories of orchestration mechanism in the control plane are introduced: single flow and multiple flow. We discuss latency and throughput-aware algorithms for flow management and study the problem of resource contention in datacenter networks. Finally, a summary is provided to indicate directions for future research.

Keywords

Network function virtualization, software-defined networks, network service chaining, orchestration layer

[1]Ming Hsieh Department of Electrical Engineering, University of Southern California, Los Angeles, CA 90089, USA
[2]Department of Electrical and Computer Engineering, National Chiao Tung University, Hsinchu 300, Taiwan

12.1 Introduction

Network functions (NFs), referred to as middleboxes, have long been used in network systems to ensure the security of applications and improve performance. NFs have traditionally been located in specific host machines, such that traffic from the users subscribing to these services are directed to those machines. However, new challenges emerge as service providers are forced to accommodate growing number of users requiring additional types of service. Current network architectures are static (i.e., difficult to change); therefore, service providers are actively seeking alternative approaches to the management of subscribers and NFs. Network function virtualization (NFV) was developed to help users and providers to manage their services and network resources flexibly [1–3].

NFV involves decoupling NFs from the hardware, and then hosting the functions on general-purpose computing machines. This approach has two advantages for network systems: (1) NFs can be hosted on any machine and (2) multiple services based on VNFs can be enabled simultaneously. Virtual machines (VMs) make it possible to mount virtualizations on any computing machine. Multiple instances of NFs can be activated on different VMs in a service network so that different flows requiring identical NFs can be routed separately; i.e., they need not be located on the same machine to avoid congestion. These two benefits enhance flexibility in the utilization of network resources and scaling up of network services.

Service providers generally try to serve as many users as possible with limited resources (e.g., the number of servers). Thus, the core problem involves developing the means by which to utilize the resources of the service network efficiently in order to accommodate more users, while taking into account the location of host machines using the respective services as constraints. Thus, the traditional solution involves mounting all services requested by a user on a single machine [4]. However, fulfilling the requirements of all users with host machines of limited capacity can be difficult. This has led to the development of service chains to extend the capacity of computing systems [5].

The use of service chains makes it possible for network operators to take advantage of all available resources in the network system, by chaining together VNFs located in different host machines. Nevertheless, this leads to the problem of utilizing network resources efficiently by steering flows among service chains in order to satisfy the user requirements. These problems have been addressed by the invention of software-defined networks (SDNs), which empower current network systems operating within an NFV infrastructure [6,7]. SDN enables flow-level management in order to make routing more effective and flexible. The SDN architecture makes it possible for network operators to centralize control in order to better utilize network resources and thereby maximize the number of users in the system.

To understand how to perform flow management for service chains of NFV in SDN, this chapter elaborates on several key elements from two major categories. Figure 12.1 illustrates the structure of this chapter. Section 12.2 overviews the basics of NFs, VNFs, service chains, and SDNs as well as the relationships among them. Meanwhile, the target problem to be solved is illustrated at the end of this section.

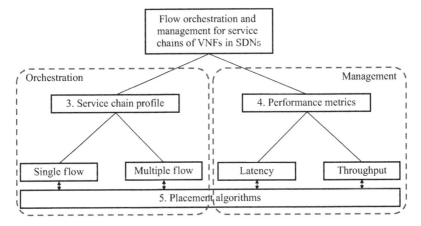

Figure 12.1   Overview of this chapter

Section 12.3 presents an investigation into one mechanism used for the orchestration of service chains of VNFs from two perspectives. One approach is the traditional method in which every user is able to allocate his service chain to only a single flow. The problem in this scenario involves orchestrating a flow among VNFs to serve the service chain of each user. However, in cases where bandwidth requirements exceed every possible flow in the network, the user cannot be served. This is unacceptable, and may increase the expenditures of network service providers. Later, the advent of the NF referred to as a "splitter/combiner" separates individual service chains into multiple flows for network operators and accommodates additional users in the network system. We also analyze methods with and without the support of SDNs when applied to these two approaches.

In Section 12.4, we summarize the flow management of VNFs in SDNs, taking two promises into consideration. One objective is the algorithm used to minimize the latency associated with users. We discuss related works that address this constraint as well as the algorithms used to achieve this goal. We also examine the expenditures of network operators. The problem of high bandwidth requirements is dealt with through the implementation of multiple flows in order to accommodate additional users. We outline the differences between various approaches to maximizing throughput and minimizing energy consumption associated with the flow management of VNFs in SDNs. A recent work (called NACHOS) is also used for case study, which makes it possible to serve the greatest number of users in Section 12.5.

Topology is the most important aspect of flow management. In Section 12.6, we highlight the issues involved in changing the topology to facilitate the management of flows with SDN. In other words, we discuss placement issues with flow management for a service chain of VNFs. In the beginning, we summarize some placement issues in the cloud-computing literature. We discuss the virtual-machine placement problem for energy saving, network traffic reduction, and resource utilization maximization. Then, we turn our focus on the placement issues while considering VNFs in NFVI. However, moving VNFs within a network system inevitably incurs costs. Thus, we point out problems that also take migration costs into account. Research directions are also provided in the end of this section. Finally, the chapter is concluded in Section 12.7.

12.2   Service chains, virtualized network functions and software-defined networks

In this section, we provide background information pertaining to service chains of VNFs in SDNs. NFs are software features with specific functionality. For example, routers and switches are two main NFs on the internet, whereas packet gateway (P-GW) and mobility management equipment (MME) are functions of cellular systems. NFs are used to manage overall network systems and facilitate the operation of computer communication networks. In content delivery networks (CDNs), video decoders and quality detectors provide user-specific experience for their services. Firewalls and intrusion detectors provide security when using the internet or other network services.

In the past when NFs were first developed, network operators built specific systems to host these functions and pre-determined the routes among the NFs. Some NFs can only be executed after other NFs; i.e., their execution is expressed as a sequence referred to as a service chain of NFs [5]. The core problem addressed in this research was the determination of the routes among NFs in networks subject to a specific order of the given service chain. The routing algorithm must not compromise with the performance of the services in the networks. Most current networks services are defined by statically combining NFs in a way that can be expressed using an NF Forwarding Graph or an NF Set construct.

Unlike NFs, which are implemented as a combination of vendor-specific software and/or hardware, VNFs use software virtualization of NFs and infrastructure elements in order to expand flexibility in the deployment of NFs. This methodology is commonly referred to as NFVs. Vendors are able to deploy VNFs using commercial-off-the-shelf hardware (i.e., general-purpose computing and storage machines) to decouple software from hardware, and thereby avoid the need to host the NFs on different machines. Using VNFs in network systems requires an NFV framework with standardized behavior for every function in the system that addresses VNFs, as outlined in [2,3]. This framework includes three key functions: VNFs, NFV infrastructure (NFVI), and NFV management/orchestration. NFVI is the infrastructure used to support the execution of VNFs within the available physical resources and to define how the VNFs can be virtualized. NFV management/orchestration is the brain of the network system used to manage the lifecycles of VNFs as well as the relationship between VNFs and NFs in the system. These functions make it possible for network operators and vendors to modify the deployment of VNFs dynamically, while managing network resources (e.g., bandwidth of networks, computing resources in servers) and satisfying additional user requirements without adversely affecting other users.

The inclusion of VNFs makes the orchestration of service chains simpler than the orchestration of service chains with NFs, thanks to the elimination of routing cost

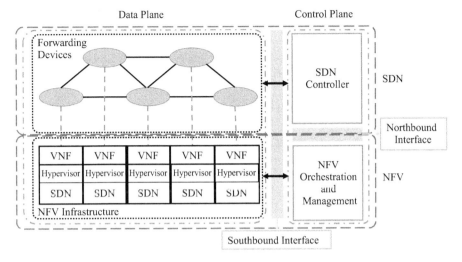

Figure 12.2   Architecture for NFV in SDN

among VNFs by locating VNFs on the same host machines and controlling the packets through the VNFs using the CPU. This also makes it possible to deploy multiple identical NFs by instantiating VNFs in the network system. This eliminates the need to route the packets of every user using the same path, by making it possible to instantiate the same NFs on different host machines. For example, network operators can deploy two firewalls with the same functions on two different host machines, such that users can access either of the machines based on their location and thereby eliminate latency between users and firewalls. In summary, NFVs apply dynamic as well as static methods in the construction and management of NF graphs or the combination of NFs. In [7], ETSI summarized these problems, including the VNF placement and VNF forwarding graph problems, within the problem of service chaining, thereby defining the process of steering the flow of traffic across a predefined set of VNFs.

Nonetheless, this does not solve the problem of orchestrating service chains of VNFs simply and efficiently [8]. SDNs are used to facilitate the smart management of network resources by partitioning the network architecture into a control plane and a data plane [9,10]. Functions on the control plane are performed by a controller, and functions on the data plane are performed by the original devices (e.g., switches and routers) on the network. The architecture of an SDN system is illustrated in Figure 12.2.

Following the integration of SDN and NFV, a controller module is used for management of the NFV and the control plane of the SDN, whereas the remaining functions of NFV (VNFs and NFV infrastructure) and the data plane of SDN are dealt with using a forwarder module. As shown in Figure 12.2, NFV management/orchestration involves the aggregation of information related to NF states and computing resources, with the SDN controller collecting information pertaining to network topology and network resources. This is then combined with information

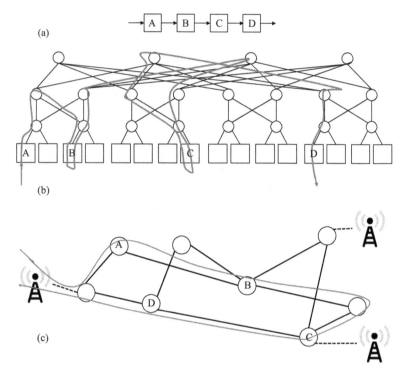

Figure 12.3   Scenarios

pertaining to user policies (i.e., quality of service, QoS, of users or SLA of network operators) in order to formulate decisions pertaining to the placement of VNFs,
the routes by which to migrate VNFs among networks, and the means by which to
steer traffic through VNFs. Communication between NFV management/orchestration
functions and SDN controllers can use the northbound interface within the SDN architecture, which was an application programming interface for application layer (above
SDN controller) to program SDN controllers. Information associated with NFV man-
agement/orchestration decisions is delivered to NFVI, whereupon the SDN controller
forwards it to SDN forwarders. Separation of the control and data planes makes it
possible for network operators to allocate network resources (e.g., bandwidth and
buffer queues) in a centralized manner. In the same manner, the orchestration of VNF
service chains can be optimized with the aid of SDNs. This makes it easier for network
operators to determine where to locate VNFs and stipulate routes for each request.

Nevertheless, designing a method to manage flow efficiently, while taking user
experience into account is not intuitive. User experience (as indicated by latency and
data rate) should be the primary objective of network operators. As to this respect,
algorithms designed to maximize throughput are no longer useful. For example,
maximizing the number of users will preclude the selection of users with large bandwidth requirements; i.e., the algorithm may disregard some individual users. In this
chapter, we review an algorithm to promote the efficient management of flow, as
illustrated by the mobile network, datacenter network, and Internet in Figure 12.3.

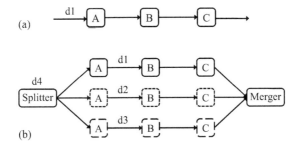

Figure 12.4   Service chain profiles of users: (a) user with single service chain (b) user with multiple service chain. User demand is d4, which exceeds d1, d2, and d3, but is smaller than their sum. Traffic demand can be split into three service chains with identical services

The objective of this architecture is to provide network as a service (NaaS) to deal with numerous types of user and numerous types of service-chain profile.

12.3   Control plane: orchestration mechanisms for service chains of VNFs

In this section, we review related studies on the service-chain orchestration mechanisms of VNFs. This discussion addresses two types of service chains: single flow and multiple flow. Typically, a service chain can be represented by a graph and its services, wherein the direction of the edges denotes the direction in which the service chain flows. Each user can be given a different service-chain profile with no loss of generality. For example, under the single-flow scenario, a user can be assigned with one service chain which executes service A, B, and C, as shown in Figure 12.4(a). But if the user's demand exceeds the capacity of a single hosting machine, multiple flows are employed. Under this situation, multiple identical instances of the same-type service are generated; therefore, the service chain profiles can be presented in Figure 12.4(b). Multiple service chains are used with particular NFs, splitter and merger, and a high-demand flow can be split into several low-demand flows in the beginning of the service chain and combined in the end of the service chain.

12.3.1   Single-flow service chain

Most previous studies on single-flow service chains for NFVs in SDNs targeted the following three problems:

(1)   Assign a flow path capable of satisfying the service chain of one user.
(2)   Design a policy by which to handle the problem of frequent packet modification as packets traverse through a service chain of VNFs.
(3)   Utilize network resources efficiently when managing the flow of service chains.

The core problem of single flow is the assignment of a flow path for the service chain associated with the user. The allocation of an appropriate service chain can give every user the best QoS. With sufficient resources, every user can be satisfied; however, when bandwidth or computing capability is scarce, users can be starved. In other words, network operators and cloud service providers must formulate a smart policy for the efficient distribution of resources among service chains. When packets traverse through a service chain, their headers need to be modified to indicate their next stops through the designated VNFs. This raises the question of how to save this information and change the headers efficiently without any error. The problem of flow mingling, referred to as intra-chain contention, should also be taken into account. In cases where flow management is not carefully implemented, intra-chain contention can cause serious congestion in the network. Every occurrence of congestion degrades the user experience and must be avoided by network operators.

MIDAS: Several studies have addressed the problem of formulating a flow path for the service chain of users. This begins with the discovery of NFs in the network. Network operators have full control over their own networks; however, they commonly need to cooperate with other network systems hosted by different network operators. This requires coordination among different network operators. MIDAS provides a framework for coordination while performing on-path flow processing [11].

Flowtags: Flowtags is the first work to address the problem of packet modification [12], based on the fact that dynamically steering traffic flows among NFs (or "middleboxes" as in the original paper) may result in intimidating network latency on modifying packet headers, which comes from dismantling the packets, extracting the information from the packets, and reforming the packet with adequate routing information. Flowtags takes advantage of the SDN architecture to retrieve contextual information from tags embedded within packet headers. SDN switches and other downstream middleboxes use tag information to facilitate routing and packet-processing.

SWF: Rather than including additional tags in packets for software-defined datacenter networks, segmented wildcard forwarding (SWF) [13] can be used to modify the fields in packet headers in order to facilitate flow migration by leveraging the symmetry of the fat-tree topology. Wildcard matching is then applied to enable to dynamic balancing of network loading. Different matching segments are checked to direct flows for SDN switches at different levels. This simple approach greatly improves the efficiency of service chaining in SDN datacenters.

SIMPLE-Fying: SIMPLE-Fying [14] provides NF policy enforcement using SDNs with centralized management as well as the decoupling of data and control planes and programmatic configuration of forwarding rules. Traffic can be steered among service chains to tame the problem of packet modification. SIMPLE-Fying also enables load balancing, which is beneficial to network operators seeking to accommodate more users. The key elements include data-plane support for chaining, unified resource management, and VNFs modification. Note that this solution is designed for service composition, which is a generalization of service chaining. The above approaches require frequent updating between NFs and SDN controllers, and may introduce issues related to scalability. SDN controllers are inapplicable for network operators seeking to provide additional services in the network systems, due to the need for back-and-forth communication.

StEERING: StEERING was proposed in [15] to facilitate the steering of traffic in large network systems. StEERING addresses the issues of scalability and mis-configuration in single-flow service chaining. This approach uses multiple tables and metadata (OpenFlow 1.1) to store information related to the routes of packets through service chains of VNFs. In this way, StEERING minimizes signaling costs between the SDN controller and switches, while storing more information in packets to enable the automatic steering of traffic.

Dynamic Chaining: There still remains the problem of NFV single-flow service chaining under the dynamic topology. In this situation, VNFs do not reside on the same host machine, even during periods of policy decision. Dynamic decision-making was proposed in [16], in a method referred to as dynamic chaining. Basically, dynamic chaining applies a finite-state machine to capture the ping-pong effect that occurs when NF states change based on decisions made in a dynamic network environment. That system, which was implemented on top of OpenStack, proved highly effective in dealing with dynamic VNF service chaining in a cloud datacenter network.

A summary of the previous work is listed in Table 12.1. As shown in the table, MIDAS, Flowtags, and SIMPLE-Fying are pioneering protocols for NFs in SDN research. These methods enhance the efficiency of single-flow service chaining among NFs in SDNs. StEERING and Dynamic Chaining put more emphasis on VNFs in SDNs, such that the relationship between the protocol and NFVI is crucial. MIDAS and Dynamic Chaining make effective decisions on assigning the route of the single flow in networks subject to the service chain of a user. Flowtags, SIMPLE-Fying, and StEERING use SDNs to deal with the problem of frequent packet modification in service chaining. SIMPLE-Fying and Dynamic Chaining, respectively, distribute traffic load among the entire network in static and dynamic network environments. However, these single-flow solutions often suffer from users with elephant flow, which consumes a huge amount of bandwidth. The appearance of this type of users in a network makes some critical NFs busier, thereby impeding flow or sacrificing the QoS. As a result, the concept of multiple flows is introduced to deal with this problem.

Table 12.1   Summary of the works on single-flow service chain

	For NFs	For VNFs	Control protocol	Packet modification	Load balancing
MIDA [11]	✓		✓		
FlowTags [12]	✓			✓	
SWF [13]	✓		✓		✓
SIMPLE-fying [14]	✓			✓	✓
StEERING [15]		✓		✓	
Dynamic chaining [16]		✓	✓		✓

12.3.2   Multiple-flow service chain

The solution based on multiple-flow service chains of NFVs in SDNs is due to the fact that the QoS requirement of users is difficult to satisfy by a single flow. Multiple flows is an ideal approach when seeking to develop an algorithm for flow management. Network operators can divide the traffic of users with elephant flows into multiple small flows to be allocated into a single flow. The following problems must be dealt with when applying multiple-flow service chains of NFVs in SDNs:

(1)   Handling of multiple flows, i.e., synchronization of packets through multiple flows for service chains.
(2)   Communication among multiple flows.
(3)   Flow mingling caused by multiple flows from the same user (i.e. "inter-chain contention") in the design of a flow-management algorithm.

When adopting the technique of multiple flows for TCP, the divided packets must be synchronized when they arrive at the destination router. Similarly, when an SDN adopts multiple-flow service chains of NFVs, the divided packets arriving at machines hosting specific NFs must be synchronized. The NF functions hosted by different machines may differ in performance, even if they are running the same software. This leads to variation in the execution time of NFs on different host machines. Given this constraint, it is difficult to divide the flow into two VNFs using traditional network systems. Take Figure 12.5(a) for example. A user subscribes to the following service chain: $A \rightarrow B \rightarrow C \rightarrow D$. Due to the scarcity of network resources for $B \rightarrow C$, the network operator may choose to divide the flows into multiple sub-flows for NF C and then combine the sub-flows in NF D, as shown in Figure 12.5(b). However, the lack of synchronization among the NF C subflows deduces that the tasks will not finish at the same time; therefore, the corresponding packets will not arrive at NF D in the correct order. This means that NF D must reorder the packets received from different sub-flows, resulting in the consumption of additional computing resources. Parallelizing tasks in B also consumes additional computing resources. Thus, the trade-off between flow-division costs and parallel-processing gains must be carefully considered when adopting this type of solution. The management of flows to avoid congestion is another burden when dealing with multiple flows. The use of SDNs makes it possible for the

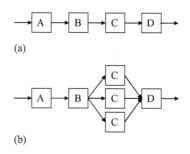

(a)

(b)

Figure 12.5   Multiple flows in service chain

centralized controller to collect the information required to reorder packets and to steer flows for the service chains of users; however, handling requests from multiple users concurrently remains a challenge. Also, for synchronizing the status of NFs in the SDN controller, communication effort and network occupation should also be taken care of in practice.

OpenNF: OpenNF [17] uses a novel control-plane architecture combining SDN and NFV to deal with the above problems. They address the problem of synchronization from the perspective of resources. As mentioned in [17], "updates to NF state due to packets may either be lost or happen out of order, violating move safety." Thus, OpenNF copies the states for each user on every NF along as well as their packets. To ensure efficient reallocation, they also address the problems of packet loss, reordering, and state consistency. Unfortunately relying on packet buffering to solve these problem induces latency and memory overhead. OpenNF reduces overhead using a flexible northbound API that controls applications capable of specifying which state to move, copy, or share.

As mentioned above, network operators should upgrade the functionality of NFs for the parallelization of tasks and packets in order to realize multiple flows; however, this approach increases the workload of each NF, thereby compromising performance. Another feasible solution involves constructing particular NFs referred to as splitters and mergers. Splitters are used for the division of flows associated with one user, whereas mergers are used to combine the flows of the same users. Splitters and mergers can communicate with the SDN controller in order to synchronize tasks among users. This makes it possible for network operators to reduce the workloads on their SDN controllers, and thereby scale up network capabilities.

Split/Merge: Split/Merge [18] is the first work to apply the split-and-merge approach to deal with the aforementioned problems. They propose state-centric, system-level abstraction to formulate elastic NFs. Taking advantage of transparent per-flow states for every NFs, Split/Merge ensures the accurate routing of flows during the splitting and merging of replicas, and reduces the latency while eliminating traffic congestion in the network. This makes it possible to scale up Split/Merge easily.

When using this type of framework, how to efficiently distribute network resources among multiple flows becomes a problem in the service chaining of VNFs in SDNs. Network operators must clearly define the efficiency of their systems, and the details of service level agreement (SLA) dealt with their customers.

Stratos: Stratos [19] applies NF migration to accommodate more users in the service network. Stratos models the determination of flow distribution as an optimization problem of linear programming, meanwhile constantly monitoring the network. When current traffic exceeds the limit of the link and thereby becomes a bottleneck, Stratos applies NF migration to dynamically eliminate congestion at the bottleneck. When Stratos detects an unacceptable packet-drop rate, the optimization algorithm is performed again to ensure quality of network service.

NACHOS: NACHOS [20] defines two types of contention problems during resource allocation for the service chaining of VNFs in SDNs with multiple flows: intra-chain contention and inter-chain contention. Intra-chain contention occurs when the service chain of one user repeatedly visits links in its flow, whereas inter-chain

Table 12.2   Summary of works related to multiple-flow service chain

	Synchronization	Architecture	Resource
OpenNF [17]	✓	✓	
Split/Merge [18]	✓	✓	
Stratos [19]			✓
NACHOS [20]			✓

contention occurs when multiple service chains from different users compete for the same network resources. NACHOS is built on the architecture of splitters and mergers in [18]; however, the focus is on handling these two contention problems within a cloud datacenter network. Details of this algorithm are elaborated in Section 12.5.

Table 12.2 presents a summary of previous works related to multiple-flow service chaining. As shown in Table 12.2, OpenNF and Split/Merge both target the synchronization problem with fundamental architectures to make multiple-flow communication possible. Stratos and NACHOS both address the issue of efficiently utilizing network resources in multiple-flow scenarios.

In summary, this section presents an overview of important related works, and provide supporting architectures on which to orchestrate the service chains of single flow and multiple flow. In the following section, we explore algorithms used for flow management in the service chaining of VNFs in SDNs. A number of these algorithms are built on the architectures introduced in this section, whereas others examine service-chaining problems from a theoretical perspective.

12.4   Data plane: flow management algorithms for service chains

In this section, we study performance metrics from two viewpoints: users and network operators. A latency-aware algorithm is preferred when user experience is the priority such that dealing with latency constraints becomes the primary objective in the design of a flow-management mechanism. A throughput-aware algorithm is preferred when utilizing network resources for maximizing profits is the priority. A number of recent flow-management works are introduced in this section.

12.4.1   Latency-aware algorithms

In this section, we begin by summarizing related works aimed at reducing latency. *MIDAS*: MIDAS [11] uses a first-fit algorithm and places all of the NFs in a service chain within the same host machine in order to reduce latency in the flow among NFs. *StEERING*: StEERING [15] applies a greedy placement heuristic to reduce flow latency in the service chain of VNFs, and generates a service dependency graph between all services. Dependency is defined as two NFs appearing consecutively in a service chain, and the degree of such dependency is determined by the amount of traffic. StEERING picks the service with the highest dependency, and locates it in the optimal location. *OpenNF*: rather than reducing the latency of single-flow service chains, OpenNF focuses on latency induced by multiple-flow assignment and processing. This method reduces latency through the use of parallel processing. More details of OpenNF can be found in [17]. *NSC*: NSC [21] addresses the issue of latency by defining a metric to represent the latency associated with a service chain using a variety of placement strategies. This work gives a detailed calculation of latency associated with packet traversal in a service chain, including the transmission time from one machine to another and the CPU time for executing functions. NSC applies a genetic algorithm (GA) for minimizing cost. *MILP*: MILP [23] uses a utility-based objective and formulates the problem as a mixed integer linear programming (MILP) problem. However, as claimed in [23], the runtime used by MILP for an optimal placement grows exponentially with the number of flows, thereby limiting its practicality.

12.4.2   Throughput-aware algorithms

MIDAS: As discussed above, MIDAS addresses the problem of latency through plac-ing all NFs within a host machine [11]. From the perspective of throughput, MIDAS employs a greedy comparison for the distribution of computational load among NF providers. Basically, this method assigns the NF task to a provider with lower load to achieve balance. *SIMPLE-Fying*: SIMPLE-Fying [14] leverages the similarity among several flows, rather than dividing the stream into chunks of fixed size. Thus, SIMPLE-fying reduces usage of links to increase the data rate to the rest of the stream. *Stratos*: Stratos [19] addresses throughput of the network through the application of multiple flows. The flow-distribution problem is formulated as a linear program-ming model, and the problem of optimization considers maximum throughput as the objective. *NSC*: NSC [21] proposes a metric to enable a comparison of bandwidths associated with different placement strategies. A GA is then used to derive the solution with the lowest cost.

    MIP: MIP [22] provides a detailed model of the application of embedded virtual networks for SDN management and orchestration. The problem is formulated as a mixed integer programming model in order to determine where to place the VNFs and how to embed flows within them. *MILP*: MILP [23] uses a utility-based objective to formulate the problem using mixed integer linear programming (MILP). However, solving the placement problem using MILP leads to exponential growth with an increase in the number of flows, which means that it cannot be applied to large scale problems. Thus, the authors proposed an alternative heuristic based on the iterative updating of constraints while taking into account issues related to latency and throughput. *VNF-OP*: VNF-OP [27] seeks to achieve higher throughput by modeling the problem within a multi-stage graph and then solving it with the Viterbi algorithm. VNF-OP provides near-optimal throughput and reduces operational costs, and can be applied to a variety of network topologies. A summary of the algorithms mentioned above and their objectives is presented in Table 12.3.

Table 12.3   Summary of works related to flow-management
algorithms

	Latency	Throughput
MIDAS [11]	✓	✓
SIMPLE-fying [14]		✓
StEERING [15]	✓	
OpenNF [17]	✓	
Stratos [19]		✓
NSC [21]	✓	✓
MIP [22]		✓
MILP [23]	✓	✓

12.5   Case study: Network-Aware CHains Orchestration Selection (NACHOS)

As discussed in Section 12.3, NACHOS takes advantage of multiple flows to deal with the problem of service-chaining for VNFs in SDNs. In this section, we look into greater detail regarding the means by which NACHOS handles the problem of resource contention in software-defined datacenter networks.

12.5.1   System architecture

In [24], NACHOS assumes that the underlying service datacenter network is a fat-tree topology. A fat-tree datacenter features great flexibility when setting up new computing machines using inexpensive commodity switches. NFs are mounted only on host machines, due to the use of commodity switches, which may not be sufficiently powerful to perform highly intensive NFs. This method allows multipath routing; therefore, a fat-tree topology possesses the characteristic of re-arrangeable non-blocking, as long as every link in the datacenter network is the same. In other words, the bottleneck is the network-card interface (i.e., the link connecting the machine and the edge switch). Generally, the complexity of a fat-tree topology also depends on the number of ports in a given switch. For example, if each switch has 6 ports, then a $k = 6$ fat-tree topology is deployed as the network for the datacenter in NACHOS.

NACHOS mainly focuses on determining which service j of type i in server s should be allocated to the service chain of a user, and which service chains can be combined to share a link. In formulating this problem, NACHOS transforms the flow management process into a graph model, referred to as a service-to-server graph. The target problem is combinatorial and difficult to analyze without a systematic approach. The service-to-server graph facilitates analysis and helps find solutions. Figure 12.6 presents an example of a service-to-server graph. Service-chain profiles can be projected onto host machines in order to meet user requirements, which makes it possible to link service instances according to their machine locations. Graph $G(V, E)$. $v_{i,j,s} \in V$ indicates whether service instance j of type i is located in server s.

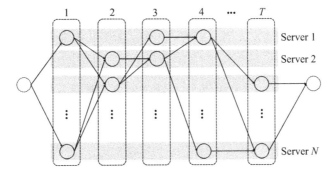

Figure 12.6   Service-to-server graph of service chains in cloud datacenter networks

$e_{(i,j,s),(i+1,\hat{j},\hat{s})} \in E$ denotes whether a link exists between server s and machine $\hat{s}$. Note that if the adjacent types of service (i.e., stages in the graph) are located in the same host machine, then only one link from $v_{i,j,s}$ to $v_{i+1,j,s}$ exists, because internal links are assumed to have unlimited bandwidth for communication within a given machine.

12.5.2   Details of NACHOS

In NACHOS, a general model is used to determine the maximum available bandwidth, given a set of services in a chain with VM locations. The optimal solution for each user is then analyzed according to the current state of the network within a fat-tree topology. The overall flow is illustrated in Figure 12.7. Details are presented as follows:

(1)   Select the service chain with the maximum available bandwidth.
(2)   Assign services to one chain of a user according to demand, and update the network status.
(3)   Repeat (1) until the SLA of the user is satisfied. In the event that no service chain is available to this user, service to this user is postponed and the original network status is restored.

   The first problem solved by NACHOS is flow mingling, which may incur intra-chain contention. Given a set of servers $\mathscr{S}$. $P_{s,\hat{s}}$ denotes the path set from server s to server $\hat{s}$, and $f_{(s,\hat{s}),p}$ represents the fraction of bandwidth assigned to this path $p \in P_{s,\hat{s}}$, wherein the bandwidth capacity of every link is B_l. $M \in E$ denotes the path selected for this service chain, and L denotes the set of all links in the datacenter network. The linear programming model used to compute maximum available bandwidth d is shown as follows:

$$\max_{d, f_{(s,\hat{s}),p}} d \tag{12.1}$$

s.t.

$$\sum_{p \in P_{s,\hat{s}}} f_{(s,\hat{s}),s} = d, \quad \forall (s,\hat{s}) \in M \tag{12.2}$$

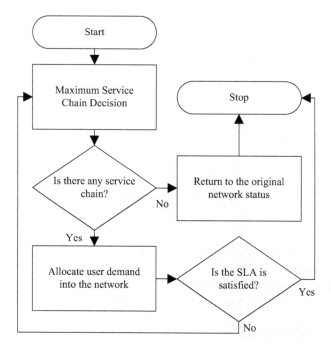

Figure 12.7   Flowchart of NACHOS

$$\sum_{s,\hat{s}\in\mathcal{S}}\sum_{p\in P_{s,\hat{s}}\wedge l\in p} f_{(s,\hat{s}),p} \leq B_l, \quad \forall l \in L \tag{12.3}$$

$$d \in \mathbb{Q} \tag{12.4}$$

$$f_{(s,\hat{s}),p} \in \mathbb{Q} \tag{12.5}$$

The rationale behind these constraints is: From (12.2) it is inferred that the sum of the fractional bandwidth assigned to each path should be equal to the maximum bandwidth available in the path set. From (12.3) it is meant that the flow in every combination from server s to server $\hat{s}$ must not exceed the bandwidth capacity of every link. Based on this model, given a set of services in a chain within their related locations in any topology, the optimal value of this model can be computed by the Simplex method. Next, considering the non-blocking characteristics of fat-tree data-center networks, an efficient solution to inter-chain mingling is proposed as follows. Given a set of services in a service chain with VM locations, the maximum demand of one service chain is $\min\left\{\dfrac{D_1}{\sum_{\forall i\in T}\sum_{\forall j\in C_i}x_{i,j,1}}, \dfrac{D_2}{\sum_{\forall i\in T}\sum_{\forall j\in C_i}x_{i,j,2}}, \ldots, \dfrac{D_s}{\sum_{\forall i\in T}\sum_{\forall j\in C_i}x_{i,j,s}}\right\}$, where $D_s \in L$ the capacity of the link from server s.

The second problem is how to apply the previous result for intra-chain mingling to the service-to-server graph. Meanwhile, taking advantage of this graph model, how to obtain an efficient solution for solving the inter-chain mingling problem. Dynamic programming is an effective approach to solving the problem of service-to-server mapping, by representing the problem within finite stages. $h(\cdot)$ denotes the objective in each stage, and S_i is the potential state in stage i. In dynamic programming, the types of services in a chain are expressed as one stage, whereupon the entire problem can be solved via reverse induction. Thus, the overall procedure is implemented according to the following formula:

$$h^*_{i,j,s}(S_i) = \max_{x_{i,j,s}} h_{i,j,s}(S_i, x_{i,j,s}), \tag{12.6}$$

$$h_{i,j,s}(S_i, x_{l,J,s}) = \min_{x_{i,j,s}} \left\{ \frac{D_s}{c_s}, h_{i+1,\hat{j},\hat{s}}(S_{i+1}, x_{i+1,\hat{j},\hat{s}}) \right\} \tag{12.7}$$

In every iteration, the algorithm examines the optimal value obtained in the next stage, and keeps c_s for every server in every candidate service chain. This algorithm possesses Markovian properties, which means that decisions in this stage depend only on the previous stage, such that the algorithm need only be performed N times during each iteration. Thus, the complexity of the algorithm is $O(N \times T)$.

After identifying the service chain with the maximum available bandwidth, NACHOS determines the amount of bandwidth that should be allocated to the service chain. According to Tso and Pezaros [25], in which flow is divided into multiple sub-flows, the technique of equal-cost multipath (ECMP) is used to enable multiple-flow service-chain assignment as follows. After using dynamic programming to compute the maximum available bandwidth for a service chain, the algorithm determines whether the current maximum available bandwidth exceeds half the residual demand of the user, but remains less than the total residual demand of the user. As long as the above criteria are satisfied, NACHOS assigns the half residual demand of the user to this flow. If the maximum available bandwidth exceeds the total residual demand of the user, then NACHOS assigns the total residual demand of the user to the service chain. ECMP makes it possible to reserve bandwidth for future demand.

12.5.3   Simulation results

NACHOS runs simulations on a fat-tree datacenter with the sizes of 6, 8, and 10 (i.e., (k), the number of ports in a switch), where the number of host servers are 54, 128, and 250, respectively. Without loss of generality, each link in the datacenter network is set to 1 Gbps. The computing capacity of each server is 3 slots; i.e., up to 3 services can be hosted on one machine. The bandwidth demand of one user averages 200 Mbps. Performance is evaluated under a variety of scenarios. First, the number of users ranges from 100 to 1,000 and the total number of satisfied users in the service network is recorded. Second, the total amount of residual bandwidth in the network is monitored in order to evaluate the utilization of network resources. Third, the number of different types of service to which one user can subscribe (a.k.a. the maximum chain length) ranges from 6 to 15. The total number of users that cannot be served by a service network can be used as an index of SLA violations to enable a comparison of NACHOS with conventional Single-Flow Service Chain (denoted "SSC"). The total

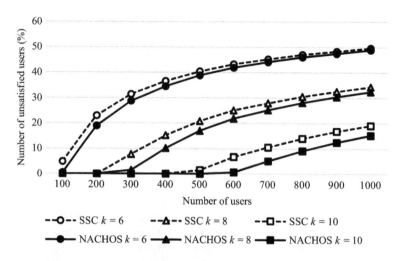

Figure 12.8   Number of unsatisfied users vs. number of users

residual bandwidth under different scenarios can also be used as an index to measure the load-balance status of the entire network.

In the first experiment, the relationship between the number of unsatisfied users and the number of total users is derived, in which the maximum length of the service chain was set to 10. Figure 12.8 presents the number of unsatisfied users along the Y-axis corresponding to the number of total users along the X-axis, under various scenarios. An increase in the total number of users led to an increase in the number of unsatisfied users. In Figure 12.8, NACHOS achieved better performance with regard to the number of unsatisfied users than did SSC, regardless of the scale of the networks. Specifically, NACHOS is able to serve 10% more users than is SSC, when the size (k) of the fat-tree service network is 10, which provides 12–20 Gbps throughput, depending on the total number of users. As the scale of the datacenter increased, the performance of NACHOS increases, due to a higher probability of allocating better service chains.

In the second experiment, the number of unsatisfied users was also monitored under various maximum chain lengths and a total of 500 users. The number of unsatisfied users was shown to increase with the maximum chain length, due to the fact that more VMs are used to provide services and chaining the services incurs higher costs. As shown in Figure 12.9, as the scale (k) of the datacenter increased, NACHOS outperformed SSC by serving 10% more users (i.e., accommodating throughput of 10 Gbps). This is a clear demonstration that the problem of flow mingling can be effectively dealt with using NACHOS.

The distribution of satisfied users is also examined in order to elucidate the influence of each demand category under various approaches. As shown in Table 12.4, NACHOS was able to satisfy 15% more high-demand users than was SSC under all scenarios. NACHOS incurs fewer SLA violations from high-demand users, which makes it possible to reduce the costs incurred by network service providers.

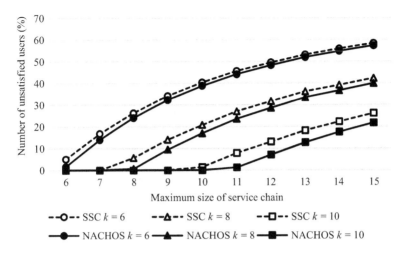

Figure 12.9   Number of unsatisfied users vs. maximum chain length

Table 12.4   Distribution of satisfied users

Demand range	Light user	Medium user	Heavy user
	Below 125 Mbps	125–275 Mbps	Above 275 Mbps
SSC $k = 6$	53.3%	31.2%	15.5%
NACHOS $k = 6$	38.1%	32.0%	29.9%
SSC $k = 8$	40.1%	42.8%	17.1%
NACHOS $k = 8$	34.3%	34.5%	31.2%
SSC $k = 10$	30.7%	43.4%	25.9%
NACHOS $k = 10$	31.8%	34.6%	33.6%

12.6   Service placement: more on flow management

Virtualization and migration provide network operators with flexibility to move NFs to the designated server, thereby facilitating the orchestration of service chains. Before the advent of service chaining of VNFs, VM placement problem (VMP) [30] has been widely studied for several years in the realm of cloud computing, especially for datacenters. In this section, we will first overview the VM placement issues in cloud datacenters, and then turn to discuss the flow management for VNFs.

12.6.1   VM placement for flow management

At the beginning, VMP was studied since cloud-service providers (CSPs) intended to reduce the energy consumption when maintaining the operation of datacenters. Consolidating multiple underutilized servers into a fewer number of servers and turning off those redundant servers can save more energy in a datacenter. Since maintaining a non-active server (i.e. not running any NFs) will consume a huge amount of power, therefore, turning off redundant servers helps CSPs save more energy in the long run. In this line of studies, how to decide which servers should be turned off, and which VMs should be migrated to which servers are the core problems. Moreover, network throughput cannot be neglected when addressing the energy minimization. How to balance this tradeoff between network throughput and energy consumption is another core issue to be dealt with. The solutions such as best-fit, first-fit, and greedy approaches are widely investigated to minimize energy consumption. The underlying idea among these solutions is bin packing, which CSPs intend to put as many VMs into fewer servers as possible and to turn off as many servers as possible.

On the other side, since communications between VMs for synchronization also occupy network resources in datacenters, how to manage these VMs and how to place these VMs to support services is the main focus for network traffic minimization. This type of communication becomes a hurdle for network throughput since some network resources are used for supporting the communication between VMs. Therefore, the intuitive solution to this problem is similar to the one for energy minimization in datacenters: placing as many VMs as possible into few servers. The only difference is the number of active servers to be used for this problem; therefore, several methods like best fit, convex optimization and some heuristics are discussed to minimize network traffic in a datacenter. Most of these solutions first place the VMs with the highest demand on communication into a server, then the VMs with the second highest demand, and so on in order.

When considering the communication between VMs, intelligent placement is not the only solution to reduce traffic among VMs located in different servers. Flow management also helps a lot in improving network performance. The concept of SDN is widely adopted since a CSP has the full control over all network links, routers, and servers in the datacenter. With the invention of SDN, steering the traffic among VMs while utilizing the fewest traffic becomes much easier than before (without SDN). As a result, new issues arise under this circumstance. For example, the CSP need to decide where to place SDN controllers and how to manage the traffic between servers, routers, and SDN controllers.

Bandwidth usage for migration also needs to be taken care of when moving VMs from a server to another. There are several reasons why VMs should be migrated. First, migration occurs when some servers are turned off for energy minimization. When turning off servers, the VMs on these servers need to be moved to other active servers. Under this situation, huge amount of traffic occupies the network resources. In addition, for reducing traffic between communicating VMs, the best way is to place them into the same hosting server, and this operation requires VM migration. Note that both VM migration and the communication between VMs demand network resources, and need to be considered concurrently for better performance. Last but not least, the workload variation should also be taken into account as well. The workload of a VM (e.g., CPU, memory, storage) may fluctuate over time. When the summation of the workload of all VMs in a server exceeds the capacity of the hosting server, VM migration is called to alleviate the workload of the over-utilized servers. The key factor to reduce the migration cost in a datacenter is the "distance" between two servers. The notion of distance here does not only refer to the distance of physical locations between two servers, but also the logical hops between them. For communication, more hops indicate more network resources used for supporting migration, also suggesting fewer resources for supporting services requested by users. When off-loading a hosting server, VMs to be migrated will be placed in a nearby server in terms of hops.

Also, the joint problem to simultaneously deal with energy consumption minimization and network traffic minimization for datacenter networks is also studied in [31]. In this work, energy-efficient with QoS-guarantee VM Placement algorithm (EQVMP) is proposed to achieve both goals at the same time. Based on the resource demands, VM traffic, and network topology, EQVMP follows the following steps: First, EQVMP partitions VMs into groups. Second, energy saving is calculated based on the best-fit decreasing principle. Then, traffic load is balanced by a dynamic-rerouting algorithm named Dynamic & Disjoint ENDIST-based (D2ENDIST) [32]. According to the simulation result, EQVMP achieves better throughput with fewer physical machines to host VMs than the other placement approaches (even with more physical machines). As a result, EQVMP successfully minimizes both energy consumption and network traffic in cloud datacenters.

12.6.2 Maximization of resource utilization

Maximizing resource utilization is another research direction studied in the literature. Cloud infrastructures are commonly composed of multiple physical and virtual resources such as CPU, memory, and network resources. In this context, efficient and balanced utilization of these resources is an important issue. To support more users while satisfying their SLAs, CSPs can either provide more physical machines or better utilize their current resources to save cost. Therefore, how to rearrange their VMs in their cloud datacenter to spare more resources is a crucial problem. The mathematical models like binary/integer programming and matrix transformation were applied to formulate this problem. Also, it is found difficult to solve such problem as the input size is huge (i.e., the datacenter with thousands of servers), and thus heuristic solutions were proposed for practicality.

Next, VMs can be not only placed in a datacenter, but also can be distributed across multiple datacenters (i.e., different cloud datacenters owned by one or many CSPs). Under this scenario, how to direct traffic from one datacenter to another is challenging. However, the traffic across different clouds is highly difficult to manage and predict when traveling through the Internet. Moreover, the topologies for some clouds can be ad-hoc, not like the common types (e.g., fat-tree, BCube, PortLand). In this context, the probability method to analyze the performance are useful, and the online learning algorithm plays an important rule for achieving better performance. The online learning algorithm is useful for making decisions based on the previous results, and the current status of networks. On the other hand, how to cooperate every CSP in this multi-cloud scenario is another issue. Since every CSP focuses on different objectives, and all of them intend to optimize their own profits, the problem to satisfy

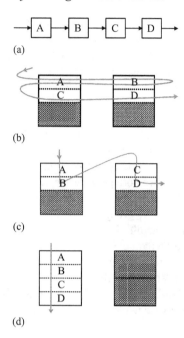

(a)

(b)

(c)

(d)

Figure 12.10   Different service placements for service chain

everyone's requirement is challenging. As a result, the solution based on the game theory is proposed for this purpose.

The VM placement in flow management for a service chain of VNFs are similar to that in network-traffic minimization. Given a service chain illustrated in Figure 12.10(a), Figure 12.10(b) presents an example for the flow-mingling problem in which the network flow of a service chain travel back and forth between two host machines due to the location of the NFs. With proper VM migration, the flow-mingling problem can be resolved as shown in Figure 12.10(c). Providing host machines with sufficient capacity to accommodate all of the VNFs in a service chain makes it possible for network operators to reduce network costs, as shown in Figure 12.10(d). Following the development of NFV techniques, flexibility becomes the dominant factor in the design of new algorithms with which to orchestrate VNFs for flow management. However, flexibility adds complexity. Thus, the research is conducted to search for efficient algorithms to facilitate the practical management and optimization of NFV orchestration with VM placement.

12.6.3   Recent placement algorithms

While taking latency into consideration, the placement strategy becomes more critical. The latency is calculated based on the number of hops through the route of service chain. To take Figure 12.10 for example, if the number of hops between these two servers is 1, the number of hops used by this service chain in Figure 12.10(b) is 3. Figure 12.10(c) shows a better result and the service chain only travels through 1 hop. Figure 12.10(d) shows the best result, in which the number of hops is 0 in this case. Considering the SLA of every user, if the latency requirement of a user is 3, the above placement solutions are all acceptable. However, if the latency requirement of a user is 1, only the solutions in Figure 12.10(c) and (d) work for this user. In summary, the SLA requirements of users need to be considered both in the VM placement and flow management to provide a feasible and low-cost solution for CSPs.

As mentioned previously, conventional solutions usually replace VMs for hosting services in chains with VNFs in SDNs. *NetVM*: NetVM proposed in [4] is a platform used to enable complex NFs with commodity hardware devices. NetVM analyzed the pros and cons of the Intel Data Plane Development Kit (DPDK), and developed a solution to deal with the drawbacks, while enabling the deployment of flexible VMs in SDNs. *VNF-P*: The authors in [26] present and evaluate a formal model for resource allocation of VNFs within NFV environments, called VNF placement (VNF-P). In this work, VNF-P focuses on a hybrid scenario where the services can be provided by dedicated physical hardware or running on VM instances. Based on their experiments with two types of service chains in a small service provider scenario, NFV can successfully expedite packet communication for a service chain.

In [19,23,27–29], hybrid placement-oriented and network-oriented solutions are investigated. *Specifying*: In [27], the authors formulated the problem of traffic congestion in a service chain, using a traffic load balancer to eliminate congestion. They focused on the deployment of traffic load balancers and VMs in the network topology to enhance system throughput. *Stratos*: In [19], the authors studied the influence of middlebox location and proposed a solution to steer flows under various traffic loading conditions. An adaptive approach was developed to satisfy the overall demand with efficiency. With Stratos, users are allowed to realize arbitrarily complex logical topologies by abstracting away the complexity of efficient VNFs composition and provisioning. *VNF-OP*: The VNFs orchestration problem (VNF-OP) was emphasized in [28]. Virtual networking embedding is used to determine where to place VNFs and how to route traffic flows with the lowest operation cost, and last the results can be transformed into a corresponding physical network. The authors in VNF-OP proposed two algorithms: one is CPLEX-based optimal solution for small networks and the other is a heuristic to deal with larger networks. The simulation results compares the performance between the heuristic and the optimal solution.

MILP: In [23], the authors outlined a joint model for service-placement and flow-steering problems. MILP aims to determine the placement of services and routing of the flows, meanwhile, minimizing the utilization of network links and network node core. The result effectively solved the proposed problem under various objectives. They also provided an incremental approach to the addition of flows, in order to deal with large systems of high tractability. Note that MILP may not be practical due to high time complexity. *ILP*: In [29], the authors studied the impact of processing-resource sharing among VNFs and scalability costs in an NFVI, when multiple service chains are deployed in the network. Placing VNFs onto NFV nodes induces two types of costs: upscaling costs and context-switching costs. Such costs lead to a trade-off of VNF sizes. To further handle this problem, the authors in [29] formulated it into integer linear programming, and run with CPLEX through simulation on a small scale of the network with various applications with different SLAs.

Placement algorithms for saving energy and reducing operational costs were studied for a long time prior to the development of NFV. However, during the orchestration of service chains of VNFs in SDNs, previous solutions may not be applicable to complex problems with multiple objectives. Thus, there remains a need for new approaches for minimizing operational costs and maximizing QoS.

12.7 Conclusion

This chapter presents an overview of flow orchestration and management for service chains of VNFs in SDNs. After an examination of previous studies, we describe an architecture based on NFV and SDNs. Two types of mechanisms (single flow and multiple flow) are then used to illustrate the orchestration of flow. Latency-aware and throughput-aware algorithms are used to illustrate flow management and design principles. We also examine important issues involved in VNF placement in operator networks. The aim was to give readers a picture of this line of research and describe several seminal works on network as a service (NaaS) and its applications.

References

[1]   B. Han, V. Gopalakrishnan, L. Ji, and S. Lee, "Network Function Virtualization Challenges and Opportunities for Innovations," *IEEE Communications Magazine*, vol. 53, no. 2, pp. 90–97, Feb. 2015.

[2]   ETSI GS NFV 001: "Network Functions Virtualization (NFV): User Cases".

[3]   ETSI GS NFV 002: "Network Functions Virtualization (NFV): Architectural Framework".

[4]   J. Hwang, K. K. Ramakrishnan, and T. Wood, "NetVM: High Performance and Flexible Networking using Virtualization on Commodity Platforms," in *USENIX NSDI'14*, Seattle, WA, USA, Apr. 2014.

[5]   H. Jeon and B. Lee, "Network Service Chaining Challenges for VNF Outsourcing in Network Function Virtualization," in *IEEE ICTC'15*, Jeju, Korea, Oct. 2015.

[6]   D. Kreutz, F. M. V. Ramos, P. Verissimo, C. E. Rothenberg, S. Azodolmolky, and S. Uhlig, "Software-Defined Networking: A Comprehensive Survey," *ArXiv e-Prints*, Oct. 2014. [Online]. Available: http://arxiv.org/pdf/1406.0440v3.pdf

[7]   J. Blendin, J. Ruchert, N. Leymann, G. Schyguda, and D. Hausheer, "Position Paper: Software-Defined Network Service Chaining," in *IEEE European Workshop on Software Defined Networks 2014*, Budapest, Hungary, Sep. 2014.

[8]   P. Quinn and T. Nadeau, "Service Function Chaining Problem Statement," *Active Internet-Draft*, IETF Secretariat, Apr. 2014.

[9] Y. Li, F. Zheng, M. Chen, and D. Lin, "A Unified Control and Optimization Framework for Dynamic Service Chaining in Software-Defined NFV System," *IEEE Wireless Communications*, vol. 22, no. 6, pp. 15–23, Dec. 2015.

[10] J. Matias, J. Garay, N. Toledo, J. Unzilla, and E. Jacob, "Toward an SDN-Enabled NFV Architecture," *IEEE Communication Magazine*, vol. 53, no. 4, pp. 187–193, Apr. 2015.

[11] A. Abujoda and P. Papadimitriou, "MIDAS: Middlebox Discovery and Selection for On-Path Flow Processing," in *IEEE COMSNET'15*, Bangalore, India, Jan. 2015.

[12] S. K. Fayazbakhsh, V. Sekar, M. Yu, and J. C. Mogul, "FlowTags: Enforcing Network-Wide Policies in the Presence of Dynamic Middlebox Actions," in *ACM HotSDN'13*, Hong Kong, China, Aug. 2013.

[13] K.-T. Kuo, C. H.-P. Wen, C. Suo, and I.-C. Tsai, "SWF: Segmented Wildcard Forwarding for Flow Migration in OpenFlow Datacenter Networks," in *IEEE ICC'15*, London, UK, May 2015.

[14] Z. A. Qazi, C.-C. Tu, L. Chiang, R. Miao, V. Sekar, and M. Yu, "SIMPLE-fying Middlebox Policy Enforcement Using SDN," in *ACM SIGCOMM'13*, Hong Kong, China, Aug. 2013.

[15] Y. Zhang, N. Beheshti, L. Beliveau, *et al.*, "StEERING: A Software-Defined Networking for Inline Service Chaining," in *IEEE ICNP'13*, Goettingen, Germany, Oct. 2013.

[16] F. Callegati, W. Cerroni, AC. Contoli, and G. Santandrea, "SDN Controller Design for Dynamic Chaining of Virtual Network Functions," in *IEEE European Workshop on Software Defined Networks 2014*, Budapest, Hungary, Sep. 2014.

[17] A. Gember-Jacobson, R. Viswanathan, C. Prakash, *et al.*, "OpenNF: Enabling Innovation in Network Function Control," in *ACM SIGCOMM'14*, Chicago, IL, USA, Aug. 2014.

[18] S. Rajagopalan, D. Williams, H. Jamjoom, and A. Warfield, "Split/Merge: System Support for Elastic Execution in Virtual Middleboxes," in *USENIX NSDI'13*, Lombard, IL, USA, Jan. 2013.

[19] A. Gember, A. Krishnamurthym S. St. John, *et al.*, "Stratos: A Network-Aware Orchestration Layer for Virtual Middelboxes in Clouds," *ArXiv e-Prints*, Mar. 2014. [Online]. Available: http://arxiv.org/pdf/1305.0209v2.pdf

[20] P.-H. Huang, K.-W. Li, and C. H.-P. Wen, "NACHOS: Network-Aware CHains Orchestration Selection for NFV in SDN Datacenter," in *IEEE CloudNet'15*, Niagara Fall, Canada, Oct. 2015.

[21] T. Kim, S. Kim, K. Lee, S. Park, "A QoS Assured Network Service Chaining Algorithm in Network Function Virtualization Architecture," in *IEEE/ACM CCGrid'15*, Shenzhen, China, May 2015.

[22] R. Guerzoni, R. Trivisonno, I. Vaishanavi, *et al.*, "A Novel Approach to Virtual Networks Embedding for SDN Management and Orchestration," in *IEEE/ IFIP NOMS'14*, Krakow, Poland, May 2014.

[23] A. Mohammadkhan, S. Ghapani, G. Liu, W. Zhang, K. K. Ramakrishnan, and T. Wood, "Virtual Function Placement and Traffic Steering in Flexible and

Dynamic Software Defined Networks," in *IEEE LANMAN'15*, Beijing, China, Apr. 2015.

[24] C. E. Leiserson, "Fat-Trees: Universal Networks for Hardware-Efficient Supercomputing," *IEEE Transactions on Computers*, vol. 34, no. 10, pp. 892–901, Oct. 1985.

[25] F. P. Tso and D. P. Pezaros, "Improving Data Center Network Utilization Using Near-Optimal Traffic Engineering," *IEEE Transactions on Parallel and Distributed Systems*, vol. 24, no. 6, pp. 1139–1148, Jun. 2013.

[26] S. Mehraghdam, M. Keller, and H. Karl, "Specifying and Placing Chains of Virtual Network Functions," in *IEEE CloudNet'14*, Luxemburg, Oct. 2014.

[27] M. F. Bari, S. R. Chowdhury, R. Ahmed, and R. Boutaba, "On Orchestrating Virtual Network Functions in NFV," *ArXiv e-Prints*, Mar. 2015. [Online]. Available: http://arxiv.org/abs/1503.06377

[28] H. Moens and F. D. Turck, "VNF-P: A Model for Efficient Placement of Virtualized Network Functions," in *IEEE CNSM'14*, Rio de Janeiro, Brazil, Nov. 2014.

[29] M. Savi, M. Tornatire, and G. Verticale, "Impact of Processing Costs on Service Chain Placement in Network Functions Virtualization," in *IEEE NFV-SDN'15*, San Francisco, CA, USA, Nov. 2015.

[30] F. L. Pires and B. Baran, "Virtual Machine Placement Literature Review," *ArXiv e-Prints*, Jun. 2015. [Online]. Available: http://arxiv.org/pdf/1506. 01509v1.pdf

[31] S.-H. Wang, P. P.-W. Huang, C. H.-P. Wen, and L.-C. Wang, "EQVMP: Energy-Efficient and QoS-Aware Virtual Machine Placement for Software Defined Datacenter Networks," in *IEEE ICOIN 2014*, Phuket, Thailand, Feb. 2014.

[32] W.-C. Lin, G.-H. Liu, K.-T. Kuo, and C. H.-P. Wen, "D2ENDIST-FM: Flow Migration in Routing of OpenFlow-based Cloud Networks," in *IEEE CLOUD 2013*, San Francisco, CA, USA, Nov. 2013.

Chapter 13

On-demand network virtualization and provisioning services in SDN-driven cloud platforms

Maha Shamseddine[1], Ali Chehab[1],
Ayman Kayssi[1] and Wassim Itani[2]

Abstract

In this chapter, we present a brief survey on the latest advancements in Network-as-a-Service (NaaS) platforms based on the software-defined networking (SDN) model. This survey sheds light on the main advantages of network virtualization and programmability in cloud computing environments that led to the provisioning of dynamic, flexible, and secure virtual network services. In spite of the great benefits provided by network virtualization in the cloud, further research is still needed to tackle a plethora of technical challenges in the fields of dynamic NaaS configuration and operation. This chapter paves the way for a better understanding of the problems hindering the NaaS adoption in current cloud architectures and represents a call for action on the need for having a dynamic NaaS configuration service to tackle these problems. Accordingly, the chapter proposes the design and implementation of a centralized cloud service for creating virtual networks (VNets) in SDN-based cloud architectures.

Keywords

Virtual networks, software-defined networks, cloud computing, Network as a Service, Network optimization, dynamic network configuration

13.1 Introduction

Computing virtualization has been the main driving force for the dramatic proliferation of the cloud computing model we are familiar with today. The main service

[1] Department of Electrical and Computer Engineering, American University of Beirut, Beirut, Lebanon
[2] Department of Electrical and Computer Engineering, Beirut Arab University, Beirut, Lebanon

execution containers in modern data centers are represented in virtual machines (VMs) that can be instantiated, managed, operated, and configured on-demand based on tenants' requirements and workload patterns. The great success achieved in the field of computing virtualization has paved the way for the emergence of networking virtualization platforms that are rapidly laying the ground for specialized Network-as-a-Service (NaaS) [1] cloud offerings. Using the NaaS model, entire virtual networks (VNets) can be created and managed instantaneously upon tenants' requests. Many cloud NaaS models exist today and the number is highly increasing with the advancements in network virtualization hypervisor design. Moreover, the emergence of a novel network architecture represented in the software-defined networking (SDN) [2] concept, where network switching units can be dynamically reprogrammed to control the various aspects of networking operations, had a sizeable positive impact on the NaaS cloud support. This is due to the fact that SDN has shed the light on the significance of coupling the network programmability features with the network virtualization counterparts to truly realize an elastic networking infrastructure.

In spite of the increasing popularity in NaaS offerings, three main issues are still hindering the wide adoption of NaaS services in the cloud, namely, (1) the NaaS model does not provide any assistance or support for the tenants in designing their network topologies or managing the network-wide services, (2) the assorted pricing schemes offered by NaaS providers, as well as the occasional price variations and deals, complicate the cost-effective selection of cloud providers, and (3) NaaS services do not support VNet partitioning, whereby the VNet is partitioned into several logical parts that can communicate seamlessly as a single unit. Network partitioning allows for cost-effective and performance-efficient VNet deployment based on the tenants' preferences and constraints.

In this chapter, we present a fundamental survey on the latest advancements in NaaS platforms based on the SDN model. This survey paves the way for a better understanding of the problems hindering the NaaS adoption in current cloud architectures and represents a call for action on the need for having a dynamic NaaS configuration service to tackle these problems. Accordingly, the chapter proposes the design and implementation of NCaaS (Network Configuration as a Service), a centralized cloud service for creating VNets in SDN-based cloud architectures. The main contribution behind the presented work is to offer an integrated service that handles the different aspects of VNet creation and management on behalf of the tenant while satisfying their requirements and constraints. NCaaS relieves tenants from the burdens and complexities of VNet creation and management by supporting dynamic provisioning operations based on Quality of Service (QoS), pricing, privacy, reliability, and energy constraints set by the tenant. NCaaS employs a flexible network partitioning mechanism that allows the hosting of the network parts on multiple cloud providers to provide a minimum VNet deployment cost. A mathematical formulation of the cost optimization problem is provided along with a set of approximation algorithms for carrying out the partitioning and topology designs. These algorithms handle the negotiation with the different SDN NaaS providers and dynamically apply the necessary VNet creation, partitioning, and migration mechanisms to ensure the satisfaction of the tenants' preferences. A test bed proof of concept implementation of the NCaaS algorithms is developed on top of the OpenVirteX network virtualization platform [3] and tested using the Mininet network emulator [4]. To the best of our knowledge, NCaaS is the first centralized cloud service for creating, managing, and provisioning VNets dynamically on behalf of tenants and based on their requirements.

The rest of this chapter is organized as follows: Section 13.2 presents a description on the main network virtualization concepts and architectures. Section 13.3 provides an introduction on SDN and demonstrates the main advantages of this novel networking model. In Section 13.4, we discuss the NaaS network virtualization model in SDN environments and stand on the latest research and industry advancements achieved on this front. In Section 13.5, we introduce the NCaaS, dynamic network creation and configuration service, present a comprehensive description of the different NCaaS algorithms, and demonstrate via an experimental simulation the performance of the NCaaS test bed implementation on the OpenVirtex network virtualization platform. Conclusions and future extensions are presented in Section 13.6.

13.2   Network virtualization: basic concepts and architecture

Network virtualization is the main enabler for the next-generation networking in the fields of telecommunication and the Internet [5]. Virtualization facilitates the feasibility of creating multiple heterogeneous networks on the same physical infrastructure. Sharing the physical resources results in "elastic" network topologies composed of virtual nodes connected via virtual links to provide dynamic end-to-end connectivity services. The network virtualization property, imposed by a network virtualization hypervisor layer, provides the spawned VNets with full isolation among each other to achieve relatively high levels of privacy and security. In such virtualization environments, VNets are characterized by elevated degrees of flexibility and ease of management, elasticity and dynamism, scalability, isolation, and heterogeneity [6].

Flexibility and ease of management: Multiple VNets can be created, deployed, and destroyed independent of the underlying physical network. Service providers (SPs) can provision their choice of network topology and forwarding protocols irrespective of the underlying physical network for the sake of delivering flexible configurations to support tenants' services.

Isolation: VNets can coexist on the same physical infrastructure and their corresponding traffic is logically segregated using the VNet hypervisor. The latter is responsible for separating the VNets address space to provide higher levels of isolation and to reduce the probability of fault propagation among the configured VNets.

Elasticity and dynamism: Using network virtualization techniques, VNets can be created, and their resources could be incrementally increased up or released down based on tenants changing demand or service request patterns. All these operations could be instantaneously executed using configurable software commands. This expedites the process of resource allocation and release without the need to restructure the underlying hardware infrastructure or network configuration.

Scalability: With network virtualization, VNets can practically scale up to the resources dedicated to the physical network with minimal performance overhead imposed by the network virtualization layer. This is considered a major property to ensure the efficient operation of the coexisting VNets on a given physical network infrastructure [7].

Heterogeneity: With network virtualization, SPs can provision VNets with their arbitrary independent topologies and forwarding protocols irrespective of those of the leased physical network infrastructure.

These technical advantages of network virtualization have been originated and evolved from the great advancements in server virtualization [8], which in turn was the main enabler for infrastructure and platform services in cloud computing [9]. Server computing virtualization has been introduced by the VM software that abstracts and decouples the software from the underlying machine hardware. Multiple VMs can coexist on the same underlying physical machine thus providing full isolation, on-demand provisioning, and flexible management.

Network virtualization has witnessed extensive attention in academia and industry due to the dynamic nature of connectivity achieved which remarkably aids in reducing the operational expenses, OPEX, and the capital expenses, CAPEX, as explained in the Network Function Virtualization work presented in [10]. The dynamism and flexibility properties provided by network virtualization push for a better utilization of the network resources and ease of management and mobility of the network components. This high level of flexibility that can be achieved by VNets, and supported lately by appealing network programmability architectures represented in the SDN networking paradigm, is the primary reason behind the proliferation of virtualization techniques in the networking research and academia projects.

Network virtualization had a highly positive impact on the cloud computing industry. Today, cloud SPs are separated into two major roles: (1) the Infrastructure Providers (InPs) that mange and offer wide varieties of physical network resources and substrates that meet the requirements and needs of tenants, and (2) the SPs which provision end-to-end network services by aggregating resources from different InPs according to tenants' requests related mainly to cost, flexibility, programmability, security, and privacy, among others.

As a result, the virtual network environment [11] greatly facilitates the utilization of a dynamic network infrastructure where multiple SP's provision multiple heterogeneous VN's that can independently coexist on the same network substrates but with full isolation and flexibility.

Figure 13.1 presents the network virtualization architecture described above and demonstrates the different hierarchical categories of providers supporting this architecture. The main important property presented by the network virtualization architecture in Figure 13.1 is the ability of the different NaaS providers to inherit the underlying network infrastructure of other NaaS providers (as if it were a physical network infrastructure) and to build on top of it an isolated set of VNets. The provider hierarchy can extend as long as the base physical network resources provided by the In Ps support it. It is worth mentioning here that the NaaS providers can run any network virtualization platform such as the virtual networking embedding model described later in this section.

In [6], the authors classify network virtualization into four main network categories, namely: Virtual Local Area Network (VLAN), Virtual Private Network (VPN),

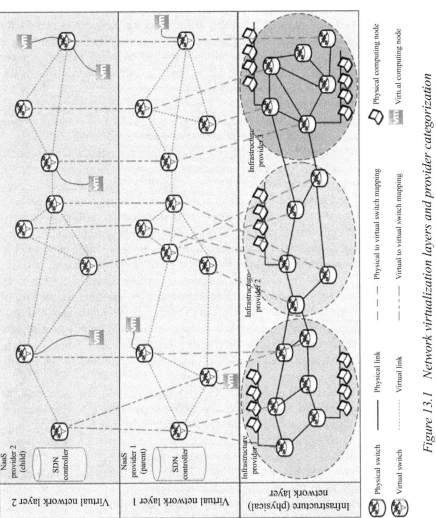

Figure 13.1   Network virtualization layers and provider categorization

active and programmable networks, and overlay networks. VLANs are networks that connect nodes and hosts together using logical addressing and have a single broadcast domain. VLANs coexist on the same physical LAN where each separate broadcast domain is enforced by the layer 2 switches to achieve traffic isolation. Broadcast domains are enforced by mapping a multiport virtual bridge on the underlying physical switch and by dedicating this bridge to a distinct VLAN. VPN is a network created on top of other networks that are located in different geographic locations. Hosts in a VPN from different network sites appear to be on the same VNet that shares the same broadcasting address. The VPN technology was initially created to ensure a secure encrypted network connection between institutional sites and their remote clients over a public internetwork such as the Internet. Many protocols currently support the VPN functionality, the most popular are IPSecurity (IPSec) [12], Point-to-Point Tunneling Protocol (PPTP) [13], and Layer 2 Tunneling Protocol (L2TP) [14]. It should be noted here that the Secure Socket Layer (SSL) and Transport Layer Security (TLS) [15] protocols can also be used to support the secure network communication of a VPN connectivity solution by employing their authentication, encryption, and integrity mechanisms. Active and programmable networks are a new generation of network technologies, protocols, and platforms that are primarily proposed to support the dynamic and active inclusion of software and hardware services into network elements. These types of networks are supposed to enhance the resiliency, security, management, and customization of the network components to ultimately enable a form of seamless plug-and-play functionality. Active and programmable networks had a great role in the demand for network virtualization to provision dynamic services decoupled from the underlying network resources. Overlay networks are networks consisting of a set of nodes with logical direct connectivity possibly over a path of multiple physical nodes. Overlay networks are mainly used to implement new features on the Internet and for experimental fixes, research, and deployment of new functions and architectures. Popular examples of overlay networks are represented in Peer-to-Peer (P2P) networks, Voice-over-IP (VoIP) networks, and content distribution networks.

Virtual network embedding [16] and network virtualization are considered the basis of the future Internet and 5G. The abstraction of network active and passive elements brings forward a more flexible and dynamic management of the network. VNE is considered of high significance in data centers for delivering a more efficient utilization of the underlying physical network resources through creating and hosting different heterogeneous VNets. VNE algorithms address the problem of mapping the VNet nodes and links to the underlying physical resources in an optimized mechanism. Moreover, VNE supports the creation of a nested level of virtual resources and VNet providers. This optimization problem is addressed in [16] where the authors consider the possible mapping of a single network node to various virtual counterparts and mapping a single virtual link to multiple physical paths on the substrate and vice versa to deliver the preset SLAs QoS including the efficient and uniform utilization of the physical links. Virtual resources are completely mapped to candidate physical resources thereby producing the embedded VNet on the physical resources. This dynamic network resource mapping is NP-hard (multi-way separator problem), which is relaxed to a simpler problem solved by heuristic and meta-heuristic approaches. In [17], the authors simplify the resource allocation problem by redesigning the physical network to enable the VNE with path splitting and node/path migration for reaching a more optimized solution. In [18], the VNE solution is divided into two major phases for node mapping and link mapping and presented the VineYard algorithm as a mixed integer problem. This algorithm is further relaxed to a linear problem to be solved using randomized and deterministic approaches. The VNE problem is addressed in different research work which consider various aspects of the resource allocation, namely: centralized or distributed, static or dynamic, and concise or redundant. In conclusion, the main objectives behind developing the VNE model are represented in (1) creating VNets that satisfy certain predefined SLA constraints and QoS parameters, (2) maximizing the utilization and economical cost efficiency of the network substrate, and (3) enhancing the resilience of the created VNets under various network states and conditions.

13.3 Software-defined networking architecture

SDN is the new-generation networking architecture that is based primarily on the concept of separating the forwarding plane from the control plane and centralizing the latter in a set of one or more controller units. This pushes the network evolution greatly by facilitating the programmability of the network functions, controls, management, and configuration. This is achieved by abstracting the operation of the switching elements in the data plane by constructing their forwarding tables and actions centrally at the control plane. Centralizing the control plane and functionally separating it from the data plane is considered a major contribution brought forward by SDN to the networking domain which is the main enabler for further advancements represented mainly in network virtualization and programmability.

According to Jain and Paul [19], the SDN architecture is targeted by all the members of the networking industry, namely: the Internet SPs, network InPs, network vendors, enterprises, and end users. This great demand is expected after the long-termed ossification of the Internet that will fend as a result of the following innovations of the SDN architecture:

1. *The separation of the control and management plane from the data plane*: In traditional IP networks, control and data plane functions are featured in the network nodes. The separation introduced by the SDN relies on introducing network nodes as forwarding elements that utilize the open flow protocol for data transfer in the network. A central controller or set of controllers represented in the control and management plane are responsible for setting the control rules adequate to transport the network messages in the data plane. The L3 routing protocols and the management functions are the responsibility of the controller in this architecture. This results in a simplified set of forwarding elements in the network that can be easily managed, maintained, and replaced.

2. *Centralizing the control plane*: The Internet architecture has been based on distributing the communication elements in order to avoid the vulnerable nature of a centralized communication system. It is only in the recent years that SDN proposed the concept of centralizing control. Accordingly, the network is divided into subnets; controlling and managing these sub-networks is aided by the statistics data that is collected from the switching elements using the Northbound API [20] of the controller that enables the efficient management and utilization of the underlying network resources. The centralization advantage lies in (1) the fast propagation of maintenance activities, (2) the introduction and enforcement of new policies and updates, (3) the flexible registration, revocation, and reconfiguration of network switching elements, and (4) and the better visibility of network topologies which facilitates enhanced internetworking and connectivity establishment.

3. *Northbound APIs for control plane programmability* [20]: The network operating system in the SDN architecture provides a set of user-friendly API's that facilitates programming the underlying network components. The controller provides a software representation of the switching hardware in order to make it vendor independent. An example is the OpenDaylight [21] SDN controller. This is analogous to the computer operating system that witnessed the programmability hype after it was hardware specific. This is the key to prevent the ossification of the Internet and provision for the exploration of network programmability virtues.

4. *Flow-based control*: This is enforced by the SDN controller on the network switching units and is governed by the specifications of the OpenFlow [22] protocol. OpenFlow defines a collection of low-level operations for updating and maintaining the switches flow tables which contain the entries that define the packet forwarding mechanisms fed to the switch by the SDN controller. Accordingly, SDN-compliant switches started emerging to support the flow-based, software-controlled forwarding mechanisms imposed by the specifications of the OpenFlow protocol. Typically, to be OpenFlow-compliant [23], a network switch must implement:

    i.   A flow table data structure to store packet forwarding rules
    ii. A secure mechanism to exchange control traffic with the SDN controller, mainly ensured using the TLS protocol [15].
    iii. The necessary software modules for consuming OpenFlow messages and executing their corresponding semantics.

The OpenFlow protocol operation in the data plane relies on three main fields comprising the entries of the switch's flow table:

1. The header field: This field identifies network flows for the purpose of packet forwarding. It is worth mentioning here that the header field can consist of a set of subfields to define flows spanning different layers in the network protocol stack, typically layers L1 to L4.

2. The flow counter field: The counter field is incremented whenever a packet header matches the respective flow table entry. This is leveraged by the SDN controller to manage the network by dynamically maintaining fresh network flow statistics.

3. The flow action field: This field specifies the actions that must be executed as a result of a match on the header field of the respective flow table entry. Three principal actions are defined, namely: forward to port, forward to controller, and drop packet. When a packet arrives at an ingress port with no matching flow table entry, the packet is forwarded as a "PacketIn" to the SDN controller over the TLS secure channel. Consequently, the controller takes the responsibility of updating the switch flow table with the suitable forwarding rules.

13.4 NaaS models in software-defined networks

The NaaS concept was firstly introduced by Costa *et al.* in [1]. With this service, tenants are granted an isolated and secure access to the underlying network resources by providing them with virtualized network views on top of the physical network infrastructure. NaaS services are mainly implemented by integrating advanced packet inspection and forwarding policies that isolate the network traffic of the respective tenants' VNet views and provide them with their own virtual routing and switching network elements. The main benefits provided by the NaaS model are represented in the following supported functionalities:

1. *Virtualized and dynamic network topologies*: with NaaS, tenants can have a virtualized logical view of the physical network that can be "elastically" up scaled and down scaled based on the tenants' application requirements. All the necessary abstractions required to implement this functionality are hidden from the tenant in the network virtualization hypervisor layer. This layer is responsible of maintaining the mapping between the logical network views and the underlying physical infrastructure.
2. *Customized packet forwarding mechanisms*: The network virtualization hypervisor implements customized packet forwarding policies and pushes it to the physical switching units to provide a traffic isolation functionality among the tenants' VNet views. It is this functionality that makes NaaS services very suitably implemented in SDN environments where the SDN controller aids the network virtualization hypervisor in pushing the forwarding rules to the physical switches to support traffic isolation.
3. *In-network processing*: the packet inspection capability of NaaS services can greatly assist the SDN controller in providing in-network processing functionality such as supporting network aggregation, opportunistic caching, packet filtering firewalls, content-based networking, among others. These functions can be better supported in NaaS due to their dependency on the application specifics. Hence, tenants directly controlling the network infrastructure via the VNet views in NaaS can provide highly efficient in-network processing functions since they are the entities aware of the application types running on the VNet views.

Extensive research work focused on providing virtualized network services in SDN-based networks. In [24], Drutskoy *et al.* presented FlowN, a NaaS architecture that utilizes SDN concepts and database mapping techniques to provide tenants with their own VNet topology, address space, and controller on top of a single physical data center. FlowN is lightweight due to the utilization of container-based virtualization which allows the system to instantiate isolated SDN controllers for the different tenant networks using one physical controller. In [25], Sherwood *et al.* presented FlowVisor, a switch virtualization architecture that maps the forwarding services of hardware switches to multiple virtual switches that can be used to construct multiple VNets with distinct switching logic. FlowVisor can run on commodity switches supporting the OpenFlow protocol, which makes it an attractive network virtualization solution for existing production data center networks.

In OpenVirteX [3], the authors followed a novel network virtualization approach that made it mimic, to a high degree, computing virtualization mechanisms and operations, such as dynamic configuration, instantiation, destruction, snapshotting, and on-demand migration. This gives network virtualization a leading edge in being capable of supporting flexible NaaS solutions whose configuration, infrastructure management, topology specification, and addressing schemes are completely under the tenants' control. It should be noted here that NCaaS relies on the OpenVirtex solution in developing the prototype test bed implementation. In CoVisor [26], the authors followed an SDN controller virtualization approach that allows a single network controller to be virtualized into multiple controller applications supporting different operational platforms and development languages. CoVisor allows for the control on shared network traffic by enabling the enforcement of multiple policies on separate controller applications such as firewall, load balancer, SSL accelerator, application gateway, router, and traffic analysis policies. The work proposed in this paper compliments the above schemes by providing a unified interface that hides the complexities of VNet creation and management from the tenant while maintaining the same levels of tenant control over the provided VNets.

OpenStack Neutron [27] is undoubtedly the most popular SDN networking project for implementing NaaS services in virtualized cloud environments. OpenStack [28] is a dedicated cloud operating system designed for controlling the compute, storage, and networking resources of big cloud data centers via a standardized Web interface known as the OpenStack dashboard. The main OpenStack projects are:

1. SWIFT and CINDER for controlling object and block storage respectively
2. NOVA for controlling compute resources
3. NEUTRON for providing virtualized networking services
4. KEYSTONE for supporting identity management services
5. GLANCE for providing image services

OpenStack Neutron (originally codenamed Quantum) provides cloud tenants with direct access on the physical network infrastructure to control the network topology and addressing schemes in multi-tenant cloud environments. Neutron is the main OpenStack component for providing network virtualization and isolation in SDN cloud platforms. Neutron supports tenants with an advanced plugin-based API for programming multiple VNets with isolated addressing mechanisms. Such APIs are the main pillar that provides tenants with the necessary tools to have better control over additional network functionality such as security, privacy, QoS, intrusion detection and prevention, packet filtering firewalls, and advanced logging and auditing.

Lately, the Open Virtual Network (OVN) project was developed by the Open-vSwitch project [29] to support Layer 2 and Layer 3 network virtualization mechanisms based on the general-purpose OpenFlow protocol in SDN networks. OVN implements logical switching units and routers and relies in its operation on tunnel-based overlay networking protocols instead of going through the complexities and overhead of VLANs and physical network management.

13.5 NCaaS: network configuration as a service

In this section, we present NCaaS, a centralized cloud service for creating VNets in SDN-based cloud architectures. The proposed service relieves tenants from the burdens and complexities of VNet creation and management by supporting dynamic provisioning operations based on QoS, pricing, privacy, reliability, and energy constraints set by the tenant. Moreover, it provides a unified interface through which tenants' network specifications and constraints are fed into the service provisioning algorithms. These algorithms, in turn, handle the negotiation with the different SDN NaaS providers and dynamically apply the necessary VNet creation, partitioning, and migration mechanisms to ensure the satisfaction of the tenants' preferences. A proof of concept test bed implementation of the proposed service will be provided on top of the OpenVirteX network virtualization platform.

NCaaS provides tenants with their predefined VNets that meet their constraints while guaranteeing a minimum cost. The tenant VNet is composed of a set of n software services that interact together to achieve the desired functionality of the tenant's business logic. These n services are to be deployed on a set of VMs distributed among t providers. NCaaS algorithms utilize provider-related information as well as tenants VNet specification and constraints to create a minimum cost VNet that complies with these specifications and constraints. Figure 13.2 demonstrates the set of NCaaS algorithms responsible for creating the VNet based on the providers'/tenants' input and specifications. This work is an extended version of an earlier work [30].[1]

13.5.1 NaaS providers offers

The NCaaS algorithms operates based on input information retrieved from a set of NaaS providers representing (1) their offered VM profile specifications and their respective cost including the hardware/software specifications of the VM, such as the number of virtual CPUs and their speed, the amount of RAM available, the storage capacity, the operating system and supporting software, and the network bandwidth, (2) the upload/download data communication rates, and (3) the energy resources

[1]This work is based on an earlier work: "NCaaS: network configuration as a service in SDN-driven cloud architectures", in proceedings of the 31st Annual ACM Symposium on Applied Computing, April 4–8, 2016, © ACM, 2016. http://dx.doi.org/10.1145/2851613.2851629.

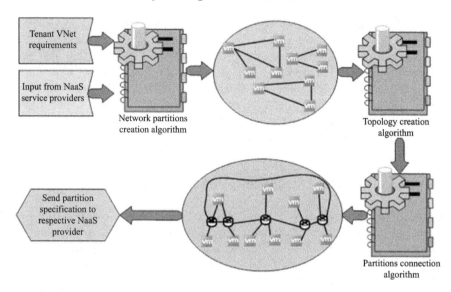

Figure 13.2   NCaaS algorithms for dynamically creating the tenants' virtual networks

used by the provider, and the privacy services offered. This information is either input from the providers, or gathered and updated by the NCaaS service. This information is further arranged in three main tables or matrices as follows:

1. The Profile-Cost Matrix (CP): this matrix (Figure 13.3(a)) shows the cost of the pre-specified VM profiles offered by different cloud providers. Each element in this matrix $CP_{a,k}$ represents the cost of a certain VM profile along with its bandwidth rates indexed by a that are offered by the cloud SP indexed by k.

2. The Upload/Download Rate Matrix (UDR): this matrix (Figure 13.3(b)) specifies the data upload/download rate offered by each of the τ cloud providers. Element UDR_i consists of the upload/download rate per data unit offered by the ith cloud SP. Without loss of generality, we assume same rates for data upload and download.

3. The Provider Profile Matrix (PP): this matrix provides three main records related to (1) the reputation rank of the provider, (2) the level of privacy supported, and (3) the eco-friendly compliance of the providers with green energy saving policies. Such metrics are mainly provided by trusted third parties as described in [31,32]. In addition to the reputation, privacy, and energy metrics, the PP matrix specifies the geographical physical location of the corresponding providers' data centers.

$$\begin{bmatrix} CP_{11} & \cdots & CP_{1\tau} \\ \vdots & \ddots & \vdots \\ CP_{\lambda 1} & \cdots & CP_{\lambda\tau} \end{bmatrix} \begin{bmatrix} UDR_1 \\ \vdots \\ UDR_\tau \end{bmatrix} \begin{bmatrix} 0 & SD_{1,2} & \cdots & SD_{1,n} \\ \vdots & 0 & & \vdots \\ SD_{n,1} & & \cdots & 0 \end{bmatrix} \begin{bmatrix} SP_1 \\ \vdots \\ SP_n \end{bmatrix} \begin{bmatrix} R_1 \\ \vdots \\ R_n \end{bmatrix}$$

(a)   (b)   (c)   (d)   (e)

Figure 13.3   *Input information to the NCaaS algorithms: (a) the profile–cost matrix; (b) the upload/download rate matrix; (c) the service dependency matrix; (d) the service profile matrix; and (e) the performance matrix*

13.5.2 Tenants preferences and constraints

The second form of input provided to the NCaaS VNet creation algorithms is a set of tenants' specifications and constraints. These are represented in three main matrices:

1. The Service Dependency matrix (*SD*): The SD matrix designates the degree of dependency among the different network services. Dependency in this context reflects the amount of data communicated between the respective services. This matrix is an *nxn* symmetric matrix for a tenant predefined set of *n* services (S_1, S_2, ..., S_n) as shown in Figure 13.3(c). Each entry $SD_{i,j}$ in the service dependency matrix indicates the amount of network traffic exchanged between services *i* and *j* per unit time. $SD_{i,j}$ of zero value indicates no linking between services *i* and *j*. Thus, obviously, the values of the matrix main diagonal are set to zeros. Moreover, assuming that $SD_{i,j}$ is the same as $SD_{j,i}$ renders the SD matrix symmetric.

2. The VNet service profile matrix (*SP*): the matrix (Figure 13.3(d)) represents the profiles and resource requirements of the underlying VMs running the network services. SP_i is the service profile of the *i*th tenant service. As indicated in Section 13.5.1, the service profile designates the hardware/software specifications of the VM such as the number of virtual CPUs and their speed, the amount of RAM available, the storage capacity, the operating system and supporting software, the network bandwidth, among others. NCaaS allocates network resources on various providers based on this matrix.

3. The performance constraint matrix *R*: this matrix corresponds to the tenant specified services that require high performance and minimum network delay. The *R* matrix entries are set in the range $1 \leq R \leq 10$ where 10 represents high-performance services and a rate of 1 indicates relatively lower service performance demands. The performance constraint matrix (Figure 13.3(e)) is used in the performance constraints and the topology creation algorithm as will be presented in Section 13.5.4.

In addition to the matrices described above, NCaaS presents a unified interface that prompts tenants to identify a set of network provider constraints as listed below:

- Location of physical provider sites according to tenant's region preferences as well as specifying tenants' black-listed provider set, if any.

- The minimum accepted provider's reputation rank (refer to [32] for more details on reputation ranking schemes in cloud computing).
- Energy requirements where the tenant may select NaaS providers with environmental friendly data centers, renewable energy options, and energy consumption limits.

The desired level of privacy on specific parts of the network (refer to [31] for more details on privacy level specifications).

13.5.3   NCaaS partitioning algorithm

NCaaS supports an adaptable VNet partitioning mechanism that provides flexibility in placing the different VNet services on different provider sites depending on the performance and cost requirements of these services. For instance, the online network services that require high-performance qualities are placed on top-notch provider sites with relatively high service pricing, while the backup and archiving components can be placed on provider sites with lower service price. Partitioning also allows for the essential VNet services to be deployed on provider sites that are geographically closer to the tenant, which plays a role in enhancing the performance of these services. Moreover, VNet partitioning enhances the reliability and fault tolerance capabilities of the network by allowing the NCaaS service to replicate, clone, and snapshot only small slices of the VNet (e.g., those with critical online operation) and hence avoiding the expensive operation on the whole VNet. The NCaaS partitioning problem aims at assigning the set of *n* services composing the tenant VNet to a set of partitions on a collection of selected cloud providers. The next subsection presents a mathematical formulation for the NCaaS partitioning optimization problem by specifying the objective cost function and the set of constraints on this objective function. This is followed by devising a greedy approximation algorithm for solving this partitioning problem.

13.5.3.1   Cost optimization formulation and objective function

In this section, we formalize the objective function representing the overall cost of deploying the tenant's VNet services on the different cloud providers. The main objective is to minimize this cost while considering the tenants' constraints and providers' specifications. Using the set S of n services ($S_1 \rightarrow S_n$) and the set PR of τ cloud providers ($PR_1 \rightarrow PR_\tau$), the cost of mapping a service S_i with VM profile P_i to provider PR_k is defined as:

$$\text{cost}(i, k) = \left(CP_{a,k} + \sum_{j=1; j \neq \text{provider } k}^{n} SD_{i,j} \times (UDR_k + UDR_{\text{prov}(j)}) \right) \qquad (13.1)$$

Equation 13.1 represents the cost of deploying service *i* with VM profile *a* on provider *k* in addition to the communication cost of service *i* with all the dependent services *j* as specified in the *SD* matrix. The function prov(*j*) returns a reference to the cloud provider hosting service *j*. Next, we define the binary constraint X_{ik} as:

$$X_{ik} = \begin{cases} 1 & \text{if } S_i \text{ is deployed on } PR_k \\ 0 & \text{otherwise} \end{cases} \qquad (13.2)$$

The objective function of the optimization problem to be minimized is presented as follows:

$$f(x) = \sum_V \text{depcost}(S_i, S_j, PR_k, PR_{m=\text{prov}(j)}) \times X_{ik} \times X_{jm} \qquad (13.3)$$

where $V = \{i, j, k, m | i, < j, i \leq n, j \leq n, k \leq \tau, m \leq \tau\}$

Minimizing $f(x)$ will result in placing the services of the VNet on providers with minimum overall deployment and communication cost in addition to the following constraints:

$$\sum_{(k,m)}^{1.....\tau} X_{ik} \times X_{jm} = 1 \forall i, j \text{ in } S \qquad (13.4)$$

where $X_{ik}, X_{jm} \in 0, 1$ and $k \neq m$

Equation 13.4 indicates that service i can be hosted by only one provider at a time. The same applies for service j.

Another essential constraint is the performance constraint directly indicated by the performance constraint matrix R and influenced by the distance between the providers PR_k, PR_m hosting the connected services S_i, S_j:

$$\text{Distance}(PR_k, PR_m) \times SD_{i,j} \leq \delta \times \max(R_i, R_j) \qquad (13.5)$$

where δ is a normalization constant that tunes the range in the R matrix to a suitable (distance × data dependency) factor.

This problem is a binary integer optimization problem, which can be reduced to a Multiple Knapsack problem [33] whose computational complexity is exponential in terms of the number of services and providers. The optimal solution of a Multiple Knapsack problem is usually obtained via branch-and-bound [34] techniques. Greedy algorithms are well-suited for approximating such type of problems in terms of computing times and storage resources. In the next section, we propose the Network Partitions Creation algorithm, which is an approximate greedy algorithm for solving this NP-complete problem.

13.5.3.2 Network partitions creation algorithm

NCaaS utilizes provider data (Section 13.5.1) along with the tenant VNet specification and constraints (Section 13.5.2) in order to achieve the minimum cost objective function presented in Equation 13.3.

To serve this goal, the NCaaS's Network Partitions Creation algorithm starts in phase 1 by short listing the set of candidate cloud providers based on the *PP* matrix and tenants requirements. The algorithm proceeds by checking providers' offers in the resource *CP* matrix against the tenant specified *SP* matrix to map each of the n VNet service to the provider of minimum cost satisfying the corresponding resource profile. At the end of this phase, all the services of the respective VNet would be arranged in partitions that are mapped to minimum cost providers. In phase 2 and to achieve minimum cost, the highly connected services are rearranged on the set of partitions to satisfy minimum connection cost among providers while ensuring tenants' specified performance criteria. Thus, the tenant-specified services' dependency matrix, along with the provider upload/download cost rate, is utilized to achieve the minimum cost objective. The pseudo code of the Network Partitions Creation algorithm is presented below:

Algorithm 1 Network Partitions Creation Algorithm.

Phase 1 Minimum cost profile selection:
Step 1
Select the candidate provider list based on the tenants' constraints on the providers' qualifications in terms of reputation, energy, physical location and privacy.
Step 2
Create the Profile-Cost Matrix (CP) with rows corresponding to n services and columns to τ providers where $CP_{i,j}$ is the cost of deploying service i on provider j.
Step 3
Create the Partitions matrix PT of dimension $\tau \times n$. Each row i in the PT matrix indicate the set of services selected to run at provider i. At this stage, this selection is merely based on the minimum cost offered by the various cloud providers for different service profiles. Effectively, each row i in the PT matrix represent a network partition to be deployed at the ith cloud provider site.
Phase 2 Achieving least inter-partition communication cost:
Step 4
Create dependency hash table $D [1 \rightarrow n]$. Each entry D_k in D represents the amount of dependency that service k has with the other $(n-1)$ services in the network. To calculate D_k we utilize the SD matrix as follows:

> **for** i from $1 \rightarrow n$
> > **for** j from $1 \rightarrow n$
> > > **if** $j \neq i$
> > > D_i += $SD_{i,j}$

Step 5
Sort D in decreasing order to start with the service of maximum connectivity influence on the other services

> **for** each service i in D
> > **Create** a cost vector $V [1 \rightarrow t]$ where each entry V_k represents the cost of deploying service i in partition k added to the cost of interconnection between service i and the rest of the services in the other partitions.
> > **for** each k in PT
> > > $V_k = CP_{i,k}$
> > **for** each service j in SD excluding those in partition k
> > > **if** Distance $(k, \text{Partition}(j)) \times SD_{i,j} \geq \delta \times \max(R_i, R_j)$
> > > > increment k (change partition)
> > > > Break
> > > > V_k += $SD_{i,j} \times UDR_k$

Step 6
Find the entry of minimum cost in V. The entry's index represents the optimum partition/provider for running service i.

13.5.4   *Topology generation*

After running the partitioning algorithm, the VNet is divided into partitions each of which contains a set of the VNet services to be deployed on a particular cloud provider. The topology creation algorithm arranges the services on the provider site in a fat-tree-based topology (popular in today's data centers), which involves a branching factor *EBR* that is inversely proportional to the number of ports involved in the topology creation. *EBR* is the ratio of the number of edge switch ports that are connected to servers to those connected to core switches. The minimum value for *EBR* is 1, which indicates that there is a dedicated link connecting the edge switch to the core switch for each node in the network. This directly contributes to minimizing the data transfer delay between the services across the edge and core switching layers. To fulfill the performance constraints specified by the tenant and represented in the performance matrix *R* (refer to Section 13.5.2), the maximum performance value R_{max} of the services in the partition is used to calculate *EBR* according to the following equation:

$$EBR = \text{trunc}(\alpha\, R_{max})$$

where α is a normalization factor that bounds the *EBR* value to a predefined range.

13.5.4.1   Topology creation algorithm

This section presents the base algorithm for creating each VNet partition topology. The parameters used in the topology creation are listed below along with the topology creation algorithm:

1. k: number of edge switches.
2. *NE*: number of ports per edge switch.
3. *NC*: number of ports per core switch.
4. *NP* is the total number of servers in a partition
5. *ES*: maximum number of edge switch ports to connect to servers deduced from *EBR* by: $ES = \text{trunc}(NE \times EBR/(1 + EBR))$
6. *EC*: number of edge switch ports to connect to core switches obtained from ES such that: $EC = NE{-}ES$
7. *B*: maximum number of ports that can connect to each edge switch where $B = ceiling(NC/k)$

13.5.5   *Partitions connection algorithm*

Using the SDN network programming features, VNet partitions can be easily joined to provide a complete VNet view to the tenant. The following example (see Figure 13.4) demonstrates how multiple VNet partitions implemented on possibly different provider sites can interact to provide the tenant with a single compositional VNet. Without loss of generality, the example demonstrates the algorithm for achieving connectivity between two VNet partitions, VNet partition 1 and VNet partition 2 consisting of two hosts each.

For functionally joining the two VNet partitions we use two forwarding hosts FH1 and FH2 at VNet partition 1 and VNet partition 2, respectively. At VNet partition 1,

---

Algorithm 2 Topology Creation Algorithm.

Phase 1 Initialization:
Compute EBR = trunc(α *Rm*)
Set Scounter = 0 //equivalent to the number of services connected
ES = trunc*(NE $\times$ EBR/(1+EBR))*
Set k =1 // (edge switch counter)
Phase 2 Creation of edge switches and their corresponding port connections:
Let i = Scounter
Create edge switch ES$_k$
While Scounter $\leq$ ES and Scounter $\leq$ NP
    if S$_i$ is not connected, connect it to ES$_k$
        increment Scounter, increment i
        for each j $\leq$ NP and j>i
            find j with maximum SD$_{i,j}$
            if S$_j$ not connected, connect it
                increment Scounter
                If Scounter < NP then increment k //
                                  //add another edge switch
        Goto Create edge switch
Phase 3 Creation of the core level
Set c=0 as the core switch counter
EC=NE-ES
B= NC div k
if B $\geq$ EC then increment c, create core switch CS$_c$ and connect each edge switch to it via EC ports
else
    let r = EC
    while r $\geq$ B
        increment c, create core switch CS$_c$ and connect each edge switch to it via B ports
        decrement r by B: r = B - r
        if r > 0
            increment c, create core switch and connect to each edge switch via r ports Redistribute edge switches over core switches by EC/c connections per edge switch

---

the SDN controller sends the rule "forward all traffic to H3 and H4 to FH1." Similarly, at VNet partition 2, the SDN controller sends the rule "forward all traffic to H1 and H2 to FH2." The FH1-to-FH2 link can be implemented as an IP tunnel as follows: All frames arriving at FH1 are encapsulated in respective IP packets and sent to FH2 and vice versa. All the packets received by FH2 are de-capsulated and the resulting frames are forwarded to their designated hosts. Analogously, all traffic received by

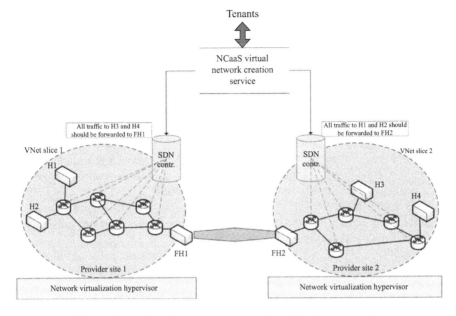

Figure 13.4   Partitions connection algorithm demonstration

FH1 is de-capsulated and forwarded to its destination. With this procedure, the tenant will view a single VNet composed of the corresponding partitions.

To prevent any performance bottlenecks due to the utilization of a single forwarding host, the Partitions Connection Algorithm can be extended by adopting more than one forwarding host per partition. Communication with other services (residing on other partitions) in the network is assigned equally among the available forwarding hosts in the source partition. In other words, the SDN controller will send the forwarding rules to the network switches to divide the outgoing load among the forwarding hosts based on the destination service receiving the traffic. In the same sense, the encapsulation algorithm on the source forwarding host is configured to forward the traffic equally among the forwarding hosts in the destination partition. This is also based on the recipient service receiving the traffic in the destination partition. That is, the source forwarding host is configured to assign the set of services deployed in the destination partition among the set of forwarding hosts available in that partition. This ensures fair distribution of workload among the forwarding hosts on the source as well as on the sink sides.

13.5.6   NCaaS implementation

The NCaaS algorithms are fully implemented using Mathworks MATLAB® 2014 [35]. To demonstrate a full-fledged functionality of the NCaaS service and to test for the fulfillment of the NCaaS objective function of minimum cost as the systems scales, we simulated four main prototype cloud configurations. Configuration 1 consists of

Table 13.1   Profile offers based on real provider pricing data

Profile offer ($/h) Range	1	2	3	4	5	6	7	8	9
Min	0.005	0.015	0.025	0.05	0.056	0.13	0.26	0.51	1.25
Max	0.013	0.026	0.053	0.104	0.126	0.252	0.504	1.008	2.53

5 providers and 10 services, Configuration 2 consists of 7 providers and 14 services, Configuration 3 consists of 14 providers and 28 services, and Configuration 4 consists of 24 providers and 36 services. The NaaS architecture utilized is the OpenVirteX network virtualization platform. This choice is due to the great flexibility provided by OpenVirteX in terms of address space isolation, topology specification, and dynamic network reconfiguration at runtime.

OpenVirteX is installed on a VirtualBox [36] VM and runs on top of the Mininet network emulator [4]. For each VNet partition, the NCaaS algorithms ultimately produce a .JSON file that formally describes the VNet partition topology. The partitions' files are fed to the OpenVirteX network embedded to automate the process of mapping the respective VNet partition onto the physical data center network. It is worth mentioning here that NCaaS may in theory rely on any available NaaS services. The only requirement here is to feed the NCaaS Topology Creation algorithm with the standard network deployment interface used by the respective NaaS services. The matrices representing the tenants and providers input (refer to Sections 13.5.1 and 13.5.2) are populated with reasonable values for each experiment run (40 runs in total for each of the four configurations). The most important matrices fed into the NCaaS algorithms are populated as follows:

1. The *CP* matrix profile values are randomly generated in the ranges specified in Table 13.1 which are based on real pricing values extracted from the websites of the top 12 cloud providers in the IaaS market.
2. The *UDR* matrix values are randomly generated in the range of $0.03/GB to $0.11/GB. This range is based on real data transfer rates extracted from the websites of six cloud providers.
3. The *SD* matrix entries linking communicating services are populated with assumed values in the range of 100 KB to 24 MB per hour.
4. The *SP* matrix is initialized with reasonable profile settings for the services in the four simulated configurations.
5. The *R* matrix is instantiated with random values in the range of 1 to 10 (this is based on the specification of the *R* matrix in Section 13.5.2).

For the performance constraint implementation, the physical distances among providers are randomly generated in the range of 1 to 20. These distances are implemented as the weights on the links connecting the partitions in the network configuration files. To achieve the required normalization level, the normalization constant δ used is 48.

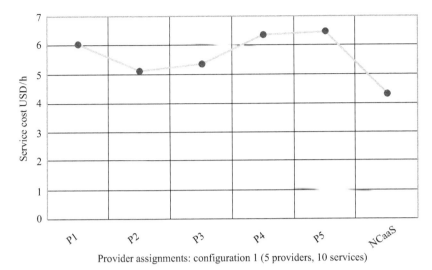

Provider assignments: configuration 1 (5 providers, 10 services)

Figure 13.5   Cost ($/h) achieved in the first configuration when deploying the VNet services on the candidate providers and when applying the NCaaS algorithms

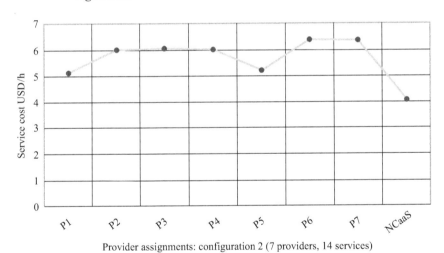

Provider assignments: configuration 2 (7 providers, 14 services)

Figure 13.6   Cost ($/h) achieved in the second configuration when deploying the VNet services on the candidate providers and when applying the NCaaS algorithms

For each of the four prototype configurations, the cost of deploying the entire tenant's VNet on a single cloud provider is computed for the whole set of candidate providers. These costs are compared to the cost of deploying the same VNet using the partitioning algorithms of the NCaaS service. It is worth mentioning here that

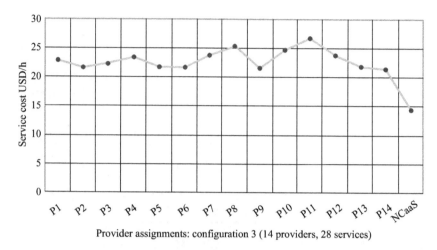

Provider assignments: configuration 3 (14 providers, 28 services)

Figure 13.7   Cost ($/h) achieved in the third configuration when deploying the VNet services on the candidate providers and when applying the NCaaS algorithms

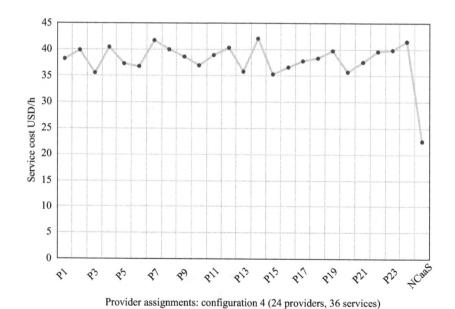

Provider assignments: configuration 4 (24 providers, 36 services)

Figure 13.8   Cost ($/h) achieved in the fourth configuration when deploying the VNet services on the candidate providers and when applying the NCaaS algorithms

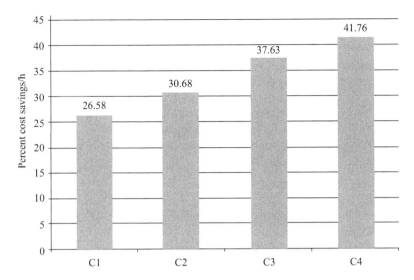

Figure 13.9   NCaaS Cost savings per hour achieved when deploying the four VNet configurations

the main focus in the implementation is directed toward the cost of services deployment on the set of available providers for a main reason represented in the fact that this cost parameter is the chief focus of the optimization objective function with the rest of the QoS-related parameters (represented in the R matrix) are designated as constraints to this optimization objective function. QoS-related parameters in the field of network virtualization are comprehensively studied in literature in the VNE work in [16]. On all the tested cloud configurations, the NCaaS service always produced the minimum cost in comparison with single NaaS provider deployment while satisfying the tenant's constraints and requirements. The results attained for configurations 1–4 are respectively presented in Figures 13.5–13.8. On average, the cost saving achieved for the first configuration (5 providers, 10 services) is 26.58%, for the second configuration (7 providers, 14 services) is 30.68%, for the third configuration (14 providers, 28 services) is 37.62%, and for the fourth configuration (24 providers, 36 services) is 41.76%. This demonstrates the scalability of the NCaaS algorithms in increasing the cost savings as the problem size increases. These results are presented in Figure 13.9.

13.6   Conclusion and future extensions

This chapter has provided a brief survey of the main technologies enabling NaaS services in the cloud, namely, network virtualization and programmability and SDN. The main advantages of NaaS offerings are represented in provisioning dynamic, flexible, and secure VNets via isolated virtual topologies, customized packet forwarding policies, and optimized in-network processing. Despite all these advantages, NaaS services suffer from limitations in supporting dynamic VNet creation and management and satisfying tenants' performance, security, and pricing requirements and constraints. An innovative network servicing approach, termed NCaaS, is presented, which combines the advantages offered by NaaS and SDN along with a set of dynamic network creation and partitioning algorithms in order to provide a centralized VNet creation and configuration service in the cloud. The chapter discussed NCaaS design and algorithms and experimentally demonstrated a proof of concept implementation on top of the OpenVirteX network virtualization platform.

In the future, several features can be augmented to the NCaaS service design and implementation. The main points include:

1. Designing an event-driven feedback mechanism at the SDN provider site to promptly inform the NCaaS cloud service with network state and status information, hence facilitating the efficient adaptation of the managed VNets to the tenants' requirements.
2. Supporting advanced virtualization operations on the VNets such as snapshotting and cloning to achieve high degrees of flexibility, reliability, and feature-rich VNet management and maintenance on behalf of the tenant.
3. Augmenting the NCaaS algorithms with energy optimization features to achieve optimal energy consumption levels.
4. Developing, deploying, and testing a real-life cloud implementation of the NCaaS service in an actual cloud computing environment.

References

[1]   Costa, Paolo, Matteo Migliavacca, Peter Pietzuch, and Alexander L. Wolf. "NaaS: Network-as-a-Service in the Cloud." *Presented as Part of the Second USENIX Workshop on Hot Topics in Management of Internet, Cloud, and Enterprise Networks and Services*. 2012

[2]   Nunes, Bruno Astuto A., Marc Mendonca, Xuan-Nam Nguyen, Katia Obraczka, and Thierry Turletti. "A survey of software-defined networking: past, present, and future of programmable networks." *IEEE Communications Surveys & Tutorials* 16, no. 3 (2014): 1617–1634

[3]   Al-Shabibi, Ali, Marc De Leenheer, Matteo Gerola, *et al.* "OpenVirteX: make your virtual SDNs programmable." In *Proceedings of the Third Workshop on Hot Topics in Software Defined Networking*, Chicago, IL, USA: ACM, 2014. pp. 25–30

[4]   De Oliveira, Rogério Leão Santos, Christiane Marie Schweitzer, Ailton Akira Shinoda, and Ligia Rodrigues Prete. "Using Mininet for emulation and prototyping software-defined networks." In *Communications and Computing (COLCOM), 2014 IEEE Colombian Conference on*, Colombia: IEEE; 2014. pp. 1–6

[5]   Hao, Fang, T. V. Lakshman, Sarit Mukherjee, and Haoyu Song. "Enhancing dynamic cloud-based services using network virtualization." In *Proceedings of the First ACM Workshop on Virtualized Infrastructure Systems and Architectures*, Barcelona, Spain: ACM, 2009. pp. 37–44

[6] Chowdhury N.M., Mosharaf Kabir, and Raouf Boutaba. "Network virtualization: state of the art and research challenges." *IEEE Communications Magazine* 47, no. 7 (2009): 20–26

[7] Drutskoy, Dmitry, Eric Keller, and Jennifer Rexford. "Scalable network virtualization in software-defined networks." *IEEE Internet Computing* 17, no. 2 (2013): 20–27

[8] Barham, Paul, Boris Dragovic, Keir Fraser, *et al.* "Xen and the art of virtualization." *In ACM SIGOPS Operating Systems Review*, vol. 37, no. 5, ACM, 2003. pp. 164–177

[9] Armbrust, Michael, Armando Fox, Rean Griffith, *et al.* "A view of cloud computing." *Communications of the ACM* 53, no. 4 (2010): 50–58

[10] Mijumbi, Rashid, Joan Serrat, Juan-Luis Gorricho, Niels Bouten, Filip De Turck, and Raouf Boutaba. "Network function virtualization: state-of-the-art and research challenges." *IEEE Communications Surveys & Tutorials* 18, no. 1 (2015): 236–262

[11] Hipp, Emily L., Yuh-yen Yeh, and Burton A. Hipp. "Virtual network environment." U.S. Patent 7,146,431, issued December 5, 2006

[12] Doraswamy, Naganand, and Dan Harkins. *IPSec: the new security standard for the Internet, intranets, and virtual private networks*. Upper Saddle River, NJ: Prentice Hall Professional, 2003

[13] Hamzeh, Kory, Grueep Pall, William Verthein, Jeff Taarud, W. Little, and Glen Zorn. *Point-to-point tunneling protocol (PPTP)*. No. RFC 2637. 1999

[14] Townsley, W., A. Valencia, Allan Rubens, G. Pall, Glen Zorn, and Bill Palter. *Layer Two Tunneling Protocol "L2TP."* No. RFC 2661. 1999

[15] Turner, Sean. "Transport layer security." *IEEE Internet Computing* 18, no. 6 (2014): 60–63

[16] Fischer, Andreas, Juan Felipe Botero, Michael Till Beck, Hermann De Meer, and Xavier Hesselbach. "Virtual network embedding: a survey." *IEEE Communications Surveys & Tutorials* 15, no. 4 (2013): 1888–1906

[17] Yu, Minlan, Yung Yi, Jennifer Rexford, and Mung Chiang. "Rethinking virtual network embedding: substrate support for path splitting and migration." *ACM SIGCOMM Computer Communication Review* 38, no. 2 (2008): 17–29

[18] Chowdhury, Mosharaf, Muntasir Raihan Rahman, and Raouf Boutaba. "Vine-Yard: virtual network embedding algorithms with coordinated node and link mapping." *IEEE/ACM Transactions on Networking (TON)* 20, no. 1 (2012): 206–219

[19] Jain, Raj, and Subharthi Paul. "Network virtualization and software defined networking for cloud computing: a survey." *IEEE Communications Magazine* 51, no. 11 (2013): 24–31

[20] Sezer, Sakir, Sandra Scott-Hayward, Pushpinder Kaur Chouhan, *et al.* "Are we ready for SDN? Implementation challenges for software-defined networks." *IEEE Communications Magazine* 51, no. 7 (2013): 36–43

[21] Medved, Jan, Robert Varga, Anton Tkacik, and Ken Gray. "OpenDaylight: towards a model-driven SDN controller architecture." In *Proceeding of IEEE International Symposium on a World of Wireless, Mobile and Multimedia Networks 2014*, Sydney, Australia. 2014

[22]   McKeown, Nick, Tom Anderson, Hari Balakrishnan, *et al.* "OpenFlow: enabling innovation in campus networks." *ACM SIGCOMM Computer Communication Review* 38, no. 2 (2008): 69–74

[23]   Wen, Heming, Prabhat Kumar Tiwary, and Tho Le-Ngoc. "Wireless virtualization." In *Wireless Virtualization*, New York City: Springer International Publishing, 2013. pp. 41–81

[24]   Drutskoy, Dmitry, Eric Keller, and Jennifer Rexford. "Scalable network virtualization in software-defined networks." *IEEE Internet Computing* 17, no. 2 (2013): 20–27

[25]   Sherwood, Rob, Glen Gibb, Kok-Kiong Yap, *et al.* "FlowVisor: a network virtualization layer." *OpenFlow Switch Consortium*, Tech. Rep. (2009): 1–13

[26]   Jin Xin, Jennifer Gossels, Jennifer Rexford, and David Walker. "CoVisor: a compositional hypervisor for software-defined networks." In *12th USENIX Symposium on Networked Systems Design and Implementation (NSDI 15)*, 2015. pp. 87–101

[27]   OpenStack Neutron home page at: http://docs.openstack.org/developer/neutron/

[28]   Sefraoui, Omar, Mohammed Aissaoui, and Mohsine Eleuldj. "OpenStack: toward an open-source solution for cloud computing." *International Journal of Computer Applications* 55, no. 3 (2012): 38–42

[29]   OpenvSwitch website: http://openvswitch.org

[30]   Shamseddine, Maha, Imad Elhajj, Ali Chehab, Ayman Kayssi, and Wassim Itani. "NCaaS: network configuration as a service in SDN-driven cloud architectures." *In Proceedings of the 31st Annual ACM Symposium on Applied Computing*, Pisa, Italy: ACM, 2016. pp. 448–454

[31]   W. Itani, A. Kayssi, and A. Chehab. " Privacy as a service: privacy-aware data storage and processing in cloud computing architectures." In *DASC'09. Eighth IEEE International Conference on Dependable, Autonomic and Secure Computing*, Chengdu, China: IEEE, 2009. pp. 711–716

[32]   W. Itani, C. Ghali, A. Kayssi, and A. Chehab. "Reputation as a service: a system for ranking service providers in cloud systems." In *Security, Privacy and Trust in Cloud Systems*. Springer: Berlin, Heidelberg, 2014. pp. 375–406

[33]   H. Kellerer, U. Pferschy, and D. Pisinger. *Introduction to NP-Completeness of knapsack problems*. Springer: Berlin, Heidelberg, 2014. pp. 483–493

[34]   S. Martello and P. Toth. "A bound and bound algorithm for the zero-one multiple knapsack problem." *Discrete Applied Mathematics* 3, no. 4 (1981): 275–288

[35]   MATLAB Mathworks home page at: www.mathworks.com/products/matlab/

[36]   VirtualBox homepage: http://www.virtualbox.org

Chapter 14

GPU-based acceleration of SDN controllers

Xuan Qi[1], Burak Kantarci[1,2] and Chen Liu[1]

14.1 Introduction

14.1.1 What is SDN?

For the last two decades, the Internet has gained a rapid development and has become an infrastructure of critical importance. Nowadays, the traditional internet structure has gradually showed its limitations in hosting emerging new applications such as social media, mobile devices, and cloud computing. For the purpose of evolving the internet to meet current and future new application scenarios, researchers have come out with various kinds of new technologies. Among them, Software-Defined Networking (SDN) is a promising direction and has become a research hot spot.

SDN is a new network operation and management solution, which decouples the control and data planes in a communication network. SDN makes it easier for network administrators to create, modify, and manage dynamic networks by abstracting low level functions and network structure [1,2]. Generally, SDN uses a centralized controller, which offers a global view of the entire network. This new feature offers the flexibility for administrators to define the strategies in terms of how the network flow is forwarded on a software level [3]. With the advancing of the research on SDN, however, distributed SDN controllers [4–6] are also introduced by researchers. One reason for the emerging of distributed SDN controller is: it can address the problems of scalability, reliability, and performance issues that a centralized SDN controller suffers from.

As shown in Figure 14.1, the framework of SDN consists of three layers: the application layer, the control layer, and the infrastructure layer. The application layer consists of numerous applications (Apps). The Apps in the application layer communicate with SDN controller through northbound Application Program Interfaces (APIs). The southbound APIs are in charge of offering the network device information to the network controller and applying the control logic to the infrastructure layer. The control layer in the middle contains the SDN controller, which is the core component

[1]Department of Electrical and Computer Engineering, Clarkson University, 8 Clarkson Avenue, Potsdam, NY 13699, United States
[2]School of Electrical Engineering and Computer Science University of Ottawa, 800 King Edward Avenue Ottawa, ON, K1K 0C4, Canada

of the whole SDN network. The SDN controller takes control of the communication between the applications and the network devices. In other words, it serves as a kind of network Operating System (OS), which provides a global view of the whole network to the applications. The infrastructure layer consists of network devices with SDN capability, for instance, the OpenFlow-based switches. Because the SDN controller decouples the control and the dataflow, SDN-based network devices are only in charge of transferring the data according to the control instructions from the SDN controller, which is different from the traditional network devices.

14.1.2 SDN, benefits, and challenges

There are many benefits that make the SDN a promising solution for future generation of computer networks. Among them, the centralized control and management, decoupling of control and data forwarding, and application centric are three most significant advantages. Besides, the benefit of distributed SDN solution also offers its own advantage.

14.1.2.1 Centralized control and management

The centralized control of SDN gives the developers and network administrators a global view of the entire network. Instead of dealing with multiple management tools or interfaces of the whole network, SDN provides users a unified and single control and management interface, which reduces the workload of administrator and improves the efficiency of network management.

14.1.2.2 Decoupling of control and data forwarding

In traditional computer networks, the control logic and data forwarding are integrated together on the network devices such as routers. Typically, different vendors have their own control hardware architectures. This situation makes developers have to adapt their control logic to different architectures. On the other hand, if SDN is adopted for network operation and management, the network device is only responsible for data forwarding; hence this solution concentrates the control logics to the SDN controller. This change eliminates the need of caring about the details of different hardware, and the developers only face the abstraction of the whole network layer provided by SDN. This benefit makes it more easy for developers to modify current control logics and implement new control software.

14.1.2.3 Application centric

When deploying a new application, since the SDN uses flow-based forwarding and gives an abstraction of the network infrastructure layer, the developers only need to focus on defining the data flow and other network behaviors of their application, and do not need to deal with the lower-level network structure and hardware. This key benefit largely reduces the difficulty of deploying a new application or modifying an existing application. Despite several benefits of SDN as mentioned above, there are associated challenges that need to be addressed. One big challenge is the computing overhead. Since SDN uses centralized controller and separates the data flow and control flow, this kind of framework could pose serious computing overhead to the controller, especially for networks of high complexity and large scale.

Besides, there are also strong needs to guarantee high throughput and low latency of the whole network. This makes the challenge of computing capability more important and inevitable.

14.1.2.4   Benefits brought by distributed controller solutions

For the distributed SDN controller, it extends the reliability and scalability of centralized solution, especially under the heavy data loads. In the work of Yazici *et al.* [6], they developed a distributed OpenFlow controller architecture with the capability of dynamic adding and removing of controllers without interrupting the network. In the work of Phemius *et al.* [5], the distributed SDN controller shows the capability of not only working in a single domain but also working adaptively to multi-domain network.

14.1.3   Summary

In this section, we introduced the basic concepts of the SDN network, the SDN controller and the benefits and challenges of applying the SDN network and controller. The following contents of this chapter is organized as follows: In Section 14.2, existing SDN solutions are summarized. We introduce one of the current approaches in dealing with the computing overhead challenge, the GPU-accelerated SDN controller in Section 14.3. Section 14.4 presents a case study on GPU-accelerated SDN in order to illustrate the benefits of this solution. The chapter is concluded in Section 14.5 with a thorough discussion of open issues and challenges in this field.

14.2   Existing SDN controller solutions

With the introduction in previous section, we know that the framework of SDN consists of three layers: the application layer, the control layer and the infrastructure layer. The northbound and southbound APIs realize the interaction of these three layers. There are three main benefits of SDN, being centralized control and management, decoupling of control and data forwarding and application centric. In this section, we are going to make some introductions about existing SDN controller solutions. As shown in Figure 14.1, the control layer plays a role as an intermediate layer between the application and infrastructure layers. The "control layer" is also referred as the "SDN controller." For the SDN controller, there are two key elements which must be introduced: OpenFlow and existing SDN controllers.

14.2.1   OpenFlow

The OpenFlow is one of the first SDN standards, which is a southbound API in the SDN framework. It serves as a communication component between the control and infrastructure layers in SDN network [7]. More specifically, the SDN maintains and updates a forwarding table and decides the packet forwarding through the OpenFlow interface as shown in Figure 14.2. The OpenFlow separates the data packet path and the control path, and offers developers a high-level abstraction of the routers and

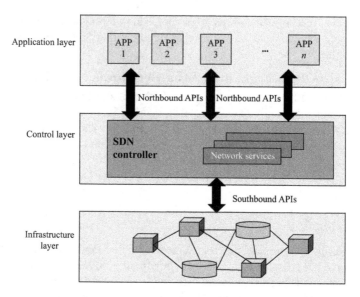

Figure 14.1   The framework of software-defined networking (SDN)

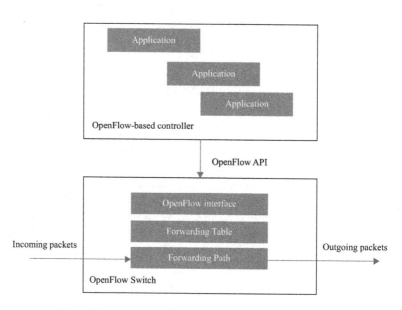

Figure 14.2   The framework of OpenFlow

switches. With the high programmability introduced by OpenFlow, developers can develop and deploy their self-defined new features easily without the knowledge of different types of network hardware. This benefit is very hard to achieve under the current network architecture as the developers need to adapt their control software to hardware with different architecture from different vendors. What is more, the centralized controller of OpenFlow gives administrators a global view of the whole network, which contributes toward easier management and security protection [8]. With all the benefits above, the OpenFlow has become a well-accepted standard and many researchers have built their own deployment on top of the OpenFlow.

Moreover, researchers are continuously working on OpenFlow and OpenFlow-based SDN controllers to overcome OpenFlow drawbacks. The main open issues of OpenFlow are the flexibility on traffic management and the scalability. First, the original design of OpenFlow has limited capability on traffic management [9]. For instance, in the case of multi-path routing, the OpenFlow forwards data flows via different routes according to a flow table. To a certain degree, this routing strategy seems "fixed," which is not adaptive to the change of the dataflow. Let us consider a simple scenario that there are more than two paths between two nodes, and in one of paths the data traffic is huge. Under this situation, a smart strategy is to split the traffic into other valid paths. However, OpenFlow does not "sense" this situation, but keeps the "fixed" forwarding strategy; hence performance is not improved. Secondly, there are also some scalability limitations especially in network functions. For example, OpenFlow has scalability limitations in topology monitoring and fault recovery functions [10], although it can utilize the Link-Layer Discovery Protocol (LLDP)-based strategy under such circumstances. The main reason is that the monitoring messages are sent over all links throughout the entire network very frequently. Thus, with the increasing network size, this strategy can cause a very huge processing and bandwidth overhead. In other words, the LLDP-based strategy in OpenFlow has a serious scalability problem, especially when the size of SDN network is large.

14.2.2 Existing SDN controller implementations

So far, there are various OpenFlow-based SDN controller implementations available. In Table 14.1, we list several representative OpenFlow-based SDN controller solutions.

The NOX controller is the first SDN controller introduced to development community by Nicira Networks [11] in 2008. The NOX is written in C++ and supports OpenFlow v1.0 standard. It now has three development branches: NOX, POX, which supports python, and the NOX Classic which has extensions like GUI (Graphical User Interface) and web services [11].

The primary components in a NOX controller-based network are illustrated in a minimalist manner in Figure 14.3. These components consist of a set of OpenFlow (OF)-based switches and the client machine as server, laptops, or routers connected to the switches. The NOXs controller can be treated as containing several different processes and a single unified network.

POX is also an open-source SDN controller solution which is a "younger sister" of NOX. The POX provides similar functions as NOX, but with Python support. POX is very suitable for rapid development of SDN controller by using python, and is also more widely used and adopted than NOX.

Table 14.1   Existing OpenFlow-based SDN controllers

Controller	Developer	Supported language	Description	Open issues
NOX [34]	Nicira	C++	NOX is the original OpenFlow controller which was first released in 2008.	Ability to deal with packet payloads, without deep packet inspection
POX [35]	Nicira	Python	POX is an OpenFlow-based controller which serves similar functions as NOX and supports Python.	Consideration of parallelism and the performance
Maestro [12]	Rice University	JAVA	Targeting improves the throughput by exploiting multi-core CPU-based parallelization.	The expansion to more accelerating devices such as GPU
Beacon [13]	Stanford University	JAVA	Beacon is a cross platform OpenFlow-based SDN controller.	Handling topologies with loops. Addressing dependency to Eclipse and OSGi
Ovs-controller [15]	Independent developer	C	Ovs-controller is an OpenFlow-based controller with the implementation of vSwitch. This controller supports any number of remote switches using OpenFlow protocol.	Issues on processing performance
Flowvisor [16]	ON.LAB	C	Flowvisor uses divided flow slices strategy to allow multiple tenants to share same physical devices.	Virtual device configuration, Virtual links, and virtual address space
RouteFlow [17]	CPQD	C++	RouteFlow is based on OpenFlow. The highlight of this controller is it can provide virtualized IP routing function.	The optimization of Protocol and scalability issues

Maestro [12] is another SDN controller solution implemented in JAVA. Since the centralized control strategy of SDN poses a great demand on computing power to the controller, the multi-threaded processing by taking advantage of multi-core processor becomes a highlight and key feature of Maestro controller.

Beacon [13] is a cross-platform OpenFlow-based SDN controller developed by Stanford University. The Beacon controller takes the develop productivity and performance as two main concerns since it was introduced to public. For productivity,

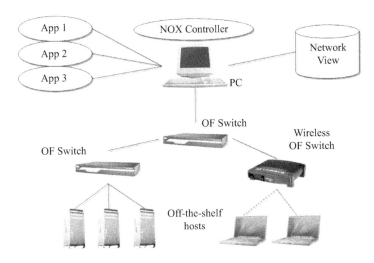

Figure 14.3   The architecture of a NOX controller-based network

Beacon uses JAVA as its development language in lieu of C/C++. As for the computing performance, Beacon also supports multi-threading. According to Erickson [14],
multi-thread-based Beacon controller can lead to linear scalability based upon using
12 cores in their experiments.

Ovs-controller [15], as its name indicates, is a controller with Open vSwitch
implementation. Open vSwitch is a virtual switch for providing network connectivity to Virtual Machines (VMs). The Ovs-controller supports any number of remote
switches based on the OpenFlow protocol.

The Flowvisor [16] supports network virtualization by dividing shared physical
devices into multiple self-defined logical networks. Also, related resources such as the
network bandwidth and flow tables are divided and partitioned to different controllers
in each virtual network.

The RouteFlow [17] is an open-source SDN controller which can provide viru-
alized IP routing based on OpenFlow-enabled network hardware. Another highlight
of this controller is it enables virtualized IP routing solutions on commodity hardware, which helps traditional network architecture evolve toward newer SDN-based
architecture.

14.3   Existing SDN controller solutions

14.3.1   GPU acceleration

The Graphics Processing Unit (GPU) is a programmable chip that specializes in image
and video rendering. The GPU can be either integrated with the CPU on the same die
or packaged on a dedicated PCI-E-based system expansion card.

CPU   GPU

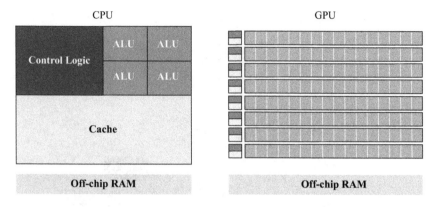

Figure 14.4   Different design philosophy of CPU and GPU

The design philosophy of GPU and CPU is very different. From Figure 14.4, we can see that the control logic and cache dominate the computing resources of a CPU chip whereas only small proportion of the resources is left for a few number of Arithmetic Logic Units (ALUs). By following this design philosophy, CPU becomes very suitable for handling workloads with complex procedures; but the downside is the parallelism of CPU is limited. For a current mainstream CPU, it can execute 8–16 threads at the same time. Inside a GPU, the majority of chip's area is allocated to Processing Elements (PEs). Furthermore, the control logic and cache, illustrated as the orange and gray areas in the figure, only occupy a small proportion of the whole area. By following this design philosophy, GPU becomes very suitable for executing massive parallel workloads. Besides, the GPU also has a much higher off-chip memory bandwidth than the CPU system does. For example, a NVIDIA GTX980 GPU has a 224 GB/s memory bandwidth [18], which is 6 times higher than an Intel Core i7 6700 CPU with a maximum bandwidth of 34.1 GB/s [19]. Hence, the GPU is also very suitable for achieving high throughput processing. Overall, the GPU can achieve much higher computing performance on the basis of high parallelism and high memory throughput [20–22]. For programming the GPU, CUDA (Compute Unified Device Architecture) and OpenCL (Open Computing Language) are the two most widely used solutions.

CUDA is a parallel programming platform with runtime API library and other acceleration libraries initially introduced by NVIDIA in 2007. The programming language of CUDA is called CUDA C/C++, which extends the standard C/C++. This key feature liberates software developers from translating their program into specific graphic programming language or API-like OpenGL, and turns GPU into General Purpose GPU (GPGPU). The OpenCL basically has the same functions as CUDA. One key difference is that OpenCL is a cross-platform language, which can be implemented not only on NVIDIAs GPU, but also on GPU from other vendor, CPU, FPGA (Field Programmable Gate Array), and DSP (Digital Signal Processor).

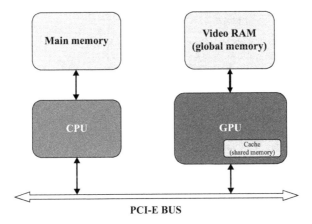

Figure 14.5   Top level architecture of CPU with GPU acceleration

14.3.2   GPU accelerated SDN controller

The top level architecture of CPU with GPU acceleration is illustrated in Figure 14.5. The CPU and GPU communicate and transfer data through the PCI-E bus. The PCI-E 3.0 has 16 GB/s bandwidth, which is the slowest data path in the whole top level of CPU-GPU hybrid computing architecture. The main memory is integrated with CPU, which is basically DDR3/DDR4 type memory, and provides from 30 to 60 GB/s bandwidth depending on the dual-channel or quad-channel configuration. On the GPU side, the off-chip video RAM, which is also called global memory in the GPU programming environment, is a type of RAM which is specifically designed for GPU. The GDDR5 (Graphics DDR 5) video RAM provides a typical bandwidth of 100–200 GB/s, which is much higher than that of the main memory connects with the CPU. The GPU chip also has a cache within the die of chip, which is also called shared memory and can provide more than 10 times bandwidth of the global memory. But the size of shared memory is much smaller than that of the global memory, which is only several megabytes in total. So, there is an unneglectable trade-off between the speed and memory size for GPU developers. In all, by understanding the whole architecture, especially the storage organization, developers can overcome the additional latency of the data transfer between CPU and GPU, and achieve a very high acceleration ratio by utilizing the high speed off-chip video RAM in GPU card, especially the super-fast shared memory within the GPU chip.

With the benefits of massive parallel processing and high memory accessing bandwidth provided by GPU computing, researchers have investigated boosting functions of an SDN controller by applying GPU acceleration and achieved higher data throughput and computing performance compared to traditional CPU-based SDN controller.

One hot research direction is accelerating the packet classification function of SDN controller. The high-level flow chart of packet classification is shown in

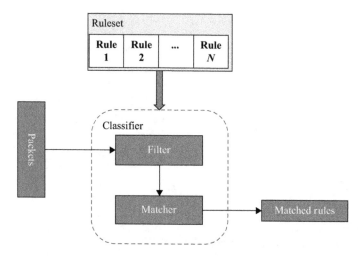

Figure 14.6   Packet classification in SDN

Figure 14.6. The purpose of packet classification is to find matched rules which include the destination for this packet and other processing activities defined by developers, and apply corresponding actions according to the matched rules for the packet. Besides, if there are multiple rules matched with the incoming packet, the priority of the rules should also be taken into consideration, and the rule with highest priority will be matched with the incoming packet.

With the consideration of improving the matching efficiency, researchers always apply a filter before the matcher inside the packet classifier. The purpose of applying a filter in advance is to verify if there is any possibility that there will be a match between the rules and the incoming packet. If the filter shows there is no possibility of a matching, the matcher will not go through the rule sets and following matching is avoided. In other words, the matching procedure is only performed if the packet passes the filter. In all, applying a filter really helps in saving time and improving the overall throughput. For the specific matching algorithms and filtering techniques, the work by Varvello *et al.* in [23] is a useful reference for the researchers and developers in this field.

Firstly, the linear search is an easily understandable and widely used matching method. This searching method applies the strategy of checking the tuple of an incoming packet against all entries in the rule set. But the linear search has some drawbacks such as the long running time and the poor scalability related to the size of rule sets (R) and the number of Tuples (T): it has the O(RT) complexity level of running time. So, the method with lower running time complexity is desired. One example is the Tuple Search introduced in [23]. The key idea of this kind of methods is using the hashing table, which has O(1) level of running time complexity. The rules belonging to the same class are stored in the same table and implemented as a hash table.

To further reduce the matching time, the pre-filtering stage is added before the searching. The key idea of pre-filtering is that the matcher can make an evaluation whether it is possible to get a match in following matching procedure or not by applying a filter. If the filter output is negative, the following matching can be avoided and the time is saved. One example to illustrate the pre-filtering technique is the Bloom filter [24], which was also applied in [23]. A Bloom filter is a data structure to check for membership of an element x in a set of n elements. Set the corresponding bits to 1 according to k hash functions, when there is an element x in the set. To check an element is in the set or not, compute it via all hash functions. If at least one bit is 0, current element is definitely not in the set.

Hsieh *et al*. presented a two-stage packet classification system accelerated by a NVIDIA K20c GPU, and achieved 356 MPPS (Million Packets Per Second) through-put for 1K 15-field rules and 213 MPPS for 100K 15-field rules [25]. From their experimental results, for the same 1K problem size, under applying the same decision-based algorithm, they achieved almost three times performance improvement over an FPGA acceleration-based work by Jiang *et al*. [26]. Also, for the same 15-field packet case, although with 3 times of problem size, their GPU-based platform achieved 4.43 times of performance improvement over a 16-core platform presented by Qu *et al*. [27]. Yun *et al*. presented a comprehensive research on packet classification by utilizing FPGA, Multi-core CPU, and GPU [28]. According to their experimental results, the Tesla K40 GPU platform leads to more than two times speedup when compared to a 16-core CPU platform. Also, the total processing latency of this platform is not longer than that of the CPU platform. From these two works, we can see that the GPU acceleration achieved encouraging performance results in packet classification and inspection function in SDN controller.

The acceleration of table matching of SDN controller is another research focus. Since the switching is flow-based in an SDN network, the OpenFlow-based switch keeps one or several flow tables. As shown in Figure 14.7, in each flow table, there is a bunch of entries for incoming data flow. For every incoming packet, the switch will pursue a matching method on them, and "guide" the packet to the matched entry of the flow table. From the flow chart in Figure 14.7, we can conclude that the whole process has obvious inner parallelism: all the flows in one flow table can be processed at the same time while all the flow tables can be processed in parallel.

For the matching algorithms, since linear search is very easy to understand and implement, it is widely used in different SDN controllers. For example, the linear search is used in switches controlling solution such as OpenvSwitch [15]. The linear search checks all the tuples in all incoming packets, and compares them by going through all entries in the rule set one by one, and get all the results of matched rules for each incoming packets. However, O(n) level time complexity of this algorithm leads to degradation in overall matching performance with the increase of the flow table size. To ensure satisfied throughput and latency, the LightFlow [29] utilized a GPU-based two-dimensional parallelization to accelerate the linear search algorithm in OpenvSwitch. Furthermore, three packet classification search algorithms: linear search, tuple search, and bloom search were applied and evaluated on CPU and GPU platform in [26]. For linear search, the rules are divided into several blocks, and each block is sent to a streaming multi-processor (SM) in GPU. Also, every multi-processor has a copy of the tuple set T. Thus, multiple rules can be compared with tuple

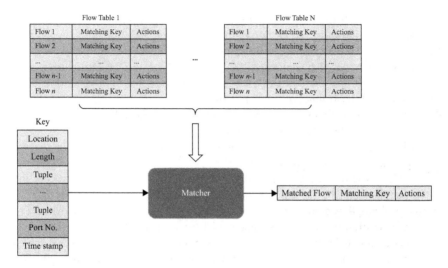

Figure 14.7   Table matching

set T at the same time by using GPU. For tuple search, rules with the same mask value are divided into same class first. Then, all classes are hashed into a hash table for efficient search. Instead of directly matching the tuple set T with the rules, in tuple search, the tuple in set T is applied hash computation first. Then, according to the hash result, the tuple is "guided" to the corresponding rule class, and only do the comparison within the sub-table consists of rules with the same mask. Thus, by applying the hashing, the matching space is reduced and the matching speed is improved. For Bloom search, a Bloom filter-based operation is applied before comparing the sub-table compared with the tuple search. The key idea of applying the Bloom filter is, if a tuple cannot "pass" the filter, the following matching to sub-table can be avoided so the overall running time or computing overhead is further reduced compared with tuple search. In both tuple search and Bloom search, the hashing and matching functions are accelerated by GPU. Overall, in their experiment results [23] the GPU acceleration achieved 7 times speedup on GSwitch and 10 times more speedup on vSwitch compared to the multi-core CPU platform.

In addition to the independent GPUs, researchers also found that the integrated GPUs can also be used for accelerating the SDN controller. First of all, nowadays, many CPUs have integrated GPU on the same die to handle the video output especially when there is no dedicated GPU in user's computer platform. Secondly, the performance of integrated GPU is rapidly increasing and already reached a unignorable level. For example, in Zhang's work [30], the integrated GPU platforms: Intel Ivy Bridge CPU, and AMD Fusion APU are compared with the dedicated GPU platform. From their experiment, the integrated GPU achieved satisfied acceleration performance and power efficiency. Lastly, integrated GPU uses faster on chip CPU–GPU interconnection instead of PIC-E bus since it is locating on the same die with the CPU. What's more, integrated GPU shares the same physical/virtual memory with

CPU, which is another advanced feature which can benefit the acceleration computing. In the work of Tseng *et al.* [31], they use Intel GT3e integrated GPUs and OpenCL programming platform to accelerate SDN packet processing. From their experiments of flow classification and IPv4/IPv6 forwarding, the overall throughput performance is accelerated by 22.5× for different workloads, compared to the high-end Xeon CPU only solutions. In another work of Zhu *et al.*, the integrated GPU is used for accelerating software routing [32]. They developed a CPU–GPU hybrid microarchitecture on the same chip with shared memory to accelerated the IP routing. In the experiments, with the acceleration of their integrated GPU solution, the total throughput is increased by five times, the average packet delay is reduced by 81.2% under best case, as well as a 72.9% reduction in delay variance [32]. To sum up, the benefits of low cost, high energy efficiency, fast on chip data bus and shared system memory, the approach of accelerating SDN controller with integrated GPU is interesting and worth to investigate furthermore.

14.4 Case study

To demonstrate the benefits brought by GPU acceleration in SDN-related field, we used packet classification as a case study and performed GPU accelerated on it. The main performance metrics shown in this case study are the overall throughput of packets processing and the processing latency for each network packet. Especially, the throughput is defined as Million Packets Per Second (MPPS), which means the number of packets can be processed by the classifier every second in the unit of millions. The latency is defined as the average latency for classifying a packet, which means the time for a single packet go through the classifier.

14.4.1 Experimental setup

The implementation of the GPU program in this case study is based on CUDA 7.0 programming environment. The hardware configurations are: Two 10-core Intel E5-2650 v3 CPUs with 2.3 GHz base frequency and 64 GB DDR4-2133 system memory. One NVIDIA Tesla K40 GPU with 2880 CUDA cores and 12 GB of GDDR5 video memory. The operating system is Ubuntu 14.04 LTS 64-bit, which is a Linux-based distribution.

14.4.2 Throughput and latency

For the experiments in this case study, we set the number of classification rules from 512 to 4096. Figure 14.8 illustrates the throughput performance under the change of the rule set number. We can observe the trend of throughput dropping with the larger rule set sizes.

However, the processing latency shown in Table 14.2 indicates that the GPU acceleration technique may introduce a longer latency for each package. The data in Table 14.1 shows a 50 times higher latency on average for GPU accelerated packet

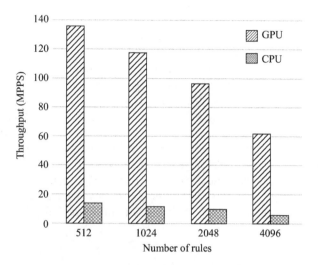

Figure 14.8   Throughput comparison

Table 14.2   Latency comparison

Number of rules		512	1024	2048	4096
Lateny (μs)	GPU	5.74	6.01	7.11	8.44
	CPU	0.11	0.12	0.14	0.17

classification compared with CPU-based solution. The longer latency is mainly caused by the following:

(1)   For achieving high processing and data transferring efficiency, GPU prefers transferring grouped data packets at one time rather than continuously transferring small data blocks. As a consequence, high bandwidth can be achieved by this pattern. But for a single packet, additional delay is caused while it is being packed with other package before transferred into GPU cores.

(2)   The GPU accelerator connects with system's main memory through PCI-E bus, which has relatively limited bandwidth. The packets need to be transferred via the PCI-E bus, and the processing result of GPU is also returned via this path. As a result, this round trip of packets also causes additional latency compared with CPU-based processing.

14.5   Open issues and challenges

There are still some open issues for GPU-accelerated SDN controller to overcome. One challenge is that there is still considerable latency caused by data transfer between

CPU and GPU accelerator. In a research by Hsieh *et al.* [25], we can see that by taking data transfer into account, the overall system performance shrinks to 70 Million Packets Per Second (MPPS) for different sized rule-sets. In other words, the data transfer causes at least 67% performance degradation when compared to the case of only considering the processing latency [25]. In another work by Eduard *et al.* [33], the authors also concluded packets transferring as one of the bottlenecks to their work. To ensure the efficiency of data transfer and GPU processing, incoming packets are grouped first, then transferred to the GPU for processing. One obvious drawback of this strategy is that it causes a longer latency for every single packet. One example in the corresponding work in [20] shows transferring 17 million packets simultaneously from CPU to GPU and getting the result back will end up with a transfer time of around 435 ms, which is about four times slower than the kernel execution in GPU.

Another situation that needs to be considered seriously is the heterogeneous packet sizes. Since the incoming packets are not identical in terms of byte count, different sized packets always coexist in the network. In some recent studies [23,25,28], one commonly applied hypothesis is that the GPU deals with incoming packets that are identical in size. This hypothesis benefits the data alignment for achieving higher throughput and better parallelism of GPU processing. But different applications can generate different sized packets, which can lower the effect of GPU acceleration. In the work of Eduard *et al.* [33], the authors considered heterogeneous packet sizes in all experiments. The experimental results in the corresponding study show that the latency can vary a lot when applying different mix portion of packets sizes.

Besides, as shown previously in Table 14.1, but not limited to it, we can see that there are various SDN controllers applying different implementation languages such as C/C++, Python, and JAVA. For now, the GPU programming environments such as CUDA and OpenCL are more C/C++ preferred, although they also have support libraries for other languages such as Python. This situation leads to a huge workload for developers, especially for those who want to translate an existing SDN controller into GPU accelerated version. Hence, there are still much more works left to do which require collaborations for SDN developers, GPU developers, and GPU technology vendors.

14.6 Conclusion

The Software-Defined Networking (SDN) controller became a research hot spot since it brings benefits such as centralized management, control-flow data-flow decoupling, and application centric. But some issues related to SDN still remaining to be solved. Among them, the performance of SDN controller is a big issue which cannot be ignored. One reason is the centralized control strategy poses a great demand on SDN controller's computation capability. The GPU becomes a natural choice for accelerating the SDN controller since it is a specialized computing unit which can handle massive parallel threads and can offer a high memory bandwidth. Many GPU-accelerated SDN controller [23,28,29,33] have been proposed and achieved good performance improvement compared with traditional CPU-based SDN controller. In this chapter, we first reviewed the background of SDN and existing SDN controller solutions. In the following sections, the background on GPU acceleration has been introduced as well as the existing GPU-accelerated SDN controller solutions. After analyzing existing research focus on GPU-based SDN controller, we presented a case study on GPU-accelerated packet classification to prove the benefits of GPU acceleration in the research field of SDN controller.

For future research related to GPU-accelerated SDN controller, one trend is to achieve higher performance and efficiency to satisfy the need of centralized control of SDN. For performance improvement, multiple-GPU approach or even GPU cluster approach is considered to handle larger network size and packets with more tuples. On the efficiency side, the power consumption and processing latency are two key points. GPU and CPU form a heterogeneous computing platform, and each component has different characteristics. While GPU has higher power efficiency, CPU has less packet processing latency. Hence, a "smart" dispatching strategy of GPU and CPU will help improve the performance of GPU-accelerated SDN controller. Also, extending the scalability of the algorithms is another concern. Since current researches are more dealing with fixed-size network packets and rule sets, and limited number of network conditions, adding more flexibility into the GPU-accelerated SDN controller will help it advance to practical implementation.

Acknowledgment

This work has been supported by the U.S. National Science Foundation (NSF) under grant numbers CNS 1464273 and CNS 1626360.

References

[1]   B. A. Nunes, M. Mendonca, X.-N. Nguyen, K. Obraczka, and T. Turletti, "A survey of software-defined networking: past, present, and future of programmable networks," *Communications Surveys & Tutorials, IEEE*, vol. 16, no. 3, pp. 1617–1634, 2014.

[2]   "Software-defined networking." https://en.wikipedia.org/wiki/Software-defined_networking.

[3]   N. Feamster, J. Rexford, and E. Zegura, "The road to SDN," *Queue*, vol. 11, no. 12, p. 20, 2013.

[4]   A. Dixit, F. Hao, S. Mukherjee, T. Lakshman, and R. Kompella, "Towards an elastic distributed SDN controller," in *ACM SIGCOMM Computer Communication Review*, vol. 43, pp. 7–12, ACM, 2013.

[5]   K. Phemius, M. Bouet, and J. Leguay, "Disco: distributed SDN controllers in a multi-domain environment," in *2014 IEEE Network Operations and Management Symposium (NOMS)*, pp. 1–2, IEEE, 2014.

[6] V. Yazici, M. O. Sunay, and A. O. Ercan, "Controlling a software-defined network via distributed controllers," *arXiv preprint arXiv:1401.7651*, 2014.

[7] N. McKeown, T. Anderson, H. Balakrishnan, *et al.*, "OpenFlow: enabling innovation in campus networks," *ACM SIGCOMM Computer Communication Review*, vol. 38, no. 2, pp. 69–74, 2008.

[8] S. J. Vaughan-Nichols, "OpenFlow: the next generation of the network?," *Computer*, no. 8, pp. 13–15, 2011.

[9] D. Tuncer, M. Charalambides, S. Clayman, and G. Pavlou, "Flexible traffic splitting in OpenFlow networks."

[10] J. Kempf, E. Bellagamba, A. Kern, D. Jocha, A. Takács, and P. Sköldström, "Scalable fault management for OpenFlow," in *2012 IEEE International Conference on Communications (ICC)*, pp. 6606–6610, IEEE, 2012.

[11] N. Gude, T. Koponen, J. Pettit, *et al.*, "Nox: towards an operating system for networks," ACM SIGCOMM *Computer Communication Review*, vol. 38, no. 3, pp. 105–110, 2008.

[12] J. Mccauley, "Pox: a python-based openflow controller." http://www.noxrepo. org/pox/about-pox/.

[13] Z. Cai, A. L. Cox, and T. E. Ng, "Maestro: a system for scalable openflow control," *Structure*, 2010.

[14] "Beacon homepage." https://openflow.stanford.edu/display/Beacon/Home, 2016.

[15] LinuxFoundation, "Open vSwitch: an open virtual switch." http://www. openvswitch.org/, Prior to Dec 30, 2010.

[16] R. Sherwood, M. Chan, A. Covington, *et al.*, "Carving research slices out of your production networks with OpenFlow," *ACM SIGCOMM Computer Communication Review*, vol. 40, no. 1, pp. 129–130, 2010.

[17] M. R. Nascimento, C. E. Rothenberg, M. R. Salvador, C. N. Corrêa, S. C. de Lucena, and M. F. Magalhães, "Virtual routers as a service: the RouteFlow approach leveraging software-defined networks," in *Proceedings of the Sixth International Conference on Future Internet Technologies*, pp. 34–37, ACM, 2011.

[18] "Nox homepage," http://www.noxrepo.org/.

[19] D. Erickson, "The beacon openflow controller," in *Proceedings of the second ACM SIGCOMM Workshop on Hot Topics in Software Defined Networking*, pp. 13–18, ACM.

[20] "GTX 980 performance specification," http://www.geforce.com/hardware/ desktop-gpus/gcforce gtx-980/specifications.

[21] "Intel CPU performance specification," http://ark.intel.com/products/88196/ Intel-Core-i7-6700-Processor-8M-Cache-up-to-4_00-GHz.

[22] S. Asano, T. Maruyama, and Y. Yamaguchi, "Performance comparison of FPGA, GPU and CPU in image processing," in *International Conference on Field Programmable Logic and Applications*, 2009. FPL 2009. pp. 126–131, IEEE.

[23] V. W. Lee, C. Kim, J. Chhugani, *et al.*, "Debunking the 100X GPU vs. CPU myth: an evaluation of throughput computing on CPU and GPU," vol. 38, pp. 451–460, ACM, 2010.

[24]   R. G. Belleman, J. Bdorf, and S. F. P. Zwart, "High performance direct gravitational n-body simulations on graphics processing units II: an implementation in CUDA," *New Astronomy*, vol. 13, no. 2, pp. 103–112, 2008.

[25]   M. Varvello, R. Laufer, F. Zhang, and T. Lakshman, "Multilayer packet classification with graphics processing units," *IEEE/ACM Transactions on Networking*, vol. 24, pp. 2728–2741, 2016.

[26]   B. H. Bloom, "Space/time trade-offs in hash coding with allowable errors," *Communications of the ACM*, vol. 13, no. 7, pp. 422–426, 1970.

[27]   C.-L. Hsieh and N. Weng, "Scalable many-field packet classification using multidimensional-cutting via selective bit-concatenation," in *Proceedings of the Eleventh ACM/IEEE Symposium on Architectures for Networking and Communications Systems*, pp. 187–188, IEEE Computer Society, 2015.

[28]   W. Jiang and V. K. Prasanna, "Scalable packet classification on FPGA," *IEEE Transactions on Very Large Scale Integration (VLSI) Systems*, vol. 20, no. 9, pp. 1668–1680, 2012.

[29]   Y. R. Qu, S. Zhou, and V. K. Prasanna, "A decomposition-based approach for scalable many-field packet classification on multi-core processors," *International Journal of Parallel Programming*, vol. 43, no. 6, pp. 965–987, 2015.

[30]   Y. R. Qu, H. H. Zhang, S. Zhou, and V. K. Prasanna, "Optimizing many-field packet classification on FPGA, multi-core general purpose processor, and GPU," in *Proceedings of the Eleventh ACM/IEEE Symposium on Architectures for Networking and Communications Systems*, pp. 87–98, IEEE Computer Society, 2015.

[31]   N. Matsumoto and M. Hayashi, "Lightflow: speeding up GPU-based flow switching and facilitating maintenance of flow table," in *2012 IEEE 13th International Conference on High Performance Switching and Routing (HPSR)*, pp. 76–81, IEEE, 2012.

[32]   Y. Zhang, M. Sinclair II, and A. A. Chien, "Improving performance portability in OpenCL programs," in *International Supercomputing Conference*, pp. 136–150, Springer, Berlin, 2013.

[33]   J. Tseng, R. Wang, J. Tsai, *et al.*, "Exploiting integrated GPUS for network packet processing workloads," in *NetSoft Conference and Workshops (NetSoft)*, 2016 IEEE, pp. 161–165, IEEE, 2016.

[34]   Y. Zhu, Y. Deng, and Y. Chen, "Hermes: an integrated CPU/GPU microarchitecture for IP routing," in *Proceedings of the 48th Design Automation Conference*, pp. 1044–1049, ACM, 2011.

[35]   E. G. Renart, E. Z. Zhang, and B. Nath, "Towards a GPU SDN controller," in *2015 International Conference and Workshops on Networked Systems (NetSys)*, pp. 1–5, IEEE, 2015.

Chapter 15

Virtualisation and management of application service networks

Indika Kumara[1], Jun Han[1], Alan Colman[1]
and Malinda Kapuruge[1]

Abstract

In a service ecosystem, organisations cooperate to achieve their goals. Some organisations provide application services while others consume these services, forming a web of interconnected services (referred to as *service network*). Multiple organisations or individuals (referred to as *tenants*) consume subsets of the services in a service network. They simultaneously share the same service network, which requires the virtualisation of the service network. The enactment and management of virtualised service networks is complex due to the heterogeneity in services, service networks, and tenants. This chapter analyses the characteristics of service networks and their virtualisation. It then presents an approach called *Software-Defined Service Networking (SDSN)* that applies the Network-as-a-Service (NaaS) model at the application level. This model is an abstraction of service network management functions that support the formation and management of virtualised service networks at runtime. We describe how a service network is virtualised and managed. We demonstrate the feasibility of our approach with a prototype implementation, and validate our support for the virtualisation and management of services networks.

15.1 Introduction

In an application service network, the application services that constitute the application are interconnected according to the capabilities provided and consumed by them [1]. The services can be computational services (e.g., weather and traffic reports) or business services, which support real-world business activities of enterprises or individuals (e.g., insurance claim handling and roadside assistance). These services exchange messages when they consume or provide their capabilities. A service

[1]School of Software and Electrical Engineering, Swinburne University of Technology, Australia

network connects a set of services, and forwards and regulates the messages between them. It is a common phenomenon in service-based businesses [1–3].

Multiple tenants consume subsets of the application services in a service network. They generally have common and variable functional and performance requirements, and thus share some of the services while also using different services as necessary. A service composition for a particular tenant in the service network needs to meet the functional and performance requirements of the tenant. Generally, the performance of a composite service depends on the performance of the individual services and the performance of the network that connects the services. The sharing and management of the services and their service network is problematic due to the heterogeneity in services, service networks, and tenants. However, the *Network-as-a-Service (NaaS)* paradigm [4], where network resources and functions are offered as services, paves the way to address this problem. The NaaS model can be employed at the application-level to abstract out the *service network management functions* that create and maintain the interconnections between the elements in a service network, route, and regulate application-level messages (instead of low-level communication network packets) between application services over the service network, and evolve the service network. These management functions can be used to virtualise and manage the services and the service network, and thus become part of the service network. The virtualisation of a service network creates tenant-specific network variants on the same service network.

Most existing works on service networks consider the modelling and analysis of service networks from specific aspects [2] such as the business value flows in a service network [5], provider–consumer relationships between services [1], and business processes in a service network [6,7]. Currently, to provide the technological realisation of service networks, their abstract models need to be converted to the configurations in existing service composition technologies such as BPMN (Business Process Management Notation) and BPEL (Business Process Execution Language) [1]. These technologies provide little or no direct support for the abstractions in a virtualised service network such as services, service interconnections, routing, regulation, and virtual service networks (VSNs), at either design time or runtime. These limitations can potentially increase the complexity of the design, enactment, and management of virtualised service networks with such technologies (similar to the impacts of such limitations on designing adaptive service collaborations [8]).

In this chapter, we present an approach called *Software-Defined Service Networking (SDSN)* that includes a range of management functions for creating and managing virtualised service networks, where a set of managed VSNs share an actual service network. A VSN represents a service composition in the service network for a particular tenant. We describe how the management functions can be used to form and manage a service network and its VSNs at runtime. These management functions are exposed as Web service operations as necessary. Conceptually analogous to the SDN (Software-Defined Networking) approach to designing virtual computer networks [9,10], SDSN similarly separates different concerns such as configuration, regulation, and dynamic management in the virtualised application service networks, but uses different design and runtime abstractions to meet the specific requirements of application service networks.

This chapter is organised as follows. Section 15.2 explores service networks, motivates their virtualisation, and identifies the requirements for an approach and framework to supporting virtualised service networks. Section 15.3 introduces our SDSN approach, and the runtime models of a service network and its VSNs. Section 15.4 describes the network management functions that support configuration, regulation, and evolution of a service network and its VSNs at runtime. It also discusses the composition and enactment of management functions. Sections 15.5 and 15.6 present the prototype implementation and evaluation of our approach, respectively. Section 15.7 reviews related work and Section 15.8 concludes the chapter.

15.2 Service networks and their virtualisation

In this section, we first provide a general overview of service networks. We then explore the *multi-tenancy* architectural principle, whereby multiple application variants can be derived from the same application to support the requirements of individual tenants [11]. Multi-tenancy can help to improve utilisation of applications and resources, and to reduce operational cost. Next, we present a set of business scenarios that highlight the sharing and runtime management of application service networks. Finally, we derive a number of general requirements for an approach that supports the enactment and runtime management of service networks and their virtualisation.

15.2.1 Service networks

A web of application services connected according to their capabilities, and the provided-required and normative relationships between them form a service network [1]. Application services are generally twofold: computational services and business services. The former encapsulate some computation objects, and the latter support real-world business capabilities of enterprises. In this research, we focus on business services, which are common in service networks [1,6]. However, our approach is also applicable to computational services. The performance properties of business services are constrained by the underlying real-world businesses, and cannot be altered by simply managing the computation resources used by the services. Business services are also generally heterogeneous concerning their capabilities and their relationships with other services.

In a service network, services collaborate to achieve the respective goals [8]. A collaboration has a structure/configuration, conversational behaviours over the structure, and regulation of such behaviours. The structure consists of a set of services and their capabilities used in the collaboration, and a set of interactions between these services. The temporal relationships between the uses of the capabilities or the interactions involved in the collaboration describe the conversational behaviours. The SLAs (service level agreements) between the service composer and the services dictate the regulation of the conversational behaviours in the collaboration by conditioning or restricting such behaviours. Since services that provide similar functionality may have differences in their capabilities, interactions, and SLAs, these services will need to be combined differently to support the same high-level requirements. These combinations of services can differ in their structures, conversational behaviours, regulations, and consequently performance achieved [12].

There are several common models of service composition: orchestration [13], choreography [13], component-based model [14], organisation-based model [15], and architecture view-based model [8]. A service orchestration models a composite service as an executable process that orders the capabilities of the component services according to the control and data flows among them. The composite service is defined, enacted, and managed from the viewpoint of the composer. A service choreography defines and orders the interactions between services from a global viewpoint. The behaviour of each participant in a collaboration is defined as its externally visible message exchanges (consumed and produced). These individual participants can be realised as service orchestrations. In the component-based model, service compositions assemble service components by wiring them together according to their provided and required interfaces. The organisation-based model represents service compositions as goal-oriented organisations. The services fulfil the positions (or roles) in an organisation, and relate to each other according to the interaction and obligation relationships (or contracts) between the corresponding roles. In the architecture view-based model, service compositions are represented, documented, and analysed using different architectural structures and views such as component-connector view, module-decomposition view, process view, and collaboration view. Thus, the other service composition models can be generally considered as specialised architecture-view based models as they use one or more architectural views, but from specific perspective(s).

The service network model can naturally support the use of the architecture view based model to describe service compositions, for example, organisation view (a network of interconnected entities) and process view (an ordered set of business activities or service capabilities). With the appropriate abstractions, the structural models can achieve better modularity [8], and improve the flexibility to manage the configuration of a composite application dynamically [16]. Furthermore, the network view or controlled message passing between entities can potentially enhance the capability to control runtime conversational behaviours in a composite application dynamically, and thus to support the management goals such as virtualisation and dynamic regulation policy enforcement (as in flexible network-based models [17,18]).

15.2.2   *Multi-tenancy in service networks*

In a service network, different tenants consume variants of service compositions provided by the same service network. Each tenant will have specific business objectives, and a specific service composition (business process) in the service network supports them. These tenant-specific service compositions may exhibit similarities and differences in the used elements of the service network, and the requirements. Thus, they can be considered as the variants of the same service composition or service network. The ability of a software application to be configured, customised, or changed in a prescribed manner to derive the variants of the application is commonly referred to as the variability in the software application [19].

There are two main multi-tenancy models for service networks (as in general cloud applications [11,20]): multi-instance multi-tenancy (MIMT) and single-instance multi-tenancy (SIMT). In the former model, each tenant uses a dedicated set of services and their service network. In the latter model, the tenants simultaneously share the same set of services and their service network. Tenants share the same services and elements of the service network for their similar requirements, and use different services and elements of the service network for their distinct requirements, *all at runtime*. Thus, the SIMT model requires the virtualisation of the service network, where a set of VSN simultaneously coexists on the same service network. In this chapter, we refer to such service networks as *multi-tenant service networks* or *virtualised service networks*.

The dedicated service networks in the MIMT model can simplify the support for tenant-specific variations and changes, while the virtualised service networks in the SIMT model can reduce operational cost and improve utilisation of services due to its higher degree of sharing. This chapter explores the enactment and management of a virtualised service network that realises the runtime variability for the possible set of tenants in the shared service network in the SIMT model.

15.2.3 Motivating scenarios

Consider a service network that offers roadside assistance to its tenants such as travel agencies and vehicle sellers by composing a number of application services such as vehicle repairers and towing providers. We will call the composite service RoSAS (Road-Side Assistance Service). Due to the benefits of the SIMT model, RoSAS employs it to manage both the roadside assistance business and the IT support. Each tenant has a VSN in RoSAS' service network to coordinate the roadside assistance for their users such as travellers and vehicle buyers.

The tenants have requirements in common but also requirements that differ. For example, repairing and towing are common to all tenants but they use either accommodation or rental vehicle. The tenants may also have similarities and differences in the required time for completing an assistance case (response time), and the number of new cases per day (throughput). For example, one tenant expects the response time of 3 days per case, and the throughput of 80 assistance cases per day, while another tenant needs the response time of 6 days per case, and the throughput of 157 assistance cases per day.

In RoSAS' service network, the services can be heterogeneous, and even services of the same type can have different capabilities, and different relationships with other services. For example, compared with the repairer MacRepair, AutoRepair needs to buy spare parts from external suppliers. The repair performances of these repairers also differ (e.g., the response times of 2 days and 3 days, and the throughputs of 200 new repairs per day and 400 new repairs per day). Some services have finite capacities or throughputs, which cannot be changed by simply managing the computation resources used by them (e.g., the aforementioned throughputs of repairers).

This commonality and variability in tenants' requirements primarily determine the commonality and variability in tenants' VSNs. Additionally, the finite throughputs of some services restrict the number of the tenants that can share them. The heterogeneity in services leads to more variations in VSNs. Consequently, VSNs can exhibit similarities and differences in their services, capabilities provided, configuration or structure, conversational behaviours, and regulation applied on such behaviours. This situation in RoSAS' service network is illustrated in Figure 15.1.

This service network needs to evolve in response to the change requests initiated by its different stakeholders such as services, RoSAS, tenants, and other external forces. A change request may concern both functional and performance objectives of the stakeholders. For example, after two months, RoSAS decides to support the taxi hire feature in its roadside assistance business. After one month, one tenant requests for 50 more assistance per day, and the towing duration of 2 hours within RoSAS' service area. From next month, the repairer AutoRepair expects to use an external assessor to estimate the cost of the repairing a damaged vehicle, which increases the repairing duration by 6 hours. To accommodate a given change request, RoSAS' service network and its VSNs need to be able to be modified at runtime. Thus, the service network should support possible types of changes that can potentially occur during its lifetime. A change needs to be realised and managed at runtime without disturbing the operations of those tenants/VSNs unaffected by the change.

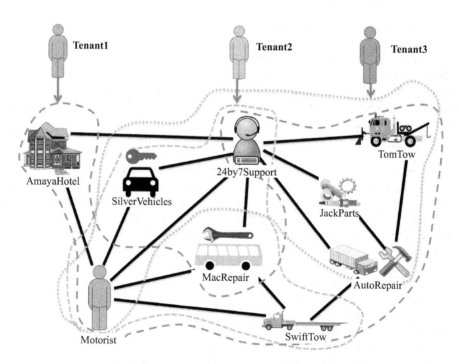

Figure 15.1   RoSAS' service network: business services and tenants

15.2.4   General requirements for an approach to virtualised service networks

The above scenarios illustrate a set of general requirements that need to be met by an approach that supports virtualised service networks. These include the ability to:

1.   Form service networks by connecting services according to the relationships between them.
2.   Route and regulate the message exchanges between the services throughout the service network that connects them.
3.   Form VSNs that share services and the service network with appropriate variations, which reflects the commonality and variability in the functional and performance requirements of the relevant tenants, and the characteristics of the services.
4.   Alter the configuration/structure of virtualised service networks at runtime.
5.   Alter the routing and regulation behaviours in virtualised service networks at runtime.

15.3   SDSN: an approach to realising and managing virtualised service networks

To address the above requirements, we present SDSN, which is an architecture-oriented approach for software engineers to design, enact, and manage virtualised service networks. In our conference paper [12], we have described how virtualised service networks can be designed with SDSN. Hence, in this chapter, we focus on how virtualised service networks can be formed and managed at runtime through a set of service network management functions.

To reduce the complexity in the design, enactment, and management of a virtualised service network, SDSN decouples the different concerns from each other, namely: structure/configuration, process/behaviour, regulation, and change management. Each concern uses its own abstractions, (de)composition strategies, and programming support. These different concerns are integrated to form a single system architecture. Moreover, SDSN introduces a number of architectural abstractions that can naturally represent a managed virtualised service network so that the semantic gaps between the domain concepts and their representations are minimised, further simplifying the software engineering activities.

In the rest of this section, we present the runtime model of a virtualised service network. In the next section, we describe the management functions that can be used to form and manage a virtualised service network at runtime.

15.3.1   Managed service networks

In SDSN, a service network is formed at runtime by architecturally connecting a set of services via a role-contract network structure. A *role* is a proxy to a service, and acts as a service network node where the messages from the other services are routed to the corresponding service via the role, and vice versa. A role includes the definitions of the used capabilities or *tasks* of the service. A message generally can be a request to provide a service or a response to that request. *Contracts* connect roles to complete the topology of the service network. A contract between the two roles defines the allowed interactions between the roles as a set of *interaction terms*, and acts as a full-duplex binary service network connection that transfer messages between two roles or service network nodes. The tasks and interaction terms define the message templates, which are used to create and validate the relevant messages.

Services interact with each other when they provide or request their tasks. The message exchanges between them over the role-contract network can be observed and regulated at a set of logical points called *regulation enforcement points (REPs)*. There are four types of REPs: synchronisation (at each role), routing (at each role), pass-through (at each contract), and coordinated-pass-through (across contracts). The synchronisation REP of a role synchronises a subset of incoming interactions from the adjacent roles and their services before invoking a task of the role's service by sending a request. The routing REP of a role receives and routes a response or request from the role's service to a subset of the adjacent roles and their services. The pass-through REP in a contract can intercept and process the interaction messages between two roles and thus their services. The coordinated-pass-through is to regulate the interactions across different pairs of roles and their services.

Each REP consists of a *knowledgebase* and a *regulation table*. The former contains a number of regulations described as event-condition-action (ECA) rules. The latter maps a message flow to a set of rules in the knowledgebase, which decides what to do with the message flow. To realise the regulation decisions, these rules use the regulation management functions such as admission control and load balance (see Section 15.4.2 for more details on these functions). A service network also has a state manager and an event manager that can record the state metrics such as measured response time and used throughput, and the events such as occurrence of an interaction or completion of a task. The events are used to coordinate the tasks or to realise the business processes in the service network (see the next section).

Figure 15.2 shows RoSAS' service network. A network of roles and contracts connect the services, and enable the passing of the messages between them. For example, the roles SC, TC1, GC1, and the contracts SC-TC1, GC1-TC1, SC-GC1 connect the services 24by7Support, SwiftTow, and MacRepair. Consider the inside of the role TC1 and its contracts with its adjacent roles as shown in Figure 15.3. The role TC1, which is played by the SwiftTow service, includes tasks such as *tPickup* and *tDeliver* while its contracts include interaction terms such as *iSendLocation*, *iNotifyDelivery*, and *iPayTow*. The synchronisation REP at the role TC1 synchronises over the interaction messages *iSendLocation* from the roles GC1 and GC2 to invoke its task *tDeliver*, which sends a request to the service SwiftTow. The routing REP at the role TC1 routes the response received from the service SwiftTow to the role SC and one of the roles GC1 and GC2 as the interaction messages *iPayTow* and *iNotifyDelivery* via the relevant contracts. The pass-through REPs at the contracts can intercept and further process these interactions, for example, generation of events or state objects. Through these events,

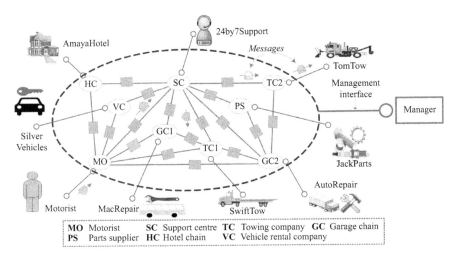

Figure 15.2   RoSAS' service network with SDSN

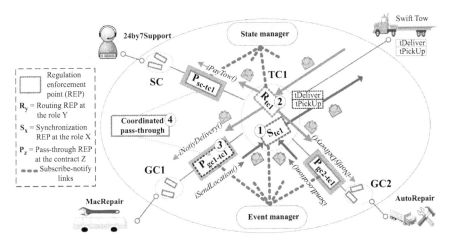

Figure 15.3   The inside of the role TC1, showing message flows and regulations

the coordinated-pass-through REP can coordinate the interactions among the roles TC1, SC, GC1, and GC2.

15.3.2   Virtual service networks

Each tenant has a VSN, which is a specific service composition in the service network that meets the functional and performance requirements of the tenant. The VSNs of different tenants simultaneously coexist on the same service network. A VSN has a set of service network paths, where each is a subset of the service network topology. Each VSN has the entries in the regulation tables at specific regulation enforcement points (REPs) in the service network paths of the VSN. A table entry at a given REP maps the messages belonging to a VSN to a subset of the regulation rules in the REP's knowledgebase. Each such rule applies a set of configured instances of management functions to the messages. Messages associated with a VSN instance are identified and kept separate so isolation of VSNs is achieved. The isolated messages are then routed and regulated on the service network path of the VSN instance, according to the configured management functions.

There exist multiple business processes in a VSN. These processes can be described and enacted as EPC (event-driven process chain) processes [12,21]. Events are used to define the temporal dependencies between the tasks used by processes. The management functions that can produce and/or consume events support the enactment of the processes (see Section 15.4.2).

VSNs of multiple tenants share some service network elements for their common requirements, and use some other service network elements for their distinctive requirements. The service network elements include roles, contracts, tasks, interaction terms, management functions, partner services, and their capabilities and throughputs/capacities. The interested reader is referred to [12] for more details on the designs of VSNs, their processes, and their sharing of the service network with appropriate functional and performance variations.

Figure 15.4 shows the service network topologies of the VSNs of two tenants in our motivating scenario. The overlaps between the topologies generally show the sharing between the two VSNs. For example, the two VSNs share some tasks of the services 24by7Support, AutoRepair, JackParts, and TomTow, the relevant roles and tasks, the contracts between these roles, the interaction terms in the contracts, and the relevant regulation rules and management functions. The sharing of a service task also implies that of its throughput. A fragment of the process in the tenant 3's VSN is shown in Figure 15.5. Once the service 24by7Support has processed the assistance request, it orders the towing of the broken-down vehicle, and rents a vehicle. When the towing is completed, the repairing of the vehicle starts.

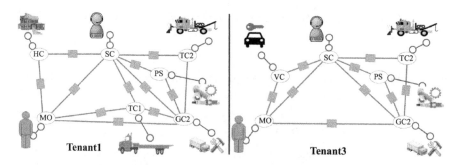

Figure 15.4   The service network topologies of two VSNs in RoSAS' service network

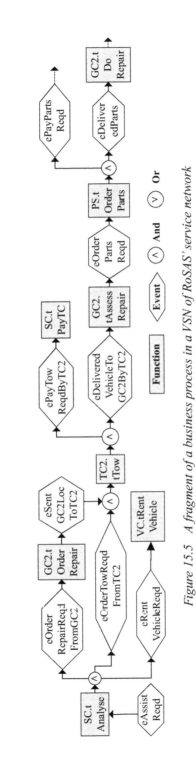

Figure 15.5 A fragment of a business process in a VSN of RoSAS' service network

15.3.3   Service network management interfaces and managers

The services in an enacted virtualised service network, and the requirements of the tenants and the service composer can change over time, requiring the changes to the configuration and/or regulation of the service network. To observe and to react to these changes, in SDSN, a service network has a number of management functions that can be used to monitor it, and modify its configuration and regulation (see Section 15.4). The service network exposes its management functions as Web service APIs (Application Program Interfaces) so that an external service network manager(s) can use these functions to realise high-level management decisions such as addition of a new service, a new tenant/VSN, or a new feature (see Section 15.4.5).

15.4   Management of virtualised service networks

In SDSN, a virtualised service network includes a number of management functions that can be used to manage the application services, the service network, and the VSNs at runtime. There are three types of such functions: configuration, regulation, and evolution. The configuration management functions can be used to create and change the interconnections between the elements in the service network (see Section 15.4.1). The regulation management functions support the routing of the messages between services through the role-contract network structure, and the observation and condition of such message flows (see Section 15.4.2). The evolution management functions can be employed to modify the configuration and regulation of the service network (see Section 15.4.3).

15.4.1   Configuration management

The main types of configuration management functions include:

- *Bind a service to a role*. This enables a role to consume the tasks of a particular service by sending request messages to the service.
- *Unbind a service from a role*. This makes a role isolated from its service. The role and the service will no longer be able to exchange messages.
- *Link roles*. This connects two roles via the contract between the two roles. Once being linked, the two roles can exchange messages between them through the contract.
- *Unlink roles*. This disconnects the two roles by isolating the contract between them. The two roles will no longer be able to exchange messages.
- *Connect a task to input and output messages*. This includes an interaction term in the inputs or outputs of a task, which indicates that the task will consume or its completion will create the interaction messages described by the term.
- *Unconnect a task from input and output messages*. This excludes an interaction term from the inputs or outputs of a task, which indicates that the task will no longer use or create the relevant interaction messages.

```
1  bind("TC1","http://swifttow.com/services/towservice");
2  link("SC","TC1");
3  connect("TC.tDeliver.inputs","GC1-TC1.iSendLocation.Req,GC2-TC1.iSendLocation.Req");
4  connect("TC.tDeliver.outputs",
5     "SC-TC1.iPayTow.Req,GC1-TC1.iNotifyDelivery.Req, GC2-TC1.iNotifyDelivery.Req");
6  registerToEvents("TC.tDeliver, "ePickedUp*(eSentGC1LocToTC1|eSentGC2LocToTC1)");
7  defineEventGeneration("TC.tDeliver,
8     "ePayTowReqdByTC1*(eDeliveredVehicleToGC1ByTC1|eDeliveredVehicleToGC2ByTC1)");
```

Figure 15.6   Some configuration management functions used in forming RoSAS'
service network

- *Register a task to events.* This makes a task being executed upon the occurrence of some events. As mentioned in Section 15.3.2, the dependencies between tasks are described using events in a loosely coupled way. Thus, this function adds or changes the dependencies of a given task on some other tasks.
- *Unregister a task from events.* A task will no longer be executed upon the occurrence of a particular set of events.
- *Define event generation from a task.* This defines the events to be generated by the completion of a particular task. Thus, this function adds or changes the dependencies of some other tasks on a given task.
- *Undefine event generation from a task.* A task will no longer generate a particular set of events upon its completion.

Figure 15.6 illustrates some uses of the configuration management functions in creating RoSAS' service network. The role TC1 is bound to the service SwiftTow, and then the role TC1 is linked with the role SC. The task *tDeliver* of the role TC1 is connected to a set of input and output messages. The task is subscribed to some events, and the events to be generated by its completion are also defined.

15.4.2   Regulation management

A regulation management function implements a unit of regulation decisions on application-level messages. We categorise them under the four types of regulation enforcement points (REPs) in a service network.

Synchronisation REP. A role stores the role–role interaction messages received from the other relevant roles (or from the corresponding services), and applies a subset of the regulation functions available in the service network. Three common types of regulation functions used at a synchronisation REP are:

- *Pull.* This pulls a set of role–role interaction messages from a message store of a role using the identifiers (internal) of those messages. The inputs and outputs in the task definitions provide these identifiers.
- *Synthesise.* This synthesises the service requests from the pulled role–role interaction messages by performing message transformations. In this research, we

use XSLT (Extensible Stylesheet Language Transformations) as the message transformation approach.

- *ExecuteTask*. This sends the created service request message to the corresponding service.

In the above three functions, the messages are first separated per each VSN instance to maintain the isolation between VSNs as well as VSN instances. The regulation rules in a synchronisation REP are generally triggered by service network events, which supports the realisation of the event-driven processes in VSNs.

Routing REP. Upon receiving a message from its service or user, the role needs to route the message to a subset from the relevant services via the roles of those services. The regulation functions used at the routing REP can realise and regulate this routing behaviour. The common types of such regulation functions include:

- *Classify*. This determines which tenant a received service message belongs to, based on the content of the message and a given policy (a set of rules). It also selects a VSN for an instantiation service message or a VSN instance for the subsequent interactions for an already created VSN instance.
- *AdmissionControl*. This decides whether to accept or reject messages from services or users. It prevents the congestion in the service network by keeping the excess service requests out of the service network in a controlled manner. The excess requests may be dropped, delayed, or served with some trade-offs (service degradation). Based on the algorithms used, the admission controller can support both response time and throughput/capacity guarantee for different tenants or classes (differentiated services).
- *Loadbalance*. This selects one interaction from a set of alternative role–role interactions for a given admitted service message to balance the load across the alternative service tasks (locally) or/and alternative processes in VSNs (globally). Different algorithms may use different constraints when selecting an alternative, for example, based on available throughput, response time, user traffic, and/or functional capabilities.
- *CreateInstance*. This creates an instance of a VSN based on a selected process from the processes in the VSN. It generates a unique identifier and attaches the generated identifier to the message. It also creates a non-executable object that records the state (active, passive, and terminated) of the VSN instance. The subsequent interaction messages for this VSN instance must carry its identifier.
- *Synthesise*. The content of the message from the service and that of an outgoing role–role interaction can be different. The synthesise function converts a service message into a role–role interaction message.
- *Forward*. This simply injects an outgoing interaction message to the pass-through REP at the contract of the destination role, and thus enforces the configuration/structure of the process of the VSN instance locally.
- *Schedule/Queue*. This determines which role–role interaction message to send next to the destination role via the relevant pass-through REP. It mainly allocates the throughput of a role–role interaction or a task (locally) or a process (globally)

among competing VSNs. However, the queuing delay can affect the response time guarantee. The scheduling policy (e.g., earliest deadline first, or priority scheduling) determines throughput and delay allocations.

- *Drop*. This drops a service interaction message and sends an error message back to the service or user indicating the reason for dropping the message.

Note that this research does not focus on the design of admission control, scheduling, and load balancing algorithms, and thus adopts some existing techniques that can achieve response time and throughput guarantees for tenants, for example [22,23].

Pass-through REP. Upon receiving a message from one of its two roles, a contract processes the received message via its pass-through REP, which can implement the regulation decisions that depend on the interactions between two services.

- *Push*. This puts a message in the message store of the destination role.
- *MonitorResponseTime*. This can measure the average response time of the capabilities (or tasks) of the destination or source services.
- *MonitorThroughput*. This can measure the used throughput of the tasks of the destination or source services.
- *PublishEvent*. This can be used to produce arbitrary events such as the indication of different states of interaction flows and tasks, and different interpretations of interaction messages.
- *SendToManager*. This can be used to communicate with the manager of the service network, for example, to send the monitored data to the manager.
- *Block*. This stops the transfer of the current message to the destination role.

Coordinated pass-through REP. This REP can listen to the events generated by the individual pass-through REPs, and implement regulation decisions based on the collected events. Thus, it can realise the regulation decisions that are relevant to multiple interactions between different pairs of roles (or services).

- *MonitorResponseTime/MonitorThroughput/PublishEvent/SendToManager*. See above for the explanations of these functions.
- *ActivateInstance/PassivateInstance/TerminateInstance*. These functions can be used to activate a new or passive instance of a VSN, to pause an active instance, and to terminate a passive instance, respectively. Only an active VSN instance can progress.

Some of the above regulation functions can be used at different types of REPs, for example, *Schedule* at synchronisation and pass-through REPs, and *SendToManager* at routing and synchronisation REPs. More regulation functions are also possible, for example, extraction and enrichment of message contents, obligation rules, and fine-grained statistics collection (see [23] for some of these functions).

Figure 15.7 illustrates some of the regulation functions used in the roadside assistance service network. These include the execution of the task *tDeliver* of the role TC1, the admission control of the assistance requests for the tenant3's VSN, the routing of a response message for the task *tDeliver*, the generation of an event for occurrence of the interaction *iSendLocation*, and the termination of a VSN instance.

```
(a)  1  ...
     2  RoleRoleInteraction[] in  =Pull("GC2-TC1.iSendLocation.Req");
     3  RoleServiceInteraction out =Synthesise(in,"DeliverToGC2.xsl");
     4  ExecuteTask(out,"tDeliver");
     5  ...
```

```
(b)  1  ...
     2  AdmissionControl("assist","157/1d",$serviceMsg);
     3  ...
```

```
(c)  1  ...
     2  Forward("GC2-TC1.iNotifyDelivery.Req",
     3              Synthesise("NotifyDeliveryGC2.xsl",$serviceMsg));
     4  ...
```

```
(d)  1  ...
     2  PublishEvent("eSentGC2LocToTC1",$interactionMsg);
     3  ...
```

```
(e)  1  ...
     2  TerminateInstance($vsnInstanceId);
     3  ...
```

Figure 15.7 *Some regulation management functions used in RoSAS' service network:*
(a) at the role TC1's synchronisation REP, (b) at the role MO's routing
REP, (c) at the role TC1's routing REP, (d) at the contract GC2-TC1's
pass-through REP, and (e) at the coordinated-pass-through

15.4.3 *Evolution management*

The evolution management functions support the runtime addition, removal, and update of the elements of the configuration and regulation designs of a virtualised service network. In general, the update function on a service network element changes its properties. The configuration evolution functions include add/remove/update of roles, contracts, tasks, and interaction terms. The relevant update functions can change the message templates of tasks and interaction terms. The regulation evolution functions include add/remove/update of regulation functions, regulation rules in knowledge-bases, regulation table entries, and service network events/states. The relevant update functions can modify the key and values of a regulation table entry, and replace the implementation of a regulation function and the content of a regulation rule. REPs (i.e., routing, synchronisation, pass-through, and coordinated-pass-through) are created or removed as part of the creation or removal of their placeholder elements (i.e., roles, contracts, and service network). Currently, an implementation of a regulation function is in the programming language used by the SDSN framework prototype (i.e., Java).

15.4.4 *Management policy: specification and enactment*

In SDSN, the management decisions are described as policies, which are essentially the composition of the instances of the three types of management functions described in the previous sections. Below, we present the policy language used by this research, as an example, to illustrate the use of policies to enact and manage virtualised service networks. We use the Drools business rule engine [24] and its language to implement the SDSN policy engine and language.

To specify the management policies, we use business rules and reactive programming concepts as in some relevant works [25,26]. A policy consists of a set of ECA (event-condition-action) rules.

- *Conditions.* A policy rule needs to be able to react to interaction messages, service network events (e.g., the completion of a task), service network state changes (e.g., the unavailability of a service), the properties of VSNs and their instances, the time of the day, and the execution state of the management policies. Thus, a condition should be a logical expression of the predicates formed by the information in these sources.
- *Actions.* The actions can include the management functions discussed in the previous sections, as well as custom functions, e.g., a performance predictor.
- *Execution.* The correct ordering of the rules as well as that of the actions within each rule are required to achieve a desired outcome. The rules are independent and are activated based on their conditions. When multiple rules are activated at the same time, the priorities of the rules can be used to resolve the conflicts. Within a rule, If-then-else conditional constructs can be used to order the actions.

The service network, its REPs, and its manager have policy engines that allow a software engineer to store, enact, and manage management policies. They can parse policies, process events, messages, state changes, and other provided information, and activate the rules in the policies as their conditions are met. When the policies are activated, the instances of the management functions used in them are executed.

15.4.5 Management policy: examples

With an example, in this section, we show how service network management functions are used. Let consider that a taxi provider service needs to be added to RoSAS' service network, and a VSN of a tenant needs to be configured to include this new service. The complete management policies are available online (see Section 15.6).

Evolution management. The role TX that represents the taxi provider service, and a contract between the role TX and the role SC that represents the 24by7Support service need to be added. The role TX includes new tasks *tOrderTaxi* and *tProvideTaxiInvoice*, and the role SC includes a new task *tPayTaxi*. Similarly, the contract SC-TX also defines new interaction terms. Figure 15.8 shows a fragment of the evolution management policy that implements these requirements.

The regulation knowledgebases and tables at the REPs of the roles TX and SC, and the contract SC-TX need to be updated. The regulation rules describe the relevant regulation decisions (see below), while the evolution management simply updates the knowledgebases with these rules. For example, the synchronisation rules *payTX* at the role SC, and *TX-SYN* at the role TX need to be added to invoke the relevant new tasks. The routing rules *analyseResponseV6* and *payTXResponse* at the role SC, and *TX-Routing* at the role TX also need to be introduced to route new service messages.

```
1  rule "conf_evol_mgt"
2    when
3      ($mgp:ManagementPolicyState(id=="conf_evol_mgt",state=="incipient"))
4    then
5      // role-level changes
6      addRole("TX");
7      addTask("TX", "tOrderTaxi");
8      addTask("TX", "tProvideTaxiInvoice");
9      ...
10     // role SC
11     addTask("SC", "tPayTX");
12     ...
13     // contract-level changes
14     addContract("SC-TX","SC","TX");
15     addTerm("SC-TX", "iOrderTaxi","AtoB");
16     addTerm("SC-TX", "iSendTaxiInvoice","BtoA");
17     ...
18     $mgp.setState("done");
19 end
```

Figure 15.8   A fragment of an (configuration) evolution management policy

```
1  rule "reg_evol_mgt"
2    when
3      ($mgp1:ManagementPolicyState(id =="conf_evol_mgt",state=="done")) and
4      ($mgp2:ManagementPolicyState(id =="reg_evol_mgt",state=="incipient"))
5    then
6      //synchronization rule changes
7      addSynchronizationRules("TX","TX_SYN.drl"); //all rules at the syn REP
8      addSynchronizationRules("SC","payTX.drl");
9      //routing rule changes
10     addRoutingRules("TX","TX_Routing.drl"); //all rules at the routing REP
11     addRoutingRules("SC","payTXResponse.drl");
12     addRoutingRules("SC","analyzeResponseV6.drl");
13     //passthrough rule changes
14     addPassthroughRules("SC-TX","SC-TX.drl");//all rules at the passth REP
15     addSynchronizationTableEntries( "Tenant2","orderTaxi:TX,payTX:SC");
16     addRoutingTableEntries( "Tenant2","analyzeResponseV6,
17           payTXResponse:SC,provideTaxiInvoice,orderTaxiResponse:TX");
18     addPassthroughableEntries( "Tenant2","orderTaxi, orderTaxiResponse,
19                       sendTaxiInvoice,sendTaxiInvoiceResponse:SC-TX");
20     ...
21     $mgp2.setState("done");
22 end
```

Figure 15.9   A fragment of a (regulation) evolution management policy

The contract SC-TX needs to include the pass-through rules to process new interaction messages. A number of regulation table entries also need to be added and modified to extend the service network path of the VSN of the tenant. A fragment of the evolution management policy that realises these requirements is in Figure 15.9.

Configuration management. The role TX needs to be bound to the service 14Cabs, the roles TX and SC need to be connected via their contract, and the relevant tasks and interactions need to be linked. For example, the outputs of the task *tAnalyse* of the role SC need to be updated so that the task's completion can initiate the interaction *iOrdertTaxi*. The inputs and outputs of the task *tOrderTaxi* of the role TX, and *tPayTX* of the role SC need to be linked to the relevant interaction terms. Figure 15.10 shows a fragment of the relevant configuration management policy.

```
1  rule "conf_mgt"
2    when
3      ($mgp1:ManagementPolicyState(id=="conf_evol_mgt",state=="done")) and
4      ($mgp2:ManagementPolicyState(id=="conf_mgt",state=="incipient"))
5    then
6      bind("TX", "http://14cabs.com/services/taxiservice");
7      link("SC","TX");
8      connect("SC.tAnalyse.outputs","SC-TX.iOrderTaxi.Req");
9      connect("TX.tOrderTaxi.inputs","SC-TX.iOrderTaxi.Req");
10     connect("TX.tOrderTaxi.outputs","SC-TX.iOrderTaxi.Res");
11     connect("SC.tPayTX.inputs","SC-TX.iSendTaxiInvoice.Req");
12     connect("SC.tPayTX.outputs,"SC-TX.iSendTaxiInvoice.Res");
13     ...
14     $mgp2.setState("done");
15 end
```

Figure 15.10   A fragment of a configuration management policy

(a)
```
1  rule "analyseResponseV6"
2    when
3      $msg : RoleServiceInteraction(opName== "analyseResponse")
4    then
5      Forward("SC-TX.iOrderTaxi.Req",Synthesise("OrderTaxReq.xsl",$msg));
6  end
```

(b)
```
1  rule "orderTaxi"
2    when
3      $msg : RoleRoleInteraction(opName == "orderTaxi")
4    then
5      PublishEvent("eOrderTaxiReqd",$msg);
6  end
```

(c)
```
1  rule "orderTaxi"
2    when
3      $e1 : Event(id == "eOrderTaxiReqd")
4    then
5      RoleRoleInteraction[] inMsgs = Pull("SC-TX.iOrderTaxi.Req");
6      RoleServiceInteraction exMsg = Synthesise(inMsgs,"OrderTaxi.xsl");
7      ExecuteTask(exMsg,"tOrderTaxi");
8  end
```

Figure 15.11   Fragments of the regulation management policies at (a) SC's routing
REP, (b) SC-TX's pass-through REP, and (c) TX's synchronisation REP

Regulation management. The regulation management functions are executed when the VSNs are enacted at runtime. Their composition is specified in the regulation rules. Figure 15.11(a) shows the use of the functions *Forward* and *Synthesise* at the role SC's routing REP (the rule *analyseResponseV6*) to route a response from the task *tAnalyse*. As shown in Figure 15.11(b), the rule *orderTaxi* in the pass-through REP at the contract SC-TX uses the function *PublishEvent* to generate an event *eOrderTaxiReqd*, indicating the occurrence of the interaction *iOrderTaxi*. Figure 15.11(c) shows the composition of the functions *Pull*, *Synthesise*, and *ExecuteTask* at the synchronisation REP of the role TX to invoke the task *tOrderTaxi*.

15.5   Prototype implementation

The prototype implementation consists of a middleware and a set of tools. Figure 15.12 shows the key elements of the prototype, namely application services, service

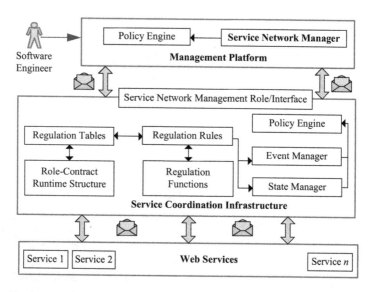

Figure 15.12   Runtime system architecture of the prototype implementation (with a single deployed service network)

coordination infrastructure, and management platform. We developed Web services to simulate the real-world business services used in our case studies. We used Apache Axis2 web service engine [27]. The service coordination infrastructure enacts the designed virtualised service networks, and enables their runtime management. It includes all service network management functions described in Section 15.4. We implemented the service coordination infrastructure by adopting and further extending the ROAD (Role-Oriented Adaptive Design)/Serendip framework [15,21], which employs an organisation-based model to support adaptive service compositions. The coordination engine is deployed on an Apache Tomcat web server [28] as an Apache Axis2 module. Each deployed service network offers its monitoring and management functions as Axis2 Web services. The management platform contains the managers of each deployed service network. We implemented the management platform as a group of Web services and consumers (deployed as an Axis2 module). A software engineer can design the service network and management policies using the provided tool support, and enact them through the relevant managers at the management platform. The tools are Eclipse plug-ins.

15.6   Evaluation

In this section, using our prototype, we evaluate the feasibility of our SDSN approach to support the virtualisation and management of service networks. The case study resources are available in https://github.com/indikakumara/SDSN_Book_Chapter.

15.6.1   Service network virtualisation

To validate the ability of our approach to realise virtualised service networks, we have fully developed the virtualised service network for the roadside assistance scenario. It consists of 9 services, 9 roles, 51 tasks, 15 contracts, 17 interaction terms, and 115 regulation rules (across all REPs). Three tenants with common as well as different functional and performance requirements share the service network, where each tenant has a VSN. The service network implements four features: *Tow*, *Repair*, *Accommodation*, and *Rental Vehicle*. Each of the first two features has two implementations with different services and different performance (throughput and response time). At least two tenants share each implementation of the four features except the feature *Accommodation*. The business processes in three VSNs were executed in parallel to validate isolation among the VSNs, and to verify that each process/VSN achieves its functional and performance requirements when they share the service network simultaneously.

In [12], we have used this roadside assistance case study to show that the significant utilisation benefits can be achieved by sharing services among tenants in a multi-tenant cloud application with our SDSN approach. We have also compared our multi-tenancy support with that in BPEL (Apache ODE [29]) for runtime performance overhead (for the same case study). Both had similar runtime overhead.

15.6.2   Service network management

To validate the realisation of different types of service network management functions and their usage in enacting and managing virtualised service networks, we have extended the original roadside assistance case study with new scenarios that change the roadside assistance service network (see Table 15.1). These scenarios consider the changes such as addition, removal, and update at the high-level features. The third scenario covers the example presented in Section 15.4.5. Each scenario has two or more sub-scenarios, which cover both functional and performance cases. For each sub-scenario, we consider both the realisation of the scenario and the rollback (removal) of the realisation. These scenarios together cover each type of management functions of the service network at least one time.

To realise a given sub-scenario, we first identify the differences between the original service network, and the target service network after the realisation of the sub-scenario, in terms of configuration and regulation changes. Then, the management policies are created to implement those changes. The created policies are enacted at runtime on the original service network. To validate the implementation of the sub-scenario, we first analyse the log traces generated when enacting the policies, to check that the expected changes have been committed. Second, we compare the responses and logs for requests to the VSNs in the changed service network with those of the target service network (manually created).

To approximate the effort for developing the management policies for the case study, we have measured the size of a policy in lines of code (LoC) (similar to [30]). Figure 15.13(a) and (b) shows the size of the management policies. The average size of the management policies for the functional change sub-scenarios is 52 LoC, and

Table 15.1   Change scenarios for the roadside assistance case study

No.	Types of changes (high-level)	Sub-scenarios
1	Add/remove a mandatory feature	*Reimbursement* feature (to be used by each tenant) Response time <30 min and max-throughput = 150 for all assistance cases
2	Add/remove an optional feature	*Accident Tow* feature (to be used by some of the existing tenants and the new tenant AsiaBus) Response time <2 d and throughput = 10 for a reimbursement
3	Add/remove feature to a feature group	*TaxiHire* feature to/from the features *RentalVehicle* and *PublicTransport* 4 d repair duration in addition to the existing 2 d and 3 d for *Repair* feature Make *Or* repair durations 2 d, 3 d, and 4 d *Alternative (XOR)* options
4	Add/remove feature dependency	The dependency *Major Repair excludes Accommodation* The dependency *Repair time = 3 d includes Tow duration = 4 h*
5	Modify feature implementation	Extend an implementation of the *Repair* feature to use an external parts supplier if the required parts are not available internally Add an external assessor to a repairing implementation, which increases repair time by 6 h

that for their rollback is 20.6 LoC. The average size of the management policies for the performance change sub-scenarios is 26.4 LoC, and that for their rollback is 14 LoC.

We have also measured the run-time change enactment time (RCET) for each scenario. RCET is the time difference between the manager of the service network receiving a management policy and the service network being ready for use after applying the policy. The framework was run on an Intel i5-2400 CPU, 3.10 GHz with 3.23 GB of RAM, and Windows 7. Figure 15.13(d) and (c) shows RCET values. The average RCET for the functional scenarios is 633.2 ms (milliseconds), and that for the rollback of the realisations is 15.2 ms. The average RCET for the performance scenarios is 307 ms, and that for the rollback is 31.2 ms.

15.7   Related work

We discuss below the existing research efforts on service networks in relation to the requirements we have identified. We also compare our SDSN for service networks with SDN (Software-Defined Networking) [9,10] for communication networks. In Section 15.2.1, we have discussed service composition models that the service network approach supports.

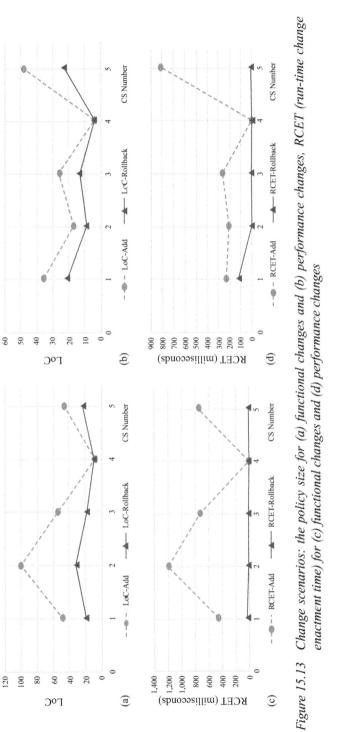

Figure 15.13 Change scenarios: the policy size for (a) functional changes and (b) performance changes, RCET (run-time change enactment time) for (c) functional changes and (d) performance changes

15.7.1   Support for service networks and their virtualisation

Different researchers use different terminologies to describe an interconnected web of services, for example, business ecosystems [3], and service value networks [2,5,31]. We call them service networks as in [1]. Most of the existing works on service networks consider the modelling and analysing of service networks from specific aspects [2] such as the business value flow in a service network [5], the business processes in a service network [6,7], and provider–consumer relationships [1]. Among these works, Allee [5] employs a graph-based notation to model value flows inside a network of entities that exchange goods, services, knowledge, and so on. Comuzzi *et al.* [6] present a framework that can maintain the monitorability of dynamic service networks, which change over time. Their focus is on the contractual agreements between the actors in a business network. To assess the completeness of a service network, Kabzeva *et al.* [7] have used a dependency model created by analysing a given process description in a service network and the service descriptions of the partner services. They provide a tool that can detect missing elements such as service descriptions and task implementations. Danylevych *et al.* [1] propose a notation to model a network of services connected according to their provider–consumer relationships. The constructs in the proposed notation was mapped to those in BPMN (Business Process Model and Notation). They observed that BPMN lacked the constructs to model the interconnections between services. Currently, to implement and enact service networks at the IT level, their abstract models need to be converted to configurations in existing service composition technologies such as BPMN and BPEL, which provide limited support for the architectural abstractions in a service network [1], limiting their utility.

Compared to these works, we consider multi-tenant or virtualised service networks, and represent domain concepts such as services, service relationships or interconnections, routing, regulation, and VSNs naturally. We propose SDSN as an enhancement to the service network model. SDSN separates multiple concerns such as configuration, conversational behaviour, regulation, and change management in a service network architecturally. These different concerns coexist in a single runtime system architecture in an isolated manner. A software engineer has the flexibility to design and implement each concern with the most appropriate abstractions and programming models.

15.7.2   Relating to SDN

To a certain degree, SDSN and SDN have similarities in their objectives and their high-level system architectures. Both SDSN and SDN primarily intend to improve the manageability of service networks or computer networks, and to simplify network management tasks, including virtualisation, enforcement of control policies, dynamic reconfiguration of enforced policies, and evolution. Both approaches pursue these goals by making the system modular and separating different concerns in the system architecture. In particular, they separate the message passing between entities in the network from the decision-making about the message passing, and allow programmability and new abstractions for each concern.

Although SDSN and SDN have similarity in their goals and their system archi-tectures, SDSN has its own specific requirements and characteristics focusing on business services. The design of the role-contract structure in SDSN naturally repre-sents services and the interaction and business relationships (e.g., obligation rules) between them. The corresponding runtime structure connects and coordinates ser-vices – roles act as proxies to services, and contracts connect roles and enforce the expected relationships between them. In a computer network, a path is an end-to-end connection between two end-hosts via network devices. The data packets are sent along a selected (best) path. In a service network, a path is a business process that can have a complex control flow among a subset of the capabilities of the services in the network. A business process is selected and instantiated. Then, the created instance is progressed by relaying the interaction messages between services that are created when the relevant service capabilities are consumed (or provided). Consequently, compared with computer networks, to regulate the message routing in service net-works, rich abstractions are needed (see synchronisation, routing, pass-through, and coordinated-pass-through regulation abstractions). Both types of networks can use different regulation mechanisms (e.g., message transformation) as well as similar regulation mechanisms (e.g., admission control and load balance). In SDN, the man-agement of a network primarily focuses on the regulation design of the network. In SDSN, the management of a service network consider both its regulation and con-figuration designs. As discussed above, the regulation design of a service network in SDSN is considerably different from that of a computer network.

15.8 Conclusion

In this chapter, we have introduced the SDSN approach to the design, enactment, and management of virtualised application service networks. In a virtualised service network, a set of managed VSNs simultaneously coexist on the same managed service network, sharing the services and their network. Each VSN comprises a subset of ser-vices in the service network to achieve the specific functional and performance goals of a tenant. We have presented the runtime models of a virtualised service network, and the network management functions that support the configuration, regulation, and evolution of a virtualised service network. We have also described the compositions of the management functions to realise management policies for service networks. We demonstrated the feasibility of our SDSN approach with a prototype implementa-tion and a related evaluation of our support for virtualised service networks and their management.

As future work, we plan to consider heterogeneous service networks, where dif-ferent types of services such as application services and communication network services collaborate or converge. This will allow us to consider in a comprehensive manner how the QoS of a service network depends on those of the application ser-vices, coordination middleware, and communication network. Issues that are also common to SDNs such as the scalability and availability of the controller, and control programming [10] also need to be explored with respect to SDSNs.

Acknowledgment

This research was partly supported by the Smart Services Cooperative Research Centre (CRC) via the Australian Government's CRC Program.

References

[1]   Danylevych O., Karastoyanova D., Leymann F. 'Service networks modelling: an SOA and BPM standpoint'. *Journal of Universal Computer Science*. 2010; **16**(13):1668–1693.

[2]   Razo-Zapata I.S., Leenheer P., Gordijn J., Akkermans H. 'Service network approaches', in Barros A., Oberle D. (eds.). *Handbook of Service Description: USDL and Its Methods*. Boston, MA: Springer US; 2012. pp. 45–74.

[3]   Barros A.P., Dumas M. 'The rise of web service ecosystems'. *IT Professional*. 2006; **8**(5):31–37.

[4]   Duan Q., Yan Y., Vasilakos A.V. 'A survey on service-oriented network virtualisation toward convergence of networking and cloud computing'. *IEEE Transactions on Network and Service Management*. 2012; **9**(4):373–392.

[5]   Allee V. 'Reconfiguring the value network'. *Journal of Business Strategy*. 2000; **21**(4):1–6.

[6]   Comuzzi M., Vonk J., Grefen P. 'Measures and mechanisms for process monitoring in evolving business networks'. *Data and Knowledge Engineering*. 2012; **71**(1):1–28.

[7]   Kabzeva A., Götze J., Müller P. 'Capturing and analysing service network models'. *Proceedings of International Conference on Information Integration and Web-based Applications & Services*; Vienna, Austria. New York: ACM; 2013. pp. 260–264.

[8]   Haesevoets R., Weyns D., Holvoet T. 'Architecture-centric support for adaptive service collaborations'. *ACM Transactions on Software Engineering Methodology*. 2014; **23**(1):1–40.

[9]   McKeown N., Anderson T., Balakrishnan H. *et al.*, 'OpenFlow: enabling innovation in campus networks'. *Computer Communication Review*. 2008; **38**(2):69–74.

[10] Kreutz D., Ramos F.M.V., Esteves V.P., Esteve R.C., Azodolmolky S., Uhlig S. 'Software-defined networking: a comprehensive survey'. *Proceedings of the IEEE*. 2015; **103**(1):14–76.

[11] Chong F., Carraro G. *Architecture Strategies for Catching the Long Tail* [online]. 2006. Available from https://msdn.microsoft.com/en-us/library/aa479069.aspx [Accessed 28 June 2016].

[12] Kumara I., Han J., Colman A., Kapuruge K. 'Software-defined service networking: runtime sharing with performance differentiation in multi-tenant SaaS applications'. *Proceedings of the 2015 IEEE International Conference on Services Computing*; New York, USA. IEEE; 2015, pp. 210–217.

[13] Peltz C. 'Web services orchestration and choreography'. *Computer.* 2003; **36**(10):46–52.

[14] *Service Component Architecture* [online]. 2007. Available from http://www.oasis-opencsa.org/sca [Accessed 28 June 2016].

[15] Colman A., Han J. 'Using role-based coordination to achieve software adaptability'. *Science of Computer Programming.* 2007; **64**(2):223–245.

[16] Kramer J., Magee J. 'Dynamic configuration for distributed systems'. *IEEE Transactions on Software Engineering.* 1985; **11**(4):424–436.

[17] Kagal L., Finin T. 'Modelling conversation policies using permissions and obligations'. *Autonomous Agents and Multi-Agent Systems.* 2007; **14**(2): 187–206.

[18] Kim H., Feamster N. 'Improving network management with software defined networking'. *IEEE Communications Magazine.* 2013; **51**(2):114–119.

[19] Van Gurp J., Bosch J., Svahnberg M. 'On the notion of variability in software product lines'. *Proceedings of the Working IEEE/IFIP Conference on Software Architecture*; Amsterdam, The Netherlands. IEEE; 2001, pp. 45–54.

[20] Guo C.J., Sun W., Huang Y., Wang Z.H., Gao B. 'A framework for native multi-tenancy application development and management'. *Proceedings of Fourth IEEE International Conference on Enterprise Computing, E-Commerce, and E-Services*; Tokyo, Japan. IEEE; 2007, pp. 551–558.

[21] Kapuruge M., Han J., Colman A. *Service Orchestration as Organization: Building Multi-tenant Service Applications*, San Francisco, CA, USA: Morgan Kaufmann; 2014.

[22] Krebs R., Momm C., Kounev S. 'Metrics and techniques for quantifying performance isolation in cloud environments'. *Science of Computer Programming.* 2014; **90**(B):116–134.

[23] Xiao X., Ni L.M. 'Internet QoS: a big picture'. *IEEE Network.* 1999; **13**(2): 8–18.

[24] *Drools – Business Rules Management System* [online]. [date unknown]. Available from http://www.drools.org [Accessed 28 June 2016].

[25] Voellmy. A, Kim H., Feamster N. 'Procera: a language for high-level reactive network control'. *Proceedings of the First Workshop on Hot Topics in Software Defined Networks*; Helsinki, Finland. ACM; 2012, pp. 43–48.

[26] Phan T., Han J., Schneider J.G., Ebringer T., Rogers T. 'A survey of policy-based management approaches for service oriented systems'. *Proceedings of 19th Australian Conference on Software Engineering*; Perth, Australia. IEEE; 2008, pp. 392–401.

[27] *Apache Axis2/Java – Next Generation Web Services* [online]. [date unknown]. Available from http://axis.apache.org/axis2/java/core/ [Accessed 28 June 2016].

[28] *Apache Tomcat* [online]. [date unknown]. Available from http://tomcat.apache.org/ [Accessed 28 June 2016].

[29] *Apache ODE* [online]. [date unknown]. Available from http://ode.apache.org/ [Accessed 28 June 2016].

[30] Hihn J., Habibagahi H. 'Cost estimation of software intensive projects: a survey of current practices'. *Proceedings of the 13th International Conference on Software Engineering*; Texas, USA. IEEE; 1991, pp. 276–287.

[31] Hamilton J. 'Service value networks: value, performance and strategy for the services industry'. *Journal of Systems Science and Systems Engineering*. 2004; **13**(4):469–489.

Chapter 16

Context-as-a-Service (CAAS) platform for smart IoT applications

Hyun Jung La[1]

Abstract

A context-aware Internet of thing (IoT) application is a software system which utilizes various contexts acquired from IoT devices in providing context-aware functionality. The key benefit of context-aware IoT applications will be the situation-specific services provided through software analytics on the rich IoT contexts. Despite of the benefits, it is not always trivial to develop high-quality IoT applications due to the runtime overhead of acquiring and analyzing the IoT contexts. In this chapter, the potential challenges of developing context-aware IoT applications are first clarified. To address the challenges, we present *Context-as-a-Service* (CAAS) platform which adopts the notion of cloud computing where resources are deployed as a unit of a service. CAAS platform plays an essential role of providing contexts to IoT applications efficiently. The key design of CAAS platform and its implementation are presented in detail.

Keywords

IoT, IoT context, context-aware application, service platform, context acquisition, context provisioning

16.1 Introduction

Internet of thing (IoT) devices are equipped with sensors, from which various contexts can be gathered. A context-aware IoT application is a software system which utilizes the IoT contexts in providing situation-specific functionality. The key benefit of context-aware IoT applications will be the appropriateness and smartness of the service provided by the application [1].

---

[1] Smarty Lab Corporation, 320 Sangdo-Ro, Dongjak-Gu, Seoul 156-725, Republic of Korea

Through our research and the literature surveys on IoT computing [2–4], we make the following observations and trends:

- An IoT device can now provide several types of contexts such as location, temperature, and proximity. Consequently, the amount of acquired contexts from the devices can be substantially large, and cannot be effectively stored in the built-in memory of the IoT device.
- A context-aware IoT application may potentially utilize a rich amount of various contexts which could be gathered from different IoT devices. That is, the IoT contexts can now be shared among applications, and hence there is a demand for effective contexts provisioning service.
- Advanced IoT applications may require analytics on an accumulated set of contexts over some time period, in addition to the current contexts gathered. That is, the applications may apply machine learning-type of analytics over a training set. Hence, there is a need for a context repository which archives acquired contexts from different devices and provides requested contexts to IoT applications efficiently.

What all the observations mean is that there is a need for a context service platform which manages a context repository on a server side and provides required contexts efficiently to smart IoT applications. This is because such a context repository and the comprehensive context service cannot be handled by each IoT application itself.

In this paper, we present a design of a context service platform for smart IoT applications, called *Context-as-a-Service* (CAAS), which uses an extensive set of shared and/or unshared contexts. The concept of CAAS is proposed by extending our previous work [5]. We address the key challenges in developing such a platform and elaborate essential parts of the design. In Section 16.2, we summarize related works. In Section 16.3, we clarify the technical challenges of IoT contexts and their provisioning to applications. In Section 16.4, we define the functional and nonfunctional requirements for the CAAS platform. In Section 16.5, we elaborate the key design of the platform. In Section 16.6, we show our PoC-level implementation of the platform and present the results of experiments with the platform.

With the proposed platform, a large amount of IoT contexts can be managed in a server-side repository and various sets of contexts can be efficiently delivered to the IoT applications. With the archived contexts, IoT applications can behave even more smartly by utilizing machine learning over an extensive training set of contexts.

16.2   Related works

There are several works on dealing with minimizing the resource consumptions by sharing contexts among multiple applications. Skorin-Kapov's work proposes cloud-based publish/subscribe middleware interfaced with a quality-driven sensor management function, applicable for building mobile IoT applications [6]. The architecture is designed so as to smartly manage and acquire sensor readings in order to satisfy global sensing coverage requirements, while obviating redundant sensor activity and consequently reducing overall system energy consumption. For the efficient data acquisition, the middleware makes intelligent decisions for choosing an optimal subset of available sensors which to keep active in order to meet subscription requirements based on sensor accuracy, level of trustworthiness, and available battery level. This work presents the architecture-level design and metrics for efficient context acquisition, which needs to be enhanced to be practically applied.

Chattopadhyay's work presents a design of algorithms and architectures for enabling a Sensing-as-a-Service paradigm in the recent era of IoTs [7]. This work raises the technical issues in realizing the concept of Sensing-as-a-Service and tries to come up with efficient mechanisms for contextual data collection, aggregation, dissemination, assimilation, and processing that can efficiently handle the data deluge resulting out of today's stupendous system. However, this work only presents the overall ideas conceptually in the architecture level, not in detail.

Lee's work presents a context monitoring platform for context aware applications in resource-limited mobile environments [8,9]. This platform supports concurrent context acquisition requests by extracting associated sensors from context monitoring queries. And, after evaluating monitoring queries, the platform chooses the minimal set of sensors to be monitored. In addition to minimizing the number of activated sensors sharing the monitoring results, other factors need to be considered.

Nath's work presents an acquisitional context engine supporting continuous context-aware applications while mitigating sensing costs [10]. The engine manages two caches for sharing contexts: one for raw contexts and the inferred contexts. And, if contexts are not in the caches, planner determines the cheapest way to acquire contexts by considering the resource consumption of each sensor and the number of sensors for the context retrieval. This work needs to consider other factors affecting efficiency and additional overhead of executing inferences.

And, some works present methods to choosing the minimal set of sensors and their monitoring frequencies and cycles (i.e., activated and inactivated durations). Wang's work presents energy efficiency mobile sensing framework [11]. The framework utilizes XML-based state descriptor given by context aware applications. By interpreting states and their transitions in the descriptor, the framework determines the minimum set of sensors and their cycles.

Han's work presents user context collection model realizing with two methods [12]. Sensor scheduling method decides the minimal number of contexts by calculating the utility function. And, dynamic sensor reconfiguration decides sampling rate and activation interval for the chosen sensor by using user current states and past history. Other factors for monitoring efficient need to be considered and the detailed description of each method needs to be elaborated.

In addition, there are several works to manage contexts, which is the broader range of acquiring contexts, in an energy-efficient way such as Roy's work [13] and Bezerra's work [14]. Roy's work deals with efficiently performing context inference by minimizing the communication overhead while maximizing accuracy. Bezerra's work presents a data reduction approach to lower the amount of data sent to the context management framework.

In summary, most of the works address the importance of minimizing costs to acquire contexts for context-aware applications. However, the solutions for the issues are too less detailed to be practically applied. And, it is demanding to have efficient context acquisition methods which are specific to IoT applications.

16.3   Challenges in acquiring and provisioning IoT contexts

Context acquisition is a prerequisite to providing the functionalities of most IoT applications. That is, IoT applications start their functionalities after acquiring contexts from IoT devices. If IoT applications depend on a larger number IoT devices, acquiring contexts from them can be the runtime overhead of the applications. In this section, the potential problems that IoT application faces with at runtime are discussed, which will lead to the motivation behind the proposed platform.

16.3.1   *Efficiency of context acquisition*

Since there is no universally accepted definition of terms about IoT computing, we define terms of an IoT application, a context, an IoT device, and their relationships in a semiformal manner.

Let APP_i be a context-aware IoT application, which provides context-specific functionality. And, $APPSET$ is defined as the set of all such applications:

$$APPSET = \{APP_i | APP_i \text{ is an application utilizing contexts of IoT devices.}\}$$

IoT applications utilize some contexts for provisioning context-specific functionality. Let CTX_j denote a context used by an application, and then the whole set of contexts used by APP_i is defined as:

$$CTXSET(APP_i) = \{CTX_x | CTX_x \text{ is used by } APP_i.\}$$

The value of some contexts is the same as a measured value of a sensor, such as temperature. The value of other contexts is the result of manipulating measured values of multiple sensors.

Let SEN_i denote a sensor equipped on an IoT device. And, let $SENSET(CTX_j)$ be a function which returns the set of sensors used for deriving a value of the given context, CTX_j. For example, $SENSET(\text{"location"})$ returns the set, $\{GPS, WiFi\}$. Here, we assume that an IoT device can embed multiple sensors to acquire different types of contexts such as Android phone.

$$SENSET(CTX_i) = \{SEN_x | SEN_x \text{ is used for acquiring } CTX_i.\}$$

Figure 16.1 shows these terms and their relationships. For "n" IoT applications, there are also "n" number of $CTXSET$ instances. For each type context, there is one instance of $SENSET(CTX_i)$. And, by combining $SENSET(CTX_i)$ for all contexts, we can figure out the total set of sensors provided by multiple IoT devices.

Efficiency is the capability of the software product to provide appropriate performance, relative to the amount of resources used, under stated conditions [15]. Context

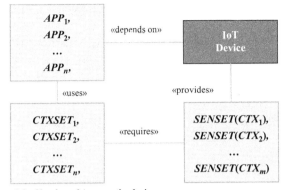

'n' : **Number of Apps on the device**
'm' : **Number of Context Types used by all the apps**

Figure 16.1   Terms and their relationships

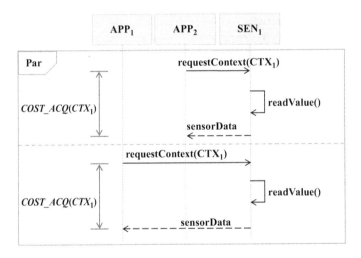

Figure 16.2   Two applications requesting the same context

acquisition is performed in two phases: reading data from IoT devices and sending the data to context-aware applications. Figure 16.2 shows a typical scenario of acquiring contexts. Two applications, APP_1 and APP_2, try to retrieve a context CTX_1, which uses a sensor SEN_1.

Then, we define a cost function for measuring the amount of resource consumption needed to acquire a context:

$$COST\_ACQ(CTX_i) = \sum_{x \in SENSET(CTX_i)} COST\_READ(x) + COST\_SEND(x)$$

That is, $COST\_ACQ(CTX_i)$ is the total cost of acquiring the given context, CTX_i. And, $COST\_READ(SEN_j)$ and $COST\_SEND(SEN_j)$ are the costs of reading data from all the context-relevant sensors and sending the context to the application respectively. Note that SEN_j is in $SENSET(CTX_i)$.

As shown in Figure 16.2, when two applications request the same context at the same time, the cost to acquire the context becomes double. That is, the total cost of the platform is evaluated by adding $COST\_ACQ(CTX_i)$ of all contexts used by all applications, represented as follows:

$$TOTAL\_COST_{PRE} = \sum_{x \in APPSET} \sum_{y \in CTXSET(x)} COST\_ACQ(y)$$

$$= \sum_{x \in APPSET} \sum_{y \in CTXSET(x)} \sum_{z \in SENSET(y)} (COST\_READ(z)$$

$$+ COST\_SEND(z))$$

This is the place where we apply an optimization of reading sensor data. That is, in our CAAS, the task of reading sensor data is performed asynchronously from the requests from the applications. More specifically, CAAS maintains a *Context Basket* which stores the most up-to-date sensor data and applications fetch contexts from the basket. More detailed description of CAAS is given in subsequent sections.

16.3.2   Incompatibility of context provisioning

IoT Applications provide their own functionalities by interacting with diverse IoT devices having different characteristics. IoT applications do not always interact with IoT devices supporting data acquisition schemes expected by the applications. That is, there is an incompatibility between IoT applications and IoT devices in terms of the data acquisition scheme. For example, an IoT application wants to read sensor data from an IoT device in a pulling manner, but an IoT device only returns the data in a pushing manner. If this incompatibility issue is not resolved, IoT applications interact with the limited number of IoT devices supporting the data acquisition schemes expected by the applications.

Before presenting the methods to resolve this incompatibility issue, we define incompatibility types between IoT applications and IoT devices as shown in Table 16.1. Drawing from the extensive surveys on available IoT devices and research articles, three representative context acquisition schemes are identified; *pulling*, *pushing*, and *notify-and-fetch* [16]. *Pulling* is very similar to *Polling*, and the difference lies in how to realize IoT applications, not the ways to send data from the devices. Hence, *Polling* is excluded from the table.

In the table, *A-2-B* indicates that an IoT application requires B scheme to get the data, but an IoT device only supports A scheme so that A scheme needs to be transformed to B scheme. For example, *Pull-2-Push* type means that an IoT device only supports a pulling scheme which does not meet the application requirements, so there is a need to mitigate a pulling scheme to a pushing scheme.

The proposed CAAS takes the responsibilities to mitigate this incompatibility on context acquisition schemes and to provide the contexts to IoT applications what

Table 16.1   Types of incompatibility of data acquisition schemes

Scheme supported by IoT device	Scheme expected by IoT application	Types of incompatibility
Pulling	Pulling	–
Pushing		*Push-2-Pull*
Notify-n-Fetch		*Notify-2-Pull*
Pushing	Pushing	–
Pulling		*Pull-2-Push*
Notify-n-Fetch		*Notify-2-Push*
Notify-n-Fetch	Notify-n-Fetch	–
Pulling		*Pull-2-Notify*
Pushing		*Push-2-Notify*

they want. As its name indicates, CAAS delivers a set of required contexts to IoT applications in the underlying technologies-neural manner.

16.4   Specification of CAAS platform

In this section, we describe the functional and nonfunctional requirements of the CAAS platform to deal with the two technical challenges addressed in the previous section.

16.4.1   Overview of CAAS platform

CAAS platform is a software system, which consists of two parts; CAAS agent running on a mobile device and CAAS server as a cloud service.

CAAS Agent: This agent runs on an IoT device or any surrogate device for an IoT device and provides an abstract layer between context-aware IoT applications and sensors embedded in IoT devices. As an intermediary, it gathers contexts from sensors and transmits the contexts to CAAS server. The key benefit of using the CAAS agent is the increased efficiency since it only reads the contexts for potential multiple IoT applications.

CAAS Server: CAAS server is a cloud service, which maintains a repository of all the contexts acquired on mobile devices for all the registered users and distributes contexts to IoT applications. The applications do not directly communicate with the sensors; rather they ask the CAAS server to provide the contexts required. As IoT devices have limited storage for storing contexts, CAAS server provides a big storage space for all the registered devices. Hence, acquired contexts will be stored safely without any loss. A value of maintaining the context repository is the potential to apply advanced analytics and machine learning on the big context.

Deployment of the Platform and Applications: The deployment of CAAS platform is shown in Figure 16.3. CAAS agent is deployed on an IoT device or any surrogate

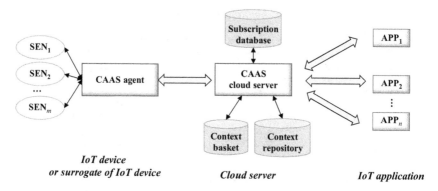

Figure 16.3　Key elements of CAAS platform

devices for an IoT device (if IoT devices does not have any computing capability), and it transmits contexts to CAAS server.

CAAS server manages a context basket which is a subset of the whole contexts and maintains up-to-date contexts required by IoT applications immediately. In addition, CAAS server manages *Context Repository* and a *Subscription Database,* which contains the whole set of contexts acquired from IoT devices and the instances of subscription of contexts by each IoT application respectively.

In Figure 16.3, multiple context-aware IoT applications are deployed on a single or multiple devices. The application communicates with the CAAS server to acquire required contexts.

In addition to the databases for contexts like *Context Basket* and *Context Repository*, the CAAS server maintains a database of registered context subscriptions. A context subscription is a tuple of (*Device, App, Context List, Context Fetch Scheme*). That is, an App registered its required subscriptions to receive the context service from the Agent.

16.4.2　Functional requirements of CAAS platform

CAAS agents and CAAS server perform their own tasks in acquiring and provisioning contexts to IoT applications by solving the technical challenges.

The CAAS agent measures and acquires the context from the sensors. The agent measures sensors values by using the sensor-specific context fetching scheme. And, it transmits the acquired contexts to the CAAS server.

In contrast, CAAS server provides a different set of tasks.

Context Repository Management: Contexts delivered from CAAS agents are stored in both *Context Basket* and *Context Repository*. The main reason of maintaining these types of the repositories is to minimize the time to retrieve contexts for IoT applications. *Context basket* only stores the latest contexts, which are immediately retrieved by IoT applications, and *Context repository* is some form of a large-scaled database, and should be securely managed.

Context Provisioning: The contexts stored in *Context Cache* are provided to IoT applications according to the subscription information. For each type of scheme specified in each IoT application, CAAS server provides the required contexts to the applications.

Context Subscription Management: A context-aware IoT application uses one or more types of contexts such as location, proximity, and gravity. To receive the contexts from a CAAS server, an application specifies the types of contexts to receive. A subscription information includes types of contexts required and fetching scheme-specific setting parameters. Subscription of contexts is essential for all the context fetching schemes except pulling scheme.

16.4.3 *Nonfunctional requirements of CAAS platform*

The first aim to design the CAAS is to minimize the value of TOTAL_COST as much as possible. Hence, maintaining the minimum values of COST_READ(SEN_i) and COST_SEND(SEN_i) is the key to realize a highly efficient context acquisition method with CAAS. Another potential problem comes only with the consideration of efficiency of acquisition. If the number of reading data from sensors is too much minimized to reduce COST_READ(SEN_i)significant sensor data can be lost. An *accuracy* of acquiring contexts should be also considered. Hence, we define the following nonfunctional requirements.

Minimizing Resource Consumption for Sensor Measurement: The platform should minimize the resource consumption for measuring sensor values. The value of the COST_READ(SEN_i) is reduced by (1) adjusting the frequency of reading sensor data and (2) adjusting the frequency of storing the acquired sensor data to the context basket. However, these frequencies should be determined not to lose any significant data. That is, the platform should minimize the measuring sensor values while maximizing the timely captures of changed contexts.

Maximizing Time Efficiency for Context Retrieval: The platform should provide the maximum time efficiency for context retrievals. That is, the time for an application to retrieve a context should be minimized.

Minimizing the Overhead of Sending Contexts to Context-aware IoT Applications: The cost of sending data is affected by the size of the data and the communication paradigm. Hence, the value of the *COST_SEND(SEN$_i$)* is minimized by (1) sending the minimum set of sensed data associated with the context and (2) choosing the right communication paradigm to send sensed data.

Maximizing Context Availability: The platform should provide the maximum context availability for applications. That is, CAAS server should maintain the optimal set of contexts in *Context Basket* to maximize the hit ratio for retrieving contexts.

In addition, CAAS platform solves the incompatibility issue between IoT devices and IoT applications, which leads to additional runtime overhead.

16.5 Design of CAAS platform

To maximize the benefits of using the context basket while minimizing the potential drawbacks, we need to design the CAAS with the consideration of the functional and nonfunctional requirements.

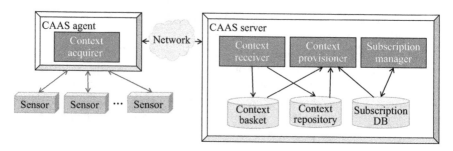

Figure 16.4   Functional view of the CAAS platform

16.5.1   Architecture design

CAAS platform consists of CAAS agents and a CAAS server as shown in Figure 16.4. By considering the functional requirements, key components comprising CAAS platform are derived.

CAAS agent consists of a component, *Context acquirer*. *Context acquirer* reads data from sensors and sends them to CAAS server.

CAAS server is a centralized manager of multiple agents and consists of three components: *Context Receiver*, *Context Provisioner*, and *Subscription Manager*. *Context Receiver* receives contexts from multiple CAAS agents and stores the contexts to both *Context Basket* and *Context Repository*. *Context Provisioner* delivers contexts to IoT applications by retrieving contexts from either *Context Basket* or *Context Repository* in the response of their requests after retrieving a set of sensor data associated with the context. And, *Subscription Manager* manages subscription information of IoT applications, which describes how the contexts are delivered to the applications.

There are "n" context-aware IoT applications and "m" sensors are equipped into multiple IoT devices. Multiple IoT applications acquire contexts from the CAAS platform, not directly from sensors. And, CAAS agent reads context data from sensors and stores them to the CAAS server. And, CAAS server maintains the context basket with the most up-to-date information in an efficient way. Since IoT devices reveal high heterogeneity in terms of their computing capabilities, this platform has diverse configurations as shown in Figure 16.5.

The left part of the figure, that is, (a) centralized CAAS, shows that both CAAS agents are CAAS server deployed on the same device. If an IoT device, like android mobile phone or Arduino board connected with various sensors, has a computing capability, CAAS agent and CAAS server can be deployed on the same device, and also IoT applications can be deployed on the same device.

The right part of the figure, that is, (b) distributed CAAS, shows the typical configuration of CAAS where CAAS agents are deployed either on IoT devices or any surrogate device for IoT devices, and CAAS server is installed on a centralized device, like cloud server. If IoT devices are not equipped with computing capabilities, like sensors connected to Arduino board, CAAS agents are installed on the designated surrogate devices. In this configuration, IoT applications, CAAS agents, and a CAAS

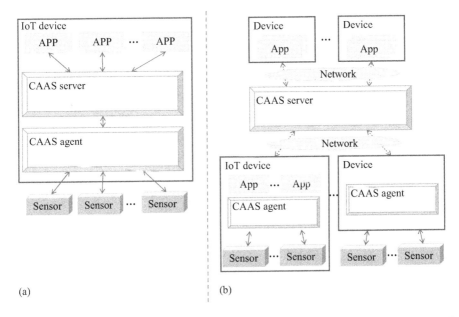

Figure 16.5   Deployment view of the CAAS platform: (a) centralized CAAS and (b) distributed CAAS

server are communicated over the network. In this chapter, we assume that CAAS platform is deployed in the distributed manner.

In the subsequent section, the detail designs showing how the technical challenges are addressed in this platform are presented.

16.5.2   Solution: CAAS and context basket for efficient sharing

To address the efficiency problem in acquiring contexts, we come up with the idea of CAAS platform itself. Before presenting the design of the platform, we first verify how CAAS with the *context basket* actually minimizes the cost to acquire context data via a metric. The following equation is used to measure the cost to acquire context by using CAAS with the *context basket*:

$$
\begin{aligned}
TOTAL\_COST_{CAAS} &= \sum_{i=1}^{n} COST\_READ(SEN_i) \\
&+ \sum_{x \in APPSET} \sum_{y \in CTXSET(x)} COST\_ACQ(y) \\
&= \sum_{i=1}^{n} COST\_READ(SEN_i) \\
&+ \sum_{x \in APPSET} \sum_{y \in CTXSET(x)} \sum_{z \in SENSET(y)} COST\_SEND(z)
\end{aligned}
$$

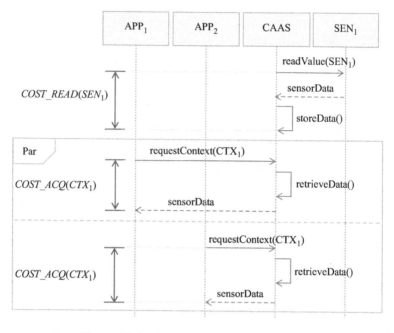

Figure 16.6   Context acquisition with CAAS

Here, *n* is the number of sensors controlled by CAAS. This equation implies that every request made by context-aware IoT applications does not result in reading sensor data from sensors. Hence, the big difference between $TOTAL\_COST_{PRE}$ and $TOTAL\_COST_{CAAS}$ lies in the cost to read data from all the sensors, which is inferred from the operands of $COST\_READ()$ of the two total costs. This is why reading sensor data and sending the data to context-aware IoT applications are always executed sequentially with a conventional context acquisition approach, while the two phases are individually performed with our CAAS.

The net benefit of deploying the CAAS is the reduced cost of context acquisition by letting applications share the acquired contexts, as shown in Figure 16.6.

In addition to the CAAS itself, *Context Basket*, a kind of cache, can maximize the runtime efficiency by leveraging fast storage and retrieval of the contexts.

16.5.3   Solution: adaptive context monitoring for efficiency

ContextAcquirer, a component installed on CAAS agent, is to read contexts as the way that IoT devices support. Hence, this component is designed to deal with applicable context acquisition schemes as shown in Figure 16.7. *ContextAcquirerInPulling*, *ContextAcquirerInPushing*, and *ContextAcquirerInNotify* are classes, which read contexts in a pulling manner, a pushing manner, and notify-n-fetch manner, respectively.

Depending on the context acquisition supported by an IoT device, the *ContextAcquisitionHandler* is specialized with one of the three classes.

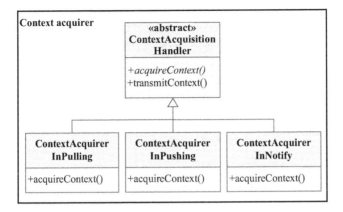

Figure 16.7   Structure of context acquirer

List 16.1   An example of context specification

```
<application id = "ca1" name = "Location Tracer">
  <contextSet>
    <context name = "User Location" frequency = "0.5">
      <sensorSet>
        <sensor type = "GPS" />
        <sensor type = "Wifi" />
      </sensorSet>
    </context>
  </contextSet>
</application>
```

And, this component, *Context Acquirer*, needs to be designed to meet the first technical challenge, which is to minimize the runtime overhead of reading data from sensors without losing any significant data. For this, we define two design tactics. These tactics are applied to designs of both CAAS agent and CAAS server.

16.5.3.1   Tactic #1. Optimizing sensing frequency

Context Acquirer first retrieves a list of contexts which will be used by an IoT application. When the context acquirer read the contexts, it has information about what contexts are required and how frequently the contexts need to be read. Before *Context Acquirer* reads the context, it needs to get the information from CAAS server. Note that *Pushing-* and *Notify-n-fetch* schemes are designed to acquire contexts whenever any context change is triggered so that it is not feasible to design an efficient context acquisition method. Hence, this tactic is only effective to *ContextAcquirerInPulling*.

As shown in List 16.1, each context-aware IoT application gives information of contexts and a set of sensors required for the context to CAAS server. Hence, CAAS can know the set of sensors needed by IoT applications by evaluating $CTXSET(APP_i)$

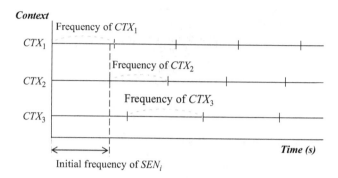

Figure 16.8   An example of setting initial frequency of a sensor (SEN$_i$)

and $\sum_{x \in CTXSET(APP_i)} SENSET(x)$. If multiple sensors are available for a single context, the smaller size of the sensor data is chosen to reduce *COST_SEND(sensor)*.

All the context-aware IoT applications deliver their context specifications to CAAS server. For each context in *CTXSET*, it contains a set of required contexts, a set of sensors associated with the context, and their desired sensing frequencies as shown in List 16.1. <contextSet>element and <sensorSet> element are equivalent to *CTXSET ()* and *SENSET ()* of the given app, respectively.

By reading the context specifications of all context-aware IoT applications, CAAS server needs to decide the optimal sensing frequencies (i.e., sampling rates) by considering efficiency and accuracy.

CAAS server, especially *Context Receiver*, sets the initial sensing frequencies of the sensors and then adjusts them. The initial frequencies are decided by reading context specifications of the IoT applications. For example, let us assume that an IoT application requires three contexts, CTX_1, CTX_2, and CTX_3, and sets their frequencies of a sensor, SEN_i, as depicted in Figure 16.8.

Since SEN_i is used for acquiring three contexts with different frequencies, the initial frequency needs to be set as the shortest frequency. Here, the frequency required by CTX_1 is set to the frequency of SEN_i.

After setting initial frequencies of sensors, the sensors start to measure contexts with their own frequencies, and those frequencies are adjusted by considering (1) configuration changes of the IoT applications and (2) variations of the sensed values. The first criterion implies that frequencies need to be changed whenever applications are installed or deleted. And, the second criterion is applied to reduce cost to acquire sensed data due to too much frequent sensing or to increase accuracy of the sensed data if there may be losses due to very scarce sensing.

When a new app, APP_{n+1}, is added in the *APPSET* or an existing app, APP_j, is removed from the *APPSET*, frequencies of sensors used by $CTXSET(APP_{n+1})$ or $CTXSET(APP_j)$ have to be calculated, respectively. Then, frequencies of relevant sensors are reconfigured.

Magnitude indicates a scale of differences among a series of measured contexts. Once a current context is measured by a sensor, the context is compared with the

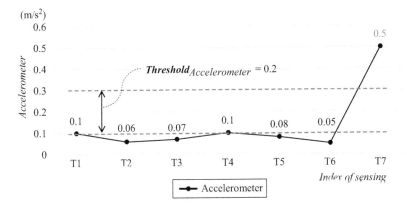

Figure 16.9   Magnitude comparison with threshold value of accelerometer

previous one to capture a difference between them. If magnitude of the difference is relatively large, then the frequency of the sensor should become smaller to increase an accuracy of the sensor.

Let CTX_k^T be a measured data of the CTX_k at the given time, T, and $Threshold_{CTX_k}$ be a threshold value. The threshold value indicates its maximum range of context changes on the given context, CTX_k. If a difference between CTX_k^T and CTX_k^{T-1} is larger than the threshold value, $Threshold_{CTX_k}$, then the frequency of the given context is adjusted to be half or double.

16.5.3.2   Tactic #2. Minimizing the overhead of storing sensor data to context basket

Compared to conventional context acquisition platforms, our CAAS platform maintains the *context basket*, which increases runtime management overhead. Hence, we need to minimize the overhead. To do this, CAAS platform is designed to reduce the number of accessing the context basket.

Storing contexts which are unchanged or very similar to previous contexts to the context basket may be waste of the time and resource. Therefore, the *Context Agent*, especially *Context Acquirer*, filters newly measured contexts by comparing them with the last updated context as shown in Figure 16.9.

Based on the latest updated context, as $CTX_{Accelerometer}^{T1} = 0.1$, newly measured context from at T2 to T7 are compared with it. If the gap between two contexts is less than the threshold value, $Threshold_{Accelerometer}$, then the newly measured context is ignored without storing it into the *Context Basket*. This tactic minimizes the number of storing contexts and also reduces the total amount of energy consumption.

16.5.4   Solution: providing contexts with context acquisition schemes required by IoT applications

Context Provisioner delivers the contexts to IoT applications and resolves incapability issues on context acquisition schemes if needed. Figure 16.10 shows the part of the

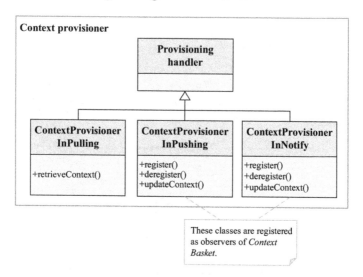

Figure 16.10   Structure of context provisioner

structural design of *Context Provisioner*. *ContextProvisionerInPulling* is to deliver contexts to applications expecting to get contexts in a pulling manner. *ContextProvisionerInPushing* and *ContextProvisionerInNotify* are classes of delivering contexts to IoT applications expecting to get contexts in a pushing manner and a notify-n-fetch manner, respectively. To deliver contexts to IoT applications in a pushing or a notify-n-fetch manner, *ContextProvisionerInPushing* and *ContextProvisionerInNotify* are registered as observers to *Context Basket* so that any context change on *Context Basket* is delivered to IoT applications.

First, a key challenge of *Push-2-Pull* and *Notify-2-Pull* mitigation methods is to transform a *pushing* scheme or *notify-n-fetch* scheme to a *pulling* scheme so that an IoT application retrieves the sensor data only when it wants without recognizing the occurrence of the incompatibility. These incompatibilities are easily tackled by the CAAS platform itself. Since *Context Acquirer* reads contexts with context acquisition schemes supported by IoT devices and *Context Receiver* stores them to *Context Basket* and *Context Repository*, *ContextProvisionerInPulling* simply returns the latest sensor data from either *Context Basket* or *Context Repository* as shown in Figure 16.11.

If the required contexts are stored in Context Basket, *ContextProvisionerInPulling* returns the contexts from *Context Basket*. If not, *ContextProvisionerInPulling* retrieves contexts from *Context Repository* and delivers them to IoT applications. Hence, pulling-relevant incompatibilities are easily handled by CAAS platform.

Second, a key challenge of *Pull-2-Push* and *Notify-2-Push* mitigation methods is to transform a *pulling* scheme or *notify-n-fetch* scheme to a *pushing* scheme so that an IoT application retrieves the sensor data from pulling-based IoT device or notify-n-fetch-based IoT devices whenever any new data is acquired. These incompatibilities are also easily tackled by the CAAS platform itself. Since *Context Acquirer* reads contexts with context acquisition schemes supported by IoT devices and *Context*

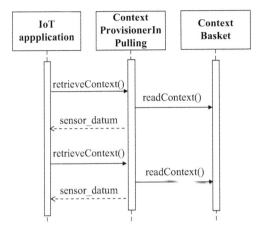

Figure 16.11 Sequence diagram for mitigating Push-2-Pull and Notify-2-Pull incompatibilities

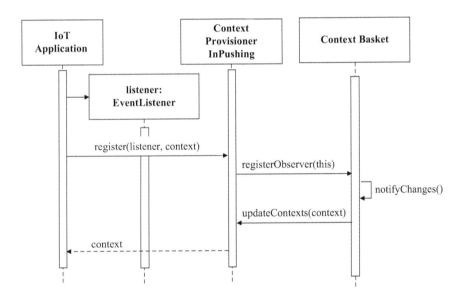

Figure 16.12 Sequence diagram for mitigating Pull-2-Push and Notify-2-Push incompatibilities

Receiver stores them to *Context Basket* and *Context Repository*, *ContextProvisioner-InPushing* simply returns contexts to registered IoT applications whenever any other change on *Context Basket* is made as shown in Figure 16.12.

An IoT application registers its interests to *ContextProvisionerInPushing* by implementing its pre-defined event listener, not to a target device because the target device does not provide the listener-based mechanisms. After the registration, *ContextProvisionerInPushing* generates events to the application whenever any new data is acquired.

Since the latest contexts are stored in *Context Basket* and *Context Basket* is designed to maintain up-to-date contexts for IoT devices, *ContextProvisionerInPushing* registers itself to *Context Basket* as an observer to get notification on context changes.

Once *Context Basket* gets a new data from the target device, it generates a notification event to *ContextProvisionerInPushing*. And, immediately, *ContextProvisionerInPushing* delivers newly generated contexts to the registered IoT applications.

Due to *ContextProvisionerInPushing*, the IoT application can get the data in a pushing manner, although the target device does not support this scheme.

Lastly, a key challenge of *Pull-2-Notify* and *Push-2-Notify* mitigation methods is to transform a *pulling* scheme or *pushing* scheme to a *notify-n-fetch* scheme so that an IoT application retrieves the sensor data from pulling-based IoT device or notify-n-fetch based IoT devices in *notify-n-fetch* manner. *Pushing* scheme and *notify-n-fetch scheme* are very similar except that an IoT device returns either changed contexts (in pulling scheme) or the fact that new contexts are generated (in notify-n-fetch scheme). Hence, *Pull-2-Notify* and *Push-2-Notify* incompatibilities are also easily addressed by CAAS platform itself, and the design for resolving these incompatibilities is derived by making minor changes to *ContextProvisionerInPushing*. Due to the page limitation, we do not present the design of the *ContextProvisionerInPushing*.

16.6 PoC implementation of CAAS and experiments

In order to evaluate applicability of our proposal, we present experiment results of energy consumptions estimated with CAAS and without CAAS.

16.6.1 Experiment settings

For the experiment, we developed three IoT healthcare applications written in python and deployed on a computer:

$$APPSET = \{APP_1, APP_2, APP_3\}$$

APP_1 computes health indexes, which tell users their current health status when a user request. APP_2 is to detect the occurrence of new contexts and retrieve the stress levels of a user when a user wants. APP_3 is to analyze the healthiness of the user's heart when a user wants. As in their functional descriptions, App_1 and App_3 expect to acquire contexts in a pulling manner, while App_2 wants to get contexts in notify-n-fetch manner.

The set of contexts used by each application is shown as:

$CTXSET(APP_1) = \{$Blood Pressure, Weight, Body Fat, Pulse, SpO2, EEG, ECG, GSR$\}$
$CTXSET(APP_2) = \{$GSR, Pulse, EEG$\}$
$CTXSET(APP_3) = \{$Blood Pressure, Pulse, ECG$\}$

Table 16.2   Types of incompatibility occurring in experiments

	Scheme expected by IoT app	Types of incompatibility
App_1	Pulling	–
App_2	Notify-n-Fetch	Pull-2-Notify
App_3	Pulling	–

Each type of context is acquired from one or more sensors. A part of the sensor sets are listed as shown here:

$SENSET(CTX_{BloodPressure}) = \{$BP Monitor by VendorA, BP Monitor by Vendor B$\}$
$SENSET(CTX_{Weight}) = \{$Body Scale$\}$
$SENSET(CTX_{BodyFact}) = \{$Body Scale$\}$
$SENSET(CTX_{EEG}) = \{$Smart Headset, Smart Hat$\}$.

All the IoT devices used for this experiment support a pulling scheme so that they return contexts when IoT applications request. Hence, App_2 has an incapability issue in utilizing the required contexts. Table 16.2 summarizes the types of the incompatibility in terms of context provisioning.

And, we develop the PoC of CAAS, which is implemented with python and executed on Django framework. CAAS has features to support efficient context acquisition and resolve incompatibilities on the context-provisioning schemes. List 16.2 shows the part of the code snippet of mitigating *PullToNotify* incompatibility.

ContextProvisionerInNotify offers sensor data of interest to IoT applications in a notify-n-fetch manner. To do this, it registers an IoT application as a listener at line 2 so that any change on the sensor data is notified to the registered IoT application. *AcquisitionThread* periodically requests to acquire data from the context basket (lines from 16 to 23). The context basket already collects the sensor data of interest in a pulling manner so that *Acquisition Thread* periodically checks the new sensor data on the context basket.

In addition to *ContextProvisionerInNotify*, CAAS is instrumented with classes resolving the other types of the incompatibilities on context-provisioning schemes.

16.6.2   Experiment results

We measured the results of the context acquisition in two cases: a case of *Direct Sensor Access* and a case of using CAAS which deploys a context basket.

Each app was set up to run the method of reading its relevant contexts with different invocation:

- Continuous Run
- Every 2 s
- Every 10 s

We have run each app for 30 minutes, and measured the efficiencies using the metrics defined in Sections 16.3.1 and 16.5.2. Figure 16.13 shows the averages of

List 16.2   Python Code of ContextProvisionerInNotify

```
1    Class ContextProvisionerInNotify:
2        def register(self, listener):
3            self.listener = listener
4            AcquisitionThread(self.device, check_event).start()
5        ...
6        def check_event(sensor_datum):
7            if is_sensor_value_changed(sensor_datum):
8                event = [Event.ValueChanged, timestamp]
9                self.data_queue[event] = sensor_datum
10               listener(event)
11       ...
12       def fetch(event):
13           sensor_datum = self.data_queue[event]
14           return sensor_datum
15
16   Class AcquisitionThread(Threading.Thread):
17       ...
18       def run(self):
19           while(True):
20               sensor_datum = basket.request_context(context)
21               self.check_event(sensor_datum)
22               ...
23               time.sleep(period)
```

the total costs measured in a graphic chart. The chart shows the average battery consumption of retrieving all the contexts required by the three IoT applications. Note that we cannot include the results of App_2 in this experiment since App_2 cannot acquire the contexts without using CAAS platform.

From the experiment results, we make the following observations:

- The more frequent applications acquire contexts, the higher gain of efficiency we have. That is, when applications access sensors directly in a continuous manner (i.e., in a loop), there is a high contention to access the shared resources, that is, sensors. If CAAS is used for the same situation, the contention is largely reduced due to the asynchronous reading of sensors and sharing among applications.
- On the contrast, if IoT applications acquire contexts with a long interval (i.e., least frequent access), then the contention is not significant and hence the efficiency gain with CAAS is low.
- We cannot get results for App_2 since there is no device supporting context acquisitions schemes that App_2 wants. That is, IoT applications can only interact with IoT devices supporting the context acquisition schemes required by the applications.

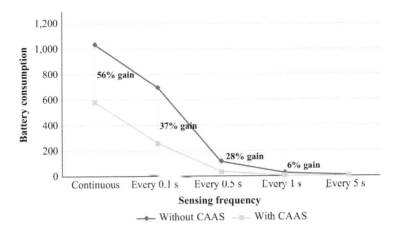

Figure 16.13   Results with battery consumption in chart

From the observations, we now derive the circumstance where the proposed CAAS yields the greatest gain of efficiency:

- When a large number of context-aware applications are deployed and managed by CAAS.
- When IoT applications need to acquire contexts at a high frequency (i.e., a short interval).
- When the battery consumption for reading sensors is considerably high.

16.7   Conclusions

A context-aware IoT application is a software system which utilizes contexts in providing situation-specific functionality. The key benefits of context-aware IoT applications will be the appropriateness and smartness of the service provided by the application. Despite the net benefits of the context-aware IoT applications, it is not quite straightforward to develop such application due to the runtime over-head of acquiring and utilizing the contexts. To clarify the potential issues in context acquisition and provisioning,

- IoT devices have limited resources so that it may have runtime overhead in read-ing and sending contexts to IoT applications. Moreover, many IoT applications request same contexts from the IoT devices, which results in inefficiency of acquiring contexts from IoT devices. If this issue is not addressed, the over-all qualities of the IoT applications may decrease. In an extreme case, the IoT applications will not be used due to their poor qualities.

- Most of the IoT devices support one context acquisition scheme, pulling, pushing, or notify-n-fetch scheme. And, IoT applications need to provide functionalities by acquiring contexts as they want. Hence, there may be incompatibility issues on the context acquisition scheme between IoT applications and IoT devices. If this problem is not solved, IoT applications interact with a limited number of IoT devices. In an extreme case, IoT applications cannot perform their functionalities due to the absence of IoT devices supporting the right context acquisition scheme.

To address the challenges, a design of CAAS platform is proposed by adopting the concept of cloud computing where any resources are deployed as a unit of a service. CAAS platform plays an essential role of providing contexts as IoT applications want. For efficiency-context acquisition, detailed designs of optimizing sensing frequency and minimizing the number of transmitting contexts to CAAS server are presented. And, for resolving incompatibility issues in context provisioning, designs of *Context Acquirer* and *Context Provisioner* are presented.

With the proposed platform, a large amount of IoT contexts can be managed in a server-side repository and various sets of contexts can be efficiently delivered to the IoT applications. With the archived contexts, IoT applications can behave even more smartly by utilizing machine learning over an extensive training set of contexts.

References

[1]   A.K. Dey, "Understanding and Using Context" *Personal and Ubiquitous Computing*, Vol. 5, No. 1, pp. 4–7 2001.

[2]   C. Perera, A. Zaslavsky, P. Christen, and D. Georgakopoulos, "Context Aware Computing for the Internet of Things: Survey," *IEEE Communications Surveys & Tutorials*, Vol. 16, No. 1, pp. 414–454, First Quarter 2014.

[3]   Y.S. Chen and Y.R. Chen, "Context-Oriented Data Acquisition and Integration Platform for Internet of Things," *In Proceedings of 2012 Conference on Technologies and Applications of Artificial Intelligence (TAAI 2012)*, pp. 103–108, Tainan, Taiwan, November 2012.

[4]   C. Perera, A. Zaslavsky, P. Christen, and D. Georgakopoulos, "CA4IOT: Context Awareness for Internet of Things," In *Proceedings of 2012 IEEE International Conference on Green Computing and Communications (GREENCOM 2012)*, pp. 775–782, Besançon, France, November 2012.

[5]   H.J. La, J.Y. Lee, Z.Z. Piao, and S.D. Kim, "An Efficient Context Acquisition for 'n' Mobile Apps," In *Proceedings of the 12th IEEE International Conference on Pervasive Intelligence and Computing (PICom 2014)*, pp. 316–321, Dalian, China, August 2014.

[6]   L. Skorin-Kapov, K. Pripuzic, M. Marjanovic, A. Antonic, and I.P. Zarko, "Energy Efficient and Quality-Driven Continuous Sensor Management for Mobile IoT Applications," In *Proceedings of 10th IEEE International Conference on Collaborative Computing: Networking, Applications and Worksharing (CollaborateCom 2014)*, pp. 397–406, Miami, USA, October 2014.

[7] S. Chattopadhyay, "Algorithmic Strategies for Sensing-As-A-Service in the Internet-of-Things Era," In *Proceedings of 2015 IEEE/ACM Eighth International Conference on Utility and Cloud Computing (UCC 2015)*, pp. 387–390, Limassol, Cyprus, December 2015.

[8] S. Kang, J. Lee, H. Jang, Y. Lee, S. Park, and J. Song, "A Scalable and Energy-Efficient Context Monitoring Framework for Mobile Personal Sensor Network," *IEEE Transactions on Mobile Computing*, Vol. 9, No. 5, pp. 686–702, May 2010.

[9] Y. Lee, S.S. Iyengar, C. Min, *et al.*, "MobiCon: A Mobile Context-Monitoring Platform," *Communications of the ACM*, Vol. 55, No. 3, pp. 54–65, March 2012.

[10] S. Nath, "ACE: Exploiting Correlation for Energy-Efficient and Continuous Context Sensing," *IEEE Transactions on Mobile Computing*, Vol. 12, No. 8, pp. 1472–1486, August 2013.

[11] Y. Wang, J. Lin, M. Annavaram, *et al.*, "A Framework of Energy Efficient Mobile Sensing for Automatic User State Recognition," In *Proceedings of the Seventh International Conference on Mobile Systems, Applications, and Services (MobiSys 2009)*, pp. 179–192, Kraków, Poland, June 2009.

[12] Y. Han, J.M. Kang, S.S. Seo, A. Mehaoua, and J.W.K. Hong, "An Energy Efficient User Context Collection Method for Smartphones," In *Proceedings of the 15th Asia-Pacific Network Operation and Management Symposium (APNOMS 2013)*, pp. 1–6, Hiroshima, Japan, September 2013.

[13] N. Roy, A. Misra, C. Julien, S.K. Das, and J. Biswas, "An Energy-Efficient Quality Adaptive Framework for Multi-Modal Sensor Context Recognition," In *Proceedings of 2011 IEEE International Conference on Pervasive Computing and Communications (PerCom 2011)*, pp. 63–73, Seattle, USA, March 2011.

[14] V. Bezerra, M.C. Junior, O. Valeria, *et al.*, "An Energy-Efficient Context Management Framework for Ubiquitous Systems," In *Proceedings of the 10th International Conference on Ubiquitous Intelligence & Computing and the 10th International Conference on Automatic & Trusted Computing (UIC/ATC 2013)*, pp. 697–702, Vietri sul Mare, Italy, December 2013.

[15] *Software Engineering – Product Quality – Part 1: Quality Model*, International Standard ISO/IEC 9126-1:2001, 2001.

[16] J.Y. Lee and S.D. Kim, "IoT Contexts Acquisition with High Accuracy and Efficiency," In *Proceedings of IEEE 11th World Congress on Services 2015 (SERVICES 2015)*, pp. 9–16, June 2015.

Index

Printed in the USA
CPSIA information can be obtained
at www.ICGtesting.com
JSHW011508221024
72173JS00005B/1241